# AMORPHOUS SILICON MATERIALS AND SOLAR CELLS

# AIP
# CONFERENCE
# PROCEEDINGS 234

# AMORPHOUS SILICON
# MATERIALS AND
# SOLAR CELLS
### DENVER, CO 1991

**EDITOR:**

**BRYON L. STAFFORD**
SOLAR ENERGY
RESEARCH INSTITUTE

**American Institute of Physics**                    **New York**

L.C. Catalog Card No. 91-55575
ISBN 0-88318-831-7
DOE CONF-910225

Printed in the United States of America.

CONTENTS

## PART I. OPENING PRESENTATIONS

## PART II. MODELING METASTABILITY

## PART III. EXPERIMENTAL DATA AND MODEL VERIFICATION

## PART IV. MATERIAL STUDIES

# PREFACE

An International Meeting on Stability of Amorphous Silicon Materials and Solar Cells was held in Denver, CO on February 20–22, 1991. The main objectives of the meeting were to bring to light–and stimulate discussion on—recent advances in (1) understanding the underlying mechanisms of light-induced instability and (2) engineering approaches to stable solar cells. This meeting followed three previous meetings with similar purposes: *International Conference on Stability of Amorphous Silicon Alloy Materials and Devices*, AIP Conference Proceedings No. 157 (Palo Alto, CA, 1987); *Optical Effects in Amorphous Semiconductors*, AIP Conference Proceedings No. 120 (Snowbird, UT, 1984); and the *Stability of Amorphous Silicon Materials Workshop* (San Diego, CA, 1982). Comparison of this proceedings with the earlier ones reveals that all of the previously identified major issues are still very active—better understood—but unresolved. Nevertheless, several of the experimental and theoretical papers presented here, particularly those regarding low-hydrogen-content materials, give cause for optimism that the phenomenon may finally be yielding to worldwide concerted efforts to understand and mitigate it.

For this proceedings, the meeting papers are organized, generally following the meeting outline, into six chapters:

> Opening Presentations
> Modeling Metastability
> Experimental Data and Model Verification
> Material Studies
> Solar Cell Studies
> Summary and Closing Remarks

The meeting cosponsors were the Amorphous Silicon Photovoltaics Program of the Solar Energy Research Institute (SERI) and the Amorphous Thin-Film Solar Cell Program of the Electric Power Research Institute (EPRI).

The meeting general co-chairmen were Mr. Byron Stafford (SERI) and Dr. Terry Peterson (EPRI). The program committee members were Prof. Walther Fuhs (University of Marburg), Prof. Yoshihiro Hamakawa (Osaka University), Prof. Craig Taylor (University of Utah), and the co-chairmen.

The meeting organizers are especially thankful for the help of the following people who served as session chairmen and discussion moderators: Dr. Warren Jackson (Xerox PARC); Prof. David Cohen (University of Oregon); Prof. Craig Taylor (University of Utah); Prof. Walther Fuhs (University of Marburg); Dr. Bill Baron (University of Delaware); Prof. Christopher Wronski (Pennsylvania State University); Dr. Pawan Bhat (Materials Research Group, Inc.); Prof. Vikram Dalal (Iowa State University); and Dr. Arun Madan (MV Associates).

A very special thanks is also owed to Prof. Peter LeComber (University of Dundee) for his excellent summary talk at the close of the meeting.

The meeting's smooth running was largely due to the expert work of Ms. Diane Christodaro, SERI's Conference Coordinator, and her staffperson, Ms. Dori Nielsen. All of the meeting attendees extended a round of applause.

Byron Stafford, SERI
Terry Peterson, EPRI

February 1991

# PART I


# OPENING
# PRESENTATIONS

# STABILIZED MODULE PERFORMANCE AS A GOAL FOR THE PHOTOVOLTAIC AMORPHOUS SILICON PROGRAM IN THE UNITED STATES.

W. Luft, B. Stafford, and B. von Roedern
Solar Energy Research Institute
1617 Cole Boulevard, Golden CO 80401

## ABSTRACT

Amorphous silicon technology offers an avenue for low-cost, thin-film photovoltaic applications. The performance of amorphous silicon-based solar cells is limited by light-induced degradation. The inadequate existing understanding of the electronic phenomena in amorphous silicon materials and devices hampers resolution of this problem. We are posing questions that should stimulate researchers to develop better descriptions for device performance and better microscopic models for defect sites. The issue of Staebler-Wronski degradation should not be addressed separately from initial performance, but research should focus on material and device properties in the stabilized state. The main focus of the a-Si:H research sponsored by the United States Department of Energy will be on improving the stabilized performance, which we anticipate to accomplish through focused development of optimized multijunction device structures combined with an improved understanding of materials.

## INTRODUCTION

In 1978, the United States Department of Energy initiated a research program on amorphous silicon materials and devices as a part of its photovoltaic research and development program. Since 1983, the Solar Energy Research Institute (SERI) has funded several cost-shared projects between the U.S. government and U. S. industry to accelerate the advancement of amorphous silicon technology. In parallel, SERI has funded fundamental research in amorphous silicon to support the government/industry projects. The progress in SERI-funded research, a large worldwide market share in photovoltaic sales, a committed worldwide industry/academia infrastructure, and extensive manufacturing experience have established amorphous silicon as a leading thin-film photovoltaic option.

As shown in Table 1, the efficiencies (air mass 1.5, 1985 American Society for Testing and Materials [ASTM] global insolation conditions) achieved under the SERI project for all-amorphous-silicon alloy, two-terminal, *different-band-gap* multijunction devices by December 1990 were 13.3% (active area) for a small-area cell and 9.3% (aperture area) for a 939-$cm^2$ module containing a-Si:H:F and a-SiGe:H:F alloy materials. Efficiencies for two-terminal, *same-band-gap*, multijunction 0.25-$cm^2$ cells and 936-$cm^2$ submodules were 11.9% (total area) and 7.2% (aperture area), respectively.   However, these are all *initial* efficiencies, before light-induced degradation, and not *stabilized* efficiencies.

**Table 1.    SERI Confirmed State-of-the-Art Initial Efficiencies for All-Amorphous Silicon PV Devices in the United States.**   Values are total-area efficiencies for cells (unless otherwise indicated) and aperture-area efficiencies for submodules and modules.

| | |
|---|---|
| **Single-Junction Technology** | |
| Small-area cells | 11.9% |
| Submodules (933 cm$^2$) | 9.8% |
| Modules (4,700 cm$^2$) | 7.8%* |
| Modules (1.2 m$^2$) | 6.2% |
| Light-induced degradation | 15% - 30% |
| | |
| **Multijunction Technology** | |
| Small-area triple-junction cells | 13.3%† |
| Small-area dual-junction cells (Si/SiGe) | 12.4%†(13.0%*†) |
| Small-area dual-junction cells (Si/Si) | 11.9%*† |
| Dual-junction submodules (936 cm$^2$) | 7.2%* |
| Triple-junction submodules (939 cm$^2$) | 9.3%‡ |
| Light-induced degradation | 10% - 15% |

* Not confirmed by SERI; † active area efficiency; ‡ Solarex submodule measured with Spire sun simulator.

It has been found that the power output of amorphous silicon cells will degrade when the cells are subjected to light exposure. This means that the initial efficiency will be reduced by a considerable fraction when subjected to operational illumination conditions. Consequently, the initial performance is irrelevant. The eventually achieved **stabilized** efficiency is the relevant parameter. Although some progress in reducing the efficiency loss resulting from illumination during normal operation has been demonstrated, further research is required to achieve stable, high-efficiency, multijunction, all-amorphous silicon modules. Thin *multijunction* amorphous silicon alloy devices have demonstrated less light-induced degradation than *single*-junction devices of comparable initial efficiencies. Further increases in stabilized efficiencies of multijunction cells and modules are needed for amorphous silicon to become cost competitive.

Same-band-gap, dual-junction submodules by the Chronar Corporation with an initial efficiency of 5% exhibited a degradation of 26% to 30%, depending on the loading conditions, after 2.7 years (961 days) of exposure to sunlight. Different-band-gap, dual-junction submodules by Solarex Inc. with initial efficiencies in the range of

5.6% to 5.7% exhibit degradation of 30% to 32% after 2 years (741 days) of exposure to sunlight (both SERI measurements).   Same-band-gap dual-junction submodules fabricated by Chronar in February 1989 having 5.3% to 5.9% initial efficiencies degraded as little as 22% and as much as 33% after 13 months of outdoor exposure (SERI measurements).  However, some recent (1990) *triple*-junction, different-band-gap cells of 9% to 10% initial efficiency have shown light-induced degradation as low as 10% to 11%  for one year's equivalent insolation [Catalano, 1990]. *Triple*-junction (Si/Si/SiGe), different-bandgap Solarex submodules have shown as little as 15% degradation from an initial efficiency of over 9% after 1,000 hours of exposure under open-circuit conditions (Solarex measurements).    Apparently, isolated defects contribute to a higher rate of degradation in large-area modules as compared to small-area cells.   Thus, the stabilized submodule efficiencies have increased from about 3.5% to more than 7.5% in the 1987 to 1990 time period.  No further *intrinsic* (that is, light-induced) degradation with illumination time is expected for these units because of thermal annealing during summer months [Jennings and Whitacker, 1990; Atmaran, Marion and Herig, 1990].

Although the intrinsic material instability in amorphous silicon films was observed as early as 1977, it was felt that the research should first concentrate on producing better materials and higher device efficiencies and deal with the instability separately. **These approaches have now been integrated into one approach.**   A third amorphous silicon research program cost-shared between the U. S. government  and industry has been initiated for the 1990 to 1992 time span.  The primary purpose of this three-year program is to concentrate research and development activities on all-amorphous-silicon alloy, two-terminal, multijunction, large-area modules (at least 900 cm$^2$), to demonstrate by FY 1993 12% stabilized, reproducible aperture-area efficiency for *different*-band-gap modules   and 10% stabilized, reproducible aperture-area efficiency for *similar*-band-gap modules.

## RESULTS OF RESEARCH ON LIGHT-INDUCED DEGRADATION

Research on the causes of light-induced degradation has been ongoing since the discovery of the effect [Staebler and Wronski, 1977], and numerous papers on the subject have been published.  This research has been addressing amorphous silicon alloy *materials* and *devices*, the latter research being mostly empirical.

To date, the *device* research has been the most successful.  The proven techniques yielding increased stabilized output are the following [Luft et al., 1990]: 1) thinner i-layers [Bennett and Rajan, 1988], 2) maximizing optical absorption, 3) better p+ and transition layers, 4) optimizing transparent conductive oxide layers, and 5) compositional grading [Guha et al., 1988].

Subsequent implementation of the above techniques (1 through 4) has resulted in 10% stable cell efficiency in 1990 [Ichikawa et al., 1990].  Device modeling to further reduce the light-induced degradation has been hampered by the lack of numerical data for materials parameters such as lifetime, mobility, and capture-rate constants.

Experimental work on solar cells has established that degradation of the i-layer poses the most serious problem to cell performance after light soaking. Degradation of doped contact layers and buffer layers may affect the open-circuit voltage and the performance of n/p tunnel junctions in multijunction cells.

The impact of the *materials* research in terms of improving the stabilized device efficiencies and/or minimizing the light-induced degradation of solar cells is more difficult to assess.

Various sources affecting the instability in amorphous silicon have been identified: light (photons), high-energy particles, quenching at high temperature, applied bias voltage, and current injection. All of these effects are reversible by annealing to a sufficiently high temperature, and all are believed to cause the same degradation mechanism [Wronski, 1990].

Several models for the light-induced degradation have been proposed, both phenomenological and microscopic/structural. Although the light-induced degradation effect has been studied in detail, total validity of any model has not been demonstrated. More importantly, there is as yet no convincing demonstration that any modification leading to an increased stability of the i-layer material can lead to increased stabilized output in solar cells. For the time being, we have to conclude that degradation and the initial performance cannot be addressed separately. So far, the optimized stabilized performance has been obtained with materials that also yield the best initial performance. It appears that when a material is produced under "non-optimum" conditions, any potential gain in stability is overwhelmed by a loss in initial performance.

Although not necessarily mutually exclusive, three models have been offered to explain the causes of instability in amorphous silicon:

1)   Weak Si-Si bonds [Stutzmann et al. 1985] or Si-H bonds may be broken by the energy released by nonradiative carrier (hole-electron) recombination.

2)   Changes of charged states of dangling bonds have been invoked. These charged states may be found in an "appropriate" local environment, such as at sites affected by C, O, or N contaminants or at sites having H-clustering or excessive local stresses [Bar-Yam et al., 1986; Branz and Silver, 1990].

3)   In *doped* a-Si:H, light-induced changes have been explained by changes in dopant coordination [Branz, 1988], by the creation of midgap states, or by a combination of both.

All the above models may involve the presence of hydrogen in the material. Experimental correlations have been established between the kinetics of annealing Staebler-Wronski defects and hydrogen motion (H diffusion coefficient). This has led some researchers to attempt to find a clue for the light-induced effect by manipulating the hydrogen content or bonding. However, it presently is not clear whether hydrogen is the *driving* force behind the effect, or whether electronic defects present may determine the mobility of hydrogen. In essence, which is the cause and which is the effect?

Redfield and Bube [1989, 1990] attempt to describe the degradation process using simple differential rate equations, based on the following considerations: (1) the number of sites at which light-induced defects can be generated is fixed for a given material, and (2) the number of convertible sites increases because of contamination and doping. The fixed number of convertible sites determines the saturated defect level at low temperatures, which appears to be temperature independent below 70°C and, thus, not determined by a competitive annealing process. While this attempt appears capable of phenomenologically describing the kinetic behavior of the degradation, it lacks the capability to lead research to clues as to the native or microscopic structure of the convertible sites.

Interesting results have been reported for heavily compensated materials [Stutzmann, 1990]. On the time scale reported (20 hours of light soaking), very few light-induced neutral dangling bond ($T_3^o$) defects are observed while the initial (native) $T_3^o$ density remains low. These results are difficult to reconcile in terms of any of the models presented above.

The majority of research to date has concentrated on optimizing the properties of the initial (as-grown) state. In many instances the initial material parameters found to be important for the optimization process were used to model solar cell performance. There has been a lack of correlating the **stabilized** materials parameters to the **stabilized** device efficiency. There has also been a tendency to dwell on the esoterics of differences in models without investigating how these differences would affect device improvements. The time has come to apply the fundamental knowledge of materials to improve the stabilized **device** performance. The research must now be narrowly focused to meet the U. S. National Photovoltaic Program goals.

## AMORPHOUS SILICON RESEARCH NEEDS

The national amorphous silicon program in the United States needs definitive answers to a number of outstanding questions, such as: 1) For device-quality materials, are there important cause-and-effect relationships between *material* quality indicators, such as ESR spin density, photoconductivity, and electron or hole life time, and *device* quality indicators, such as open-circuit voltage, short-circuit current density, and fill factor? 2) Provided there is a correlation, what are the limits of these main material quality indicators at which they cease to be of significance for device performance?  3) Is the degradation inherent and intrinsic to amorphous silicon, and if so, can the degree of instability be reduced?  4) Is the *change* in material quality indicators resulting from light soaking reflected in a corresponding change in device quality indicators?  Are there correlations of importance that have been overlooked?  5) What are the *preferred* instability indicators?  Is there only one or are there several?  Should they be *device* quality indicators?  If so, should they be relative changes or absolute changes.  6) How do light-induced changes in doped layers or at interfaces affect the stabilized device performance?  How can the influence of doped layers and interfaces in a device be separated from changes in the intrinsic layers?  7) Are the currently used optimization procedures for best initial material properties necessary or sufficient for optimized stabilized device performance?

In addition to these fundamental questions there are a number of specific questions for cases where there is inconclusive evidence about the answers offered [Luft et al., 1990].

### RESEARCH DIRECTION IN THE UNITED STATES

While the directions presented below relate specifically to the DOE/SERI-supported amorphous silicon research program, these directions should be applicable to all researchers interested in advancing the amorphous silicon photovoltaic technology.

- It is in the interest of the U.S. amorphous silicon research program to achieve the highest possible **stabilized** efficiency for amorphous silicon modules. Here, the term "stabilized" refers to the performance level reached after at least 600 hours of exposure to 1 sun's illumination or its equivalent at an operating temperature of about 50°C.

- The research in the near term will be directed toward satisfying the program goals for the mid-1990s: stabilized module efficiencies of 10% at a cost of $90/m^2$ with life expectancies of up to 30 years, to reach a levelized electricity cost target of 12 cents/kWh (1986 dollars).

- The research should be directed toward answering the questions enumerated in the previous section.

- All research should consider that its contribution will ultimately be used for multijunction device applications, because devices with thin i-layers degrade at a lower rate and by a lesser amount than those with thicker i-layers.

- Investigations of degradation phenomena should concentrate on light-induced instability and set aside a fundamental investigation of the other degradation inducers, such as high-energy particles, quenching, or bias voltage, unless such work can help to understand light-induced degradation.

- Regarding the Staebler-Wronski effect, the **fundamental** research should be directed primarily toward a-Si:H instability, leaving light-induced degradation of a-SiGe:H for later consideration. This direction in no way detracts from the importance of improving low-band-gap a-SiGe:H alloy properties.

- Any materials investigations should be complemented with device investigations. Researchers are encouraged to establish collaborations to accomplish this goal. Claims about good-quality materials with increased stability have little impact unless the materials have been tested in a device configuration.

- Groups that work simultaneously on materials and device improvements should attempt to better separate the contributions to improved stabilized output as being due to either the material or device optimization. If these contributions are confused, other workers could be led to conduct needless research.

## CONCLUSIONS

While the stabilized module efficiencies have increased from about 3.5% to more than 6% between 1987 and 1990, higher stabilized efficiencies are required. There are a number of important questions concerning the light-induced instability that must be addressed in the near future. Especially urgent are answers regarding preferred instability indicators and regarding the relationships (if any) between material and device quality indicators. Another question relates to the optimization methodology that should be used in amorphous silicon research. Resolution of many of these questions can be accomplished through focused research. For that purpose the U.S. amorphous silicon research program has defined specific research directions for the next three years. Among them are stabilized efficiency goals, multijunction devices, complementary materials and device investigations, and a fundamental instability research focused on a-Si:H. Mitigation of the light-induced instability will have a significant impact on increasing the stabilized performance of amorphous silicon-based modules and will improve the probability of meeting the cost goals for thin-film flat-plate modules.

## ACKNOWLEDGEMENT

This work was supported by the U.S. Department of Energy under contract DE-AC02-83CH10093.

## REFERENCES

Atmaran, G. H., B. Marion, and C. Herig, "Performance and Reliability of a 15 kWp Amorphous Silicon Photovoltaic System", Proc. 21st IEEE PV Specialists Conference, 1990, pp. 821-830.

Bar-Yam, Y., D. Adler, and J. D. Joannopoulos, "Structure and Electronic States in Disordered Systems", Phys. Rev. Letters, **57**, (1986) 467.

Bennett, M. S., and K. Rajan, "The Stability of Multijunction a-Si Solar Cells", Proc. 20th IEEE Photovoltaic Specialists Conference, IEEE, New York, 1988, p. 67.

Branz, H. M., "Charge Trapping Model of Metastability in Doped Hydrogenated Amorphous Silicon", Phys. Rev. B, **38** (1988) 7474.

Branz, H. M., and M. Silver, "Potential Fluctuations Due to Inhomogeneity in Hydrogenated Amorphous Silicon and the Resulting Charged Dangling Bond Defects", Phys. Rev. **B**, **42** (1990) 7420.

Guha, S., J. Yang, A. Pawlikiewicz, T. Glatfelter, R. Ross, and J. R. Ovshinsky, "A Novel Design for Amorphous Silicon Alloy Solar Cells", Proc. 20th IEEE Photovoltaic Specialists Conference, IEEE, New York, 1988, p. 79.

Ichikawa, Y., S. Fujikake, T. Yoshida, T. Hama, and H. Sakai, "A Stable 10% Solar Cell with a-Si/a-Si Double-Junction Structure", Proc. 21st IEEE PV Specialists Conference, 1990, pp. 1475-1480.

Jennings, C., and C. Whitacker, "PV Module Performance Outdoors at PG&E", Proc. 21st IEEE PV Specialists Conference, 1990, pp. 1023-1029.

Luft, W., B. Stafford, and B. von Roedern, "The Amorphous Silicon Program Needs with Respect to Stabilized Module Performance", SERI PV AR&D Meeting, October 1990, also in *Solar Cells*, to be published.

Redfield, D., and R. H. Bube, "Interpretation of Degradation Kinetics of Amorphous Silicon", Appl. Phys. Lett., **54** (1989) 1037 and "Identification of Defects in Amorphous Silicon", Phys. Rev. Lett. **65** (1990) 464.

Stutzmann, M., W. B. Jackson, and C. C. Tsai, "Light-Induced Metastable Defects in Hydrogenated Amorphous Silicon: A Systematic Study", Phys. Rev **B**, **32** (1985) 23.

Stutzmann, M., "Light-Induced Defect Creation in Amorphous Silicon: Single Carrier vs. Excitonic Mechanisms", Appl. Phys. Lett., **56** (1990) 2313.

Wronski, C. R., "Instabilities in a-Si:H Solar Cells: Materials and Device Issues", Proc. 21st IEEE PV Specialists Conference, 1990, pp.1487-1492.

# RESEARCH ON THE STABILITY OF A-SI:H BASED SOLAR CELLS BY SMART

C. R. Wronski
The Pennsylvania State University, University Park, PA,  16802

N. Maley
Coordinated Science Lab, University of Illinois, IL, 61801

## ABSTRACT

Stable Materials Advisory Research Team (SMART) was established to develop a unified approach to address the stability problem in a-Si:H based solar cells.  Goal of the coordinated research effort by industrial laboratory and research institution members is to resolve whether a-Si:H based materials are intrinsically unstable and if high efficiency cells can have a 20 year lifetime.  This paper reviews ongoing research which addresses both material and device issues in the effort to improve the material properties and solar cell performance.  Results are presented for materials and device structures obtained using several deposition techniques with hydrogen content varying from about 20% to 8%.  Also several issues are discussed which arose from the wide range of measurements carried out on the same materials in different laboratories.

## INTRODUCTION

Fundamental studies on the SWE have improved the understanding of the metastability in a-Si:H.  The development of both new materials and to a large extent improvements in cell design have led to a significant increase in the stability of solar cells.[1]  However, there are still two key questions related to the long term stability of high efficiency cells and their technology First, are a-Si:H based materials intrinsically unstable?  Second, can light induced degradation of high efficiency cells be held to less than the tolerable limits necessary for a 20 year lifetime?  An approach for addressing this was formulated by members of the Stable Materials Advisory Research Team (SMART).  SMART is an EPRI-organized volunteer group of researchers from 12 United States photovoltaic industry and universities who are jointly addressing these questions.  There is currently interest and organization of similar programs in Europe and Japan.  Based on the current understanding of a-Si:H based materials a consensus was reached at SMART workshops that the possible causes of metastability are:  hydrogen; impurities; inhomogeneities ($\sim 20$Å scale); and strained bonds.  It was concluded that there are still material issues which have to be resolved in order to understand the origins of these defects to a degree which can allow systematic improvements to be made in material and solar cell properties. In addition it was recognized that there is an absence of quantitative correlations between the results on materials and those on the long term stability of solar cells.

A unified approach was established on how to address these issues in a coordinated research effort. In this approach emphasis is placed on carrying out a wide range of different measurements on identical materials and establishing self consistency between the results. In addition emphasis is placed on investigating both materials and devices in order to obtain quantitative correlations between material properties and solar cell parameters and performance. It was decided to address the question of hydrogen first by investigating a-Si:H materials and devices deposited using a range of different deposition techniques to reduce hydrogen concentrations in good quality a-Si:H. Several experimental techniques are used to characterize the intrinsic and light induced defects in the materials and device structures. Correlation is also being established between results on the intrinsic materials and the devices by the appropriate modeling of the device parameters.

These collaborative studies have led to several new insights about hydrogen in a-Si:H and measurements commonly used to characterize the quality and stability of materials. Hydrogen between 15 and 8 atomic % shows interesting and beneficial trends and it is important to investigate materials with lower concentrations. Measurements of only photoconductivity are not reliable criteria for the quality and stability of material properties related to solar cells. It is also found that the extraction of the densities of the midgap and valence band tail states from sub-bandgap photoconductivity has to be improved and standardized.

## MATERIALS AND MEASUREMENTS

The a-Si:H films and device structures were deposited using PACVD (RF and DC), photo CVD and reactive sputtering. The hydrogen content has thus far been varied between 20% to 8%. Because of the focus on hydrogen it was important to establish an accurate and self consistent methodology for characterizing the hydrogen in the materials. The methodology developed by the University of Illinois was chosen for this purpose[2]. One $\mu$m thick films were codeposited on c-Si and Corning 7059 substrates for measurements of: IR absorption; optical absorption; spectroscopic ellipsometry; SAXS; photoconductivities; sub-bandgap absorption (CPM, PDS, dual beam photoconductivity); ESR; and diffusion lengths. Saturation studies of the light induced changes in these films are based on the methodology developed by Princeton University[3]. In addition, $n^+$ - intrinsic and p-i-n device structures with the same intrinsic layers were also codeposited on specular tin oxide. The p-i-n structures <u>did not</u> include any technologies developed to optimize their efficiencies, but consisted of the simplest possible p-i-n devices in order to correlate the material properties with cell parameters.    The device characteristics of the metal n-i Schottky barrier structures were characterized by I-V and quantum efficiency measurements (QE)[4]; the p-i-n cells by I-V, QE and cell parameter measurements.

The procedure agreed on for establishing the annealed state of the materials and devices consisted of an anneal at 170°C for 10 hours followed by slow cooling to room temperature.    The light induced changes were measured using both AM1, or equivalent red light, illuminations for 10 and 100 hours as well as higher intensities to obtain accelerated degradation. Several of the commonly used measurements were duplicated in different laboratories in order to establish a baseline for the uncertainties that can be expected in reported values.    Detailed numerical modeling was undertaken in order to correlate the measured material properties with solar cell parameters.    Although there have been many results reported on the effects of hydrogen on a-Si:H and its solar cells, there have been no studies on materials prepared using different deposition techniques where concurrent measurements were carried out on the same materials and devices in different laboratories.

## HYDROGEN

The hydrogen content of the films was determined by infrared absorption.    Thermal evolution measurements on randomly chosen films yielded consistent results. The infrared data analysis incorporated a) thickness dependent corrections where necessary[2] b) the use of analytic expressions to remove interference fringes and c) oscillator strengths measured by Langford et al.[5] to determine the total hydrogen content (640 cm$^{-1}$ mode) and the amount in the isolated and clustered phases (2000 and 2100 cm$^{-1}$ modes, respectively).    Figure 1 shows the hydrogen content of the 2000 and 2100 cm$^{-1}$ modes as a function of the total hydrogen content for sputtered and PACVD films.    The total hydrogen content determined from the wagging and stretching modes is in good agreement except for a few films which were deposited on single side polished c-Si substrates.

The improvements in IR analysis have increased the sensitivity to observe the 2100 cm$^{-1}$ mode.    Figure 1 shows that there is a measurable amount of hydrogen in the clustered phase even for $C_H$ < 10%.    In fact, rather than vary linearly above a threshold, the amount of hydrogen in the clustered phase seems to vary as the square of the total hydrogen content.    It is also important note that despite the wide range of deposition conditions and methods used to produce these films, the amount of hydrogen in the isolated and clustered phases seems to depend only on the total hydrogen content.    It appears that by reducing the total hydrogen to about 3-4% or less will further reduce the clustered phase which is considered desirable for improving the light-induced degradation.

## OPTICAL PROPERTIES

The optical gaps, $E_{opt}$, were measured using optical transmission and reflection, and PDS. As expected, the optical bandgaps decrease as the hydrogen content is lowered but the values for $E_{opt}$ obtained in the different

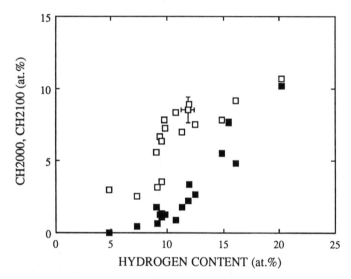

Figure 1. The content 2000 cm$^{-1}$ (open symbols) and 2100 cm$^{-1}$ (full symbols) hydrogen modes versus total hydrogen content for wide range of a-Si:H films.

laboratories had a spread of up to 30meV. This can be attributed to the problems associated with the averaging of the interference fringes in the optical measurements and whether $E_{opt}$ is determined from reflection and transmission or transmission alone. The results though clearly showed that the relation between hydrogen and the optical gap is not unique but depends on the deposition conditions, implying that the optical properties are determined by both hydrogen as well as the changes in the Si-Si network when the deposition conditions are modified.

## MICROSTRUCTURE

The microstructure of the a-Si:H films was characterized using spectroscopic ellipsometry[6]. Because of its sensitivity to detect microvoid differences of 1%, this was felt to be a useful method of quantifying even small changes in the structure of a-Si:H which are related to the Si-Si packing densities in high quality materials. Systematic changes were observed in the microstructure of the a-Si:H films with hydrogen content where this is illustrated in Fig. 2 with results obtained for films deposited by RF and DC plasma CVD and reactive sputtering. The volume void fraction is given relative to that of a PACVD a-Si:H film (deposited at 300° C) which had a volume void fraction similar to that of low pressure CVD a-Si.[6] There are clear trends in Fig. 2 which reflect an increase in the a-Si:H density with decrease in hydrogen but it is also important to note that there are differences in the microstructure for films having the same hydrogen contents. Care must therefore be taken in separating the role of hydrogen from that of the changes in deposition conditions, particularly for materials with less than 10 atomic % of hydrogen.

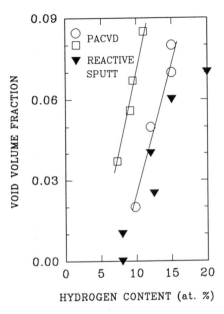

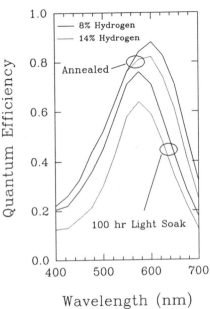

Figure 2. The volume void fraction of PACVD and reactively sputtered films as of hydrogen content.

Figure 3. The short circuit internal quantum efficiency of Photo CVD Pd-i-n structures before and after soaking for 100 hrs with 100 mw/cm$^2$ light.

## DEVICE STRUCTURES

The studies on device structures fabricated using the same intrinsic layers have so far been not as extensive as those on the thin films and are still in progress. However results on Schottky barrier, metal -i-n structures on both PACVD and Photo CVD materials showed that their stability is affected by the hydrogen content. This is clearly indicated by the light induced degradation in their carrier collection efficiencies such as illustrated in Fig. 3 for Pd Schottky barrier cell structures formed with photo CVD materials having 14 and 8 atomic % hydrogen. The results in Fig. 3 clearly show that degradation in the short-circuit quantum efficiencies, after illumination with 100 mw/cm$^2$ white light for 100 hours, is significantly smaller in the case of the lower hydrogen material. This encouraging trend and the direct correlation between the results on the device structures and those on the thin film materials is currently being investigated further. Using detailed modeling of device parameters[4], progress is being made in understanding and quantifying the relationship between intrinsic material and solar cell properties.

## PHOTOCONDUCTIVITIES

Photo and dark conductivities with their ratio are extensively used to characterize both the quality of a-Si:H as well as its and stability. Photoconductivity measurements from different laboratories showed very good agreement when the same generation rates of carriers, f, were used. The films however exhibit a wide range of $\gamma$ values. There were also large differences in the values of dark conductivities, $\sigma_D$, particularly for values in the $10^{-11}$ $\Omega$ cm$^{-1}$ range which did not allow the ratio $\sigma_L/\sigma_D$ to be accurately quantified. Because of the differences in the optical absorption the mobility-lifetime ($\mu\tau$) products must be used to characterize the lifetimes. No unique trends were found in the changes of $\mu\tau$ with hydrogen. In a series of PACVD films $\mu\tau$ (f = $10^{20}$ cm$^{-3}$ s$^{-1}$ ) increased from 5 x $10^{-8}$ to 1 x $10^{-6}$ cm$^2$/V as the hydrogen concentration decreased from 19 to 9%. However, such an increase of $\mu\tau$ with decrease in hydrogen is not a general characteristic. In direct contrast the reactively sputtered films exhibit a systematic decrease over the same range of $\mu\tau$ and hydrogen values.

Although no extensive studies were carried out on the kinetics of the changes in $\sigma_L$ with light soaking, very good agreement was obtained between the values of $\sigma$(annealed))/$\sigma$ (soaked) using 100 mw/cm$^2$ ELH white light and volume absorbed red light. Comparisons were also made between accelerated degradation (400 mw cm$^2$) and standard illuminations based on kinetics having an (intensity)$^2$ dependence. Although agreement was obtained in some cases, there was a significant scatter (up to a factor of 2) between the results. Thus far it has not been possible to directly relate the light induced changes in the photoconductivities to the hydrogen content or the densities of midgap states. The studies, however, do indicate that photoconductivity results by themselves cannot adequately characterize the quality and stability of different a-Si:H material

## DENSITIES OF STATES

Midgap densities of states were measured primarily using sub-bandgap photoconductivities normalized to the optical absorption ,$\alpha$, and by the integration of $\alpha(h\nu)$ between about 0.8 and 1.4eV (using an integration constant[7] 1.9 x $10^{16}$. Photothermal deflection spectroscopy (PDS) and ESR measurements were also carried out on a number of these films. The PDS results on annealed films exhibited subgap absorption up to an order to magnitude higher than that obtained from sub-bandgap photoconductivity. This clearly showed that surface states dominate these results when the midgap densities of states are low ($< 10^{16}$ cm$^{-3}$). Similarly large contributions of surface states were also found in the measurements of dangling bonds using ESR. Sub bandgap photoconductivity measurements, which reflect the bulk properties, can be used for characterizing a-Si:H both in the annealed and light soaked states. However, their interpretation needs to be improved as discussed later.

The densities of midgap states was in the range 1 x $10^{15}$ to 6 x $10^{16}$ cm$^{-3}$. For a given hydrogen content the sputtered films had lower bandgap and higher densities of states. After 100 AM1 or equivalent light soaking and upon saturation there was no clear dependence of DOS on the hydrogen content or bandgap.

Comparison of sub-gap absorption data from different laboratories obtained by CPM and dual beam photoconductivity (DBP)[8] revealed three important features. First, interference fringes in optical absorption spectra in the photon energy range where CPM and DBP data are normalized to them can lead to large errors in the estimated mid-gap defect density (factor of 3). When the same normalization procedure is used, CPM and DBP spectra obtained in different labs showed very good agreement. Nevertheless the differences in the deconvolution led to a 3-5 meV variation in $E_o$ and factor of 2 variation in densities of midgap states. With light soaking most of the samples also showed a small increase in $E_o$, but this result is not definite at this stage. All these factors have a strong influence on the reliability of sub-gap absorption spectra to evaluate film quality and stability. The SMART group is now attempting to establish voluntary guidelines to make these measurements more reliable.

## SUMMARY

Studies carried out by SMART have investigated the role of hydrogen in the stability of a-Si:H materials and their solar cells. Results obtained on materials fabricated using various deposition conditions indicate that the reduction of hydrogen between about 15 and 8 atomic % has beneficial effects. Because the changes in the hydrogen content are achieved by modifying the deposition techniques from those normally used to obtain high efficiency solar cells, it is not easy to establish systematic and quantitative correlations which are directly related to hydrogen. Nevertheless, some systematic changes in the properties are observed and further reduction of hydrogen in a-Si:H is being investigated. It appears, however, that in these studies it is very important to distinguish between the role of hydrogen per se and those of changes in the deposition conditions.

The multilaboratory studies have raised several issues related to the current methodologies used to characterize the quality and stability of a-Si:H materials. It is found that it is not possible to rely on only photoconductivity (dark conductivity) measurements for quantifying the quality and stability of different materials. It is also found that it is necessary to improve and standardize the methodologies used to measure the densities of midgap states in different a-Si:H materials. This issue as well as the direct correlation of the material properties to the performance and degradation of solar cells is being further addressed by SMART.

The efforts to improve the material properties and stability of a-Si:H and alloys are worldwide. It is thus important to establish a set of measurements to facilitate easy and reliable comparison of samples grown and characterized in different laboratories. To this end it is desirable to measure as many parameters as possible, through collaborative efforts, if necessary. Structural properties required to evaluate materials include not only hydrogen content, its clustering but also the microstructure. The quality and stability of materials has to be characterized not only with $\mu\tau$ products, but also the midgap and band tail states densities of states for after annealing and prolonged (saturated) degradation. In addition, more emphasis should be placed on characterizing the corresponding device structures which are not optimized for high efficiency.

We wish to acknowledge the contributions from the various members of the SMART program. This work was supported by the EPRI Thin Film Solar Cell Program.

## REFERENCES

1.    C. R. Wronski, Conf. Record of the 21st IEEE Photovoltaic Specialists Conf. (IEEE 1990).

2.    N. Maley and I. Szafranek, MRS Proceedings, Vol. 192, 673 (1990).

3.    H. R. Park, J. Z. Liu, and S. Wagner, Appl. Phys. Lett., 55, 2658 (1989).

4.    P. J. McElheny, S. Nag, S. J. Fonash, and C. R. Wronski, Conf. Record of the 21st IEEE Photovoltaic Specialists Conf. (IEEE 1990).

5.    A. A. Langford, B. P. Nelson, M. H. Flect, W. A. Lanford, and N. Maley (to be published).

6.    R. W. Collins, "Amorphous Silicon and Related Materials", H. Fritzche Ed., (World  Scientific Press, Singapore) p. 1003.

7.    Z. E. Smith, V. Chu, K. Shepard, S. Aljashi, D. Slobodin, J. Kolodzy, S. Wagner, and T. L. Chu, App. Phy. Lett., 50, 1521, 1987.

8.    S. Lee, S. Kumar, and C. R. Wronski, J. Non-Cyrst. Solids, 114, 316 (1989).

# PART II

# MODELING
# METASTABILITY

# METASTABILITY AND THE HYDROGEN DISTRIBUTION IN A-SI:H

R. A. Street
Xerox Palo Alto Research Center, Palo Alto, CA 94304

## ABSTRACT

Studies of metastability relate defect creation to the motion of hydrogen between different bonding sites. The mechanism is discussed in terms of a hydrogen density of states distribution, whose general features are obtained from simple chemical models of hydrogen bonding and weak Si-Si bonds. The model relates metastability to the equilibrium defect density and to the hydrogen diffusion.

## INTRODUCTION

The phenomena of defect creation in a-Si:H by illumination has been studied since 1977.[1] The phenomenology of the effect is well documented and several models have been proposed to account for the observations.[2-7] So far, however, none of this work has resulted in a successful strategy for eliminating the effects in solar cell devices. Indeed, some models attribute the defect creation to the intrinsic structure of the film, suggesting that the effects cannot be removed without a radical change in the growth process.

Light-induced defects are one of a broad class of metastable defect creation processes.[8] Other examples include defects induced by quenching from high temperature, charge accumulation at interfaces, doping, electric current, etc.[9-11] The effects are so pervasive that they involve almost every aspect of the material, and all a-Si:H electronic devices. Metastability results from a thermally activated recovery, such that the relaxation rate,

$$\tau_R = \omega_0^{-1} \exp(E/kT) \qquad (1)$$

is very long at room temperature. Thus, the defects remain when the creation source is removed, but are annihilated by annealing at elevated temperatures (typically 100-200°C). Fig. 1 summarizes defect creation and annihilation processes which can be by internal thermal excitation, or by an external excitation such as recombination, trapping or a shift of the Fermi energy.

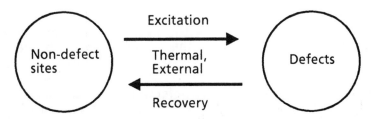

FIGURE 1. Illustration of the reversible transfer between non-defect and defect sites by different excitation mechanisms.

The metastable effects may be grouped according to their type as follows.

REVERSIBLE NON-EQUILIBRIUM EFFECTS. These occur when there is a non-thermal source of energy to create defects, such as trapping or recombination of electrons and holes. There is a low thermal activation energy for defect creation because part of the energy arises from the recombination. Examples are light-induced defects and current-induced defects.[1,2,10] These have a slow defect creation rate, and it is not yet resolved whether stretched exponential or a power law is the more appropriate description of the creation kinetics.[2,6] Saturation of the defect density occurs at long excitation times.

An example of this type of metastability is given in Fig 2 which shows recent room temperature defect creation in p-i-n devices induced by a forward bias current.[12] The defect creation rate exhibits approximately

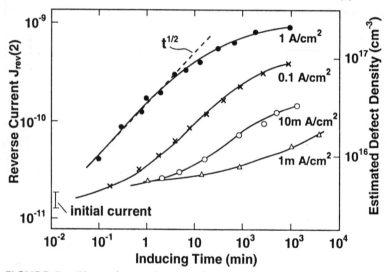

FIGURE 2. Time dependence of the current-induced defect density (right axis) for different inducing currents as indicated.

$t^{1/2}$ kinetics with a rate dependent on the current, and saturation at long inducing times. Recombination is the likely mechanism, since both electrons and holes contribute to the current. Space-charge limited currents in p-i-p structures induce defects presumably by the trapping of holes.[10]

REVERSIBLE EQUILIBRIUM EFFECTS. Defects are created thermally in an equilibrium process. Examples are thermally- and bias-induced defects, and defects in doped a-Si:H.[4,9,11] There is a high activation energy (~1 eV) for creation and annealing, and both rates have stretched exponential kinetics. Quenching from high temperature freezes in a high defect density, which is reduced by annealing at a lower temperature. A shift of the Fermi energy associated with charge accumulation at an interface or doping, reduced the defect formation energy and creates defects.

Dopants also have metastable properties and a temperature dependent doping efficiency.

IRREVERSIBLE CHANGES. There are thermally-induced changes in the defect density associated with substantial structural or compositional changes in the a-Si:H film. Examples are annealing of low substrate temperature films and hydrogen evolution and rehydrogenation. The exact relation of these effects to the reversible changes remains unclear.

## HYDROGEN MEDIATED METASTABILITY

The three most often cited models for metastability are that it results from a change of electron occupancy of defects which are always present;[5] that it is associated with structural changes of either silicon or impurities;[7] and that it is a result of the redistribution of bonded hydrogen.[3,4,8] The remainder of this paper discusses the last of these models.

There is strong circumstantial evidence for the role of hydrogen, although a complete proof is still lacking. The evidence stems primarily from the observations of hydrogen motion at the same temperature as the metastability effects. The activation energy of defect annealing is comparable with that of hydrogen diffusion. Furthermore, the doping trends are the same - dopants which result in a larger hydrogen diffusion coefficient also lead to faster defect relaxation. In addition, the stretched exponential form of the relaxation is in quantitative agreement with the dispersive time dependence of the hydrogen diffusion. Finally, the observation of hydrogen diffusion implies that the hydrogen distribution is in equilibrium, and therefore that any defect associated with the breaking of a Si-H bond are also in equilibrium. Thus the defect equilibration is a predicted consequence of the hydrogen diffusion.

## HYDROGEN CHEMICAL REACTIONS

The model assumes that defects are created by hydogen migration between alternative bonding sites, specifically Si-H and Si-Si bonds. The reaction,

$$\text{Si-H} + \text{Si-Si} \leftrightarrows \text{Si}\bullet + \text{Si-H} \bullet \text{Si} \qquad (2)$$

corresponds to the transfer of one hydrogen and the creation of two defects, both of which are dangling bonds (denoted Si$\bullet$) but in different local environments. A second reaction,

$$\text{Si-H} + \text{Si-H} \bullet \text{Si} \leftrightarrows \text{Si}\bullet + \text{Si-HH-Si} \qquad (3)$$

results in both defects at Si-H sites and the Si-Si bond broken but fully passivated by the two hydrogen atoms. The energy to insert hydrogen into a Si-Si bond is quite large and is accompanied by a significant relaxation of the silicon atoms. The energy is reduced if the bond is suitably distorted, leading to the idea of preferential hydrogen insertion into weak bonds. Reaction 2, when applied to a distribution of weak bonds is one manifestation of the weak bond-dangling bond conversion model. Street and Winer show that the equilibrum defect density depends on the details of the reactions, and find that reaction 2 agrees well with the observed equilibrium defect density in undoped a-Si:H.[9]

Reaction 3 shows how an Si-Si bond can satisfy its bonding requirements by incorporating two H atoms. The combined reaction,

$$2\,\text{Si-H}\bullet\text{Si} \leftrightarrows \text{Si-Si} + \text{Si-HH-Si} \tag{4}$$

may be exothermic, because the configurations on the right side are low energy states containing no unsatisfied bonds, provided that the network is sufficiently flexible to incorporate the two hydrogen atoms. This confers the Si-Si bond with the property of a negative hydrogen correlation energy (negative $U_H$), analogous with negative U electronic defects sometimes found in the presence of a large lattice relaxation.[13,14]

### THE HYDROGEN DENSITY OF STATES

Chemical reactions become inconvenient when the energy states are broadened by disorder. We have recently suggested that the hydrogen density of states (HDOS) is a useful representation of the chemical reactions, and is illustrated in Fig. 3a. The vertical axis is the energy of

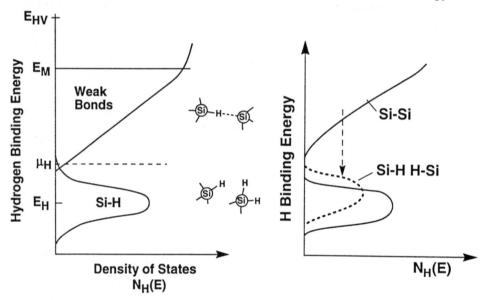

Fig. 3. (a) Illustration of the HDOS distribution showing the Si-H bonding states centered at energy $E_H$, the migration enrgy $E_M$, the chemical potential, $\mu_H$ and the energy $E_{HV}$ of hydrogen in vacuum. (b) Illustration of the HDOS for two-hydrogen states with large structural relaxation.

hydrogen in the various configurations, and the horizontal axis is the number $N_H(E)$ of states at energy E. Fig. 3a shows Si-H states which are occupied by hydrogen apart from the small density of Si dangling bonds, and the Si-Si bonds which are empty of hydrogen except for the few defects of the Si-H $\bullet$Si type. The HDOS is referenced to the energy $E_{HV}$ of atomic hydrogen in a vacuum, and the energy of hydrogen migration at $E_M$ is estimated to be 0.5-1 eV below $E_{HV}$. Negative $U_H$ states can be included in the HDOS, as illustrated in Fig 3b, and consists of a band of Si-HH-Si below the singly occupied weak bond states.

The equilibrium hydrogen distribution is described by a chemical potential $\mu_H$ which separates the occupied and unoccupied states. The hydrogen difusion and the defect density are both given in terms of $\mu_H$ and $N_H(E)$. The defect density is minimized when $\mu_H$ lies in the minimum $N_H(E)$ between the Si-H bonds and the Si-Si bonds. It seems plausible that chemical reactions during growth cause $\mu_H$ to be positioned such that the defect density is minimized. A calculation of the temperature dependence of the defect density, making this assumption, and agree closely with the analysis based on reaction 2, as expected because the model is essentially the same.

The HDOS is a convenient way of comparing different models. For example, Fig. 4 illustrates the models proposed recently by Zafar and Schiff and by Jackson.[13,17] Both models invoke negative $U_H$ weak bonds, but

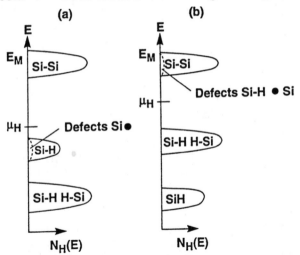

Fig. 4. Alternative models for the ordering of Si-H and weak bond states, proposed by (a) Zafar and Schiff and (b) Jackson.[13,17]

differ in the assumptions made of the ordering of the states. Jackson places the Si-H bonds below Si-HH-Si, with the result that the Si-H bonds do not contribute to the dangling bonds, which are instead the singly occupied Si-H •Si states. In contrast, Zafar and Schiff place Si-H between the singly and doubly occupied weak bond states, and attribute the defects to liberation of the hydrogen from Si-H. It is important to understand the bonding distribution better and its implications to diffusion and defects. Experiments are needed to establish $N_H(E)$.

## LIGHT INDUCED DEFECTS

The HDOS approach leads to a new description of metastability which focusses attention on the hydrogen distribution. The equilibrium defect density is given by the Fermi distribution of hydrogen and is the sum of the unoccupied Si-H states and the singly occupied weak bonds. Non-equilibrium metastability follows from the excitation of hydogen into a non-equilibrium distribution, as is illustrated in Fig. 5. The recombination of an electron-hole pair is assumed to excite hydrogen off a Si-H bond into the distribution of weak bonds. Deep trapping of the hydrogen at the weak bonds corresponds to metastable defect creation, and annealing allows the hydrogen to migrate and return to its original distribution. These processes are analogous to electronic photoconductivity and trapping of carriers in deep states. Thus we anticipate describing the process in terms of excitation, trapping and recombination steps, but of hydrogen atoms instead of electronic carriers.

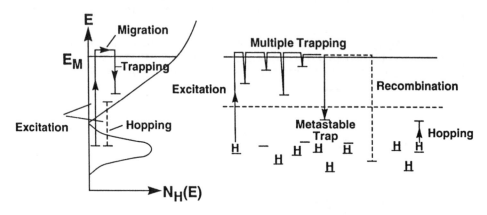

Fig. 5. illustration of possible hydrogen excitation, trapping and recombination mechanisms leading to light-induced defect creation, as described in the text.

### Excitation

There is good evidence that defect creation results from the energy released by the recombination of an electron-hole pair or the trapping of a single carrier,[2] but the details of the mechanism are not yet completely clear. A phonon mechanism is usually assumed in which the recombination energy is transferred to vibrational excitations. Here it is assumed that the energy excites hydrogen from an Si-H bond to a weak bond state. Additional thermal excitation may be needed if the recombination energy is insufficient to break the bond. For example the hydrogen might be excited out of a state which already has considerable vibrational excitation. The additional thermal energy should be observed in the activation energy of the excitation rate. Light-induced defect creation has a low activation energy, indicating that the recombination energy is almost sufficient to release hydrogen, as might be expected since the hydrogen diffusion energy (1.5 eV) is less than the band gap energy. Defect creation by trapping of a single carrier releases a smaller energy and is expected to have a larger activation energy.

### Hydrogen trapping and defect formation

The excited hydrogen must reach a trapping site of high binding energy to form a metastable defect. Figure 5 shows two possible mechanisms of hydrogen trapping. The hydrogen is free to migrate away if it is excited into the mobile state at $E_M$, provided that immediate recombination does not occur. We anticipate that the subsequent migration follows a multiple trapping mechanism within the distribution of weak bonds, until it reaches a trap which is sufficiently deep that it is not released within the time scale, t, of the experiment. A demarcation energy, $E_D$, can be defined, as in the multiple trapping model of electronic carriers,

$$E_D = E_M - kT \ln(\omega_0 t) \tag{5}$$

$E_D$ defines the minimum trap depth for which release does not occur within time t. $E_D$ is about $E_M - 1$ eV at room temperature and a time of $10^3$-$10^4$ sec.

The multiple trapping model suggests that hydrogen migrates a substantial distance before finding a metastable site. Assuming an exponential tail to the distribution of weak bonds of slope $E_0 = 0.1$ eV,[9,15] then the fraction of states further from $E_M$ than $E_D$ is $f \sim \exp(-E_D/E_0)$. The average number of migration steps is $1/f$ for which the diffusion distance, $R_D$, is

$$R_D = af^{-1/2} = a \exp(E_D/2E_0) \sim 500\text{Å} \tag{6}$$

where a is the mean free path for hydrogen capture, taken to be 3Å.

An alternative trapping mechanism aplies when hydrogen makes a direct hop to a weak bond site, as illustrated in Fig. 5. The Si-H bond is assumed to be immediately adjacent to the trapping site; an example is the model proposed by Stutzmann, Jackson and Tsai, in which a hydrogen atom remains attached to the same Si atom, but rotates into a weak bond.[2] This type of mechanism has the advantage of probably requiring a lower excitation energy (although there must be a barrier to prevent immediate recombination), but presumably can only occur at a small fraction of all the sites. The observation of light-induced hydrogen migration would distinguish between these two mechanisms.

### Recombination

The analogy between this metastability model and electronic photoconductivity leads us to expect that a steady state defect density is eventually reached when excitation and recombination rates balance. It is not necessary that all the available defect sites are occupied, and indeed the occupancy fraction might be very low. Recombination is illustrated in Fig. 5, and occurs when the migrating hydrogen binds to a dangling bond. The steady state concentration of defects depends on the relative capture cross-sections for trapping at dangling bonds and weak bonds, neither of which are known, and the generation rate. This approach suggests that the rate equation for the defect density, $N_D$, is of the form,

$$dN_D/dt = \beta_{cr} N_H R - \beta_r N_H^* N_D \tag{7}$$

where R is the electronic recombination rate (presumably proportiona to electron and hole concentration), $N_H^*$ is the concentration of excited mobile hydrogen, and $\beta_{cr}$ and $\beta_r$ are excitation and recombination rates, which probably contain time dependent dispersive factors.

The model also indicates that there is an upper limit on the metastable defect density which can be derived without knowledge of the rate constants. Only those weak bond states deeper than $E_D$ are stable, because hydrogen in shallower states can migrate and presumably can recombine relatively quickly. Thus the upper limit on the metastable defect density is $\sim E_0 N_H(E_D)$, which is about $10^{17}$ cm$^{-3}$ at room temperature.[15]

## SUMMARY

Hydrogen chemical reactions which lead to defect formation are conveniently described by a hydrogen density of states distribution. The general features of the distribution are obtained from simple models of chemical bonding, but further experiments are needed to establish the actual distribution. The defect density, hydrogen diffusion, metastability effects and the structure formed during growth can be related to the density of states.

Reversible metastable changes are described by the rearrrangement of hydrogen within the HDOS. Equilbrium defects are given by the Fermi distribution of hydrogen, while the non-equilibrium metastability is described by the excitation of hydrogen into non-equilibrium states in which the hydrogen remains trapped. Irreversible metastable changes are accompanied by structural changes which modify the HDOS and so change the hydrogen distribution. These effects are discussed further elsewhere.[15]

## ACKNOWLEDGEMENTS

Many helpful conversations with W. Jackson and P. Santos are gratefully acknowledged. The research is supported by the Solar Energy Research Institute.

## REFERENCES

1.  D. L. Staebler and C. R. Wronski, Appl. Phys. Lett., 31, 292 (1977).
2.  M. Stutzmann, W. B. Jackson and C. C. Tsai, Phys. Rev., B32, 23 (1985).
3.  D. E. Carlson, Appl. Phys., A41, 305 (1986)
4.  S. Zafar and E. A. Schiff Phys. Rev., B40, 5235 (1989).
5.  H. M. Branz and M. Silver, MRS Symp. Proc., 192, 261 (1990).
6.  D. Redfield and R. H. Bube, Appl. Phys. Lett., 54, 1037 (1989).
7.  D. Redfield, Appl. Phys. Lett., 52, 493 (1989).
8.  R. A. Street, Solar Cells, 24, 211 (1988).
9.  R. A. Street and K. Winer, Phys. Rev. B40, 6236 (1989).
10. W. Kruhler, H. Pfleiderer, R. Plattner and W. Stetter, AIP Conf. Proc. 120, 311 (1984).
11. W. B. Jackson and M. D. Moyer, Phys. Rev. B36, 6217 (1987).
12. R. A. Street, submitted for publication.
13. S. Zafar and E. Schiff, Phys. Rev. B40, 5235 (1989).
14. R. A. Street, M. Hack and W. B. Jackson, Phys. Rev. B37, 4209 (1988)
15. R. A. Street, Solar Cells, in press.
16. R. A. Street Phys. Rev. B43, 2454 (1991).
17. W. B. Jackson, Phys. Rev. B41, 10257 (1990).

# METASTABILITY IN HYDROGENATED AMORPHOUS SILICON: THE ADLER MODEL REVISITED

Howard M. Branz and Richard S. Crandall
Solar Energy Research Institute, Golden, CO 80401

Marvin Silver
University of North Carolina, Chapel Hill, NC 27599

## ABSTRACT

Ten years after it was first proposed, the Adler model of the Staebler-Wronski effect in hydrogenated amorphous silicon (a-Si:H) remains plausible and elegant. Adler suggested that during illumination, charged threefold-coordinated silicon dangling-bond defects capture photogenerated electrons and holes and reconfigure into metastable neutral dangling bonds. We present recent refinements to this model and review the considerable experimental evidence that a-Si:H has important metastable charge-trapping defects. Problems with this and other models of the Staebler-Wronski effect are discussed.

## INTRODUCTION

The metastable degradation of hydrogenated amorphous silicon (a-Si:H) material and devices due to light exposure has been an important research topic since its discovery by Staebler and Wronski[1] in 1977. The degradation of a-Si:H films involves increased carrier recombination, decreased photoconductivity, a shifted Fermi level, and increased neutral dangling bond ($T_3^0$) spin density. In a-Si:H solar cells, light soaking reduces nearly all parameters that determine the energy conversion efficiency. Several models of the Staebler-Wronski effect have been proposed, but none is universally accepted.

In this paper, we reexamine the model of the Staebler-Wronski effect originally proposed by Adler in 1981. Adler[2, 3] suggested that during illumination, charged threefold-coordinated silicon defects ($T_3^+$ and $T_3^-$) capture photogenerated electrons and holes and reconfigure into metastable neutral dangling bonds. In the 10 years since Adler published his model, significant supporting theoretical and experimental work has accumulated. We find that the Adler model remains a plausible and elegant explanation of the experimental data.

Numerous recent studies of crystalline semiconductor defects that trap carriers and metastably reconfigure[4] demonstrate the charge-trapping mechanism upon which Adler based his model. For example, illumination of n-type InP:Fe supplies holes that induce transformation of the MFe defect to a metastable configuration.[5] The metastable configuration anneals out in the presence of electrons with an activation energy of 0.35 eV. In boron-doped crystalline silicon, there is a hole-induced metastability that is thermally activated in both production (0.92 eV) and annealing (0.74 eV).[6] These phenomena in crystals resemble the metastabilities observed in a-Si:H and suggest that charge-trapping models of light-induced degradation in a-Si:H should be carefully considered.

## THE ADLER MODEL

Adler proposed[3] that because amorphous silicon is inhomogeneous it includes regions where the effective correlation energy ($U_{eff}$) of the dangling-bond defect is negative. In these regions, spinless charged dangling-bond defects form. The $T_3^+$ ($T_3^-$) defects are in their equilibrium $sp^2$ ($s^2p^3$) hybridizations, and form 120° (90°) bond angles. The defects capture photogenerated carriers and reconfigure according to

$$T_3^+(sp^2) + e^- \longrightarrow T_3^0(sp^3) \tag{1}$$

and

$$T_3^- (s^2p^3) + h^+ \longrightarrow T_3^0(sp^3). \tag{2}$$

Reconfiguration into the $sp^3$ hybridization involves bond angle shifts of 10 to 20 degrees and an accomodation by the surrounding atoms. The resulting $T_3^0$ defects are indistinguishable from the native $T_3^0$ defects, but are metastable. To return to equilibrium during annealing, they must reconfigure over a barrier. This may be an enthalpy barrier or simply the requirement that neighboring atoms cooperate to enable reconfiguration.

It is important to recognize that the Adler model does not require a negative $U_{eff}$ for the dangling bond, only the presence of equilibrium $T_3^+$ and $T_3^-$ defects. Recent theoretical work on thermodynamic models of a-Si:H suggests that copious $T_3^+$ and $T_3^-$ defects are also found in both films[7, 8] and solar cells[9] even when $U_{eff}$ is everywhere positive. There is also significant experimental evidence for these defects.[8]

The charged dangling-bond defects form in a-Si:H films due to inhomogeneity and equilibrium statistics.[8] Inhomogeneities (for example H clusters, microvoids, and impurities) create short-range potential fluctuations (V) of about 0.3 eV full-width, which are observed as the defect-band width in optical absorption and other experiments. The charged dangling-bond defects are frozen-in at about 200°C and V(x) is one of the terms in their local formation energies. Therefore, large densities of $T_3^+$ or $T_3^-$ defects form wherever the potential fluctuation charges a dangling bond in electronic equilibrium, i.e. in regions with $|V(x)| > U_{eff}/2$. Charged dangling-bond densities are estimated[8] to be above $10^{17}$ cm$^{-3}$, comparable to the observed saturation density of the Staebler-Wronski effect.[10]

In solar cells and other devices, equilibration of defect populations with the Fermi sea creates numerous charged dangling bonds in the undoped regions adjacent to a doped layer, as in the doped layers themselves. Branz and Crandall[9] calculate defect densities in a p-i-n solar cell *of homogeneous a-Si:H with positive $U_{eff}$* and find that the density of $T_3^-$ defects varies from $10^{18}$ cm$^{-3}$ at the n-i interface to more than $10^{16}$ cm$^{-3}$ at 200Å from the interface. Similarly, more than $10^{17}$ cm$^{-3}$ form near the p-i interface. The observed effect on spectral quantum efficiency of blue- and red-light soaking p-i-n and n-i-p solar cells from either side are explained[9] by assuming that charged dangling-bond defects near the interfaces undergo the Adler metastability.

Figure 1 is a schematic diagram of the density of states for dangling-bond optical transitions in the Adler model of metastability. Photons excite electrons from states below $E_F$ to extended states above the conduction-band mobility edge ($E_c$) and holes from states above $E_F$ to extended states below the valence-band mobility edge ($E_v$). In both films and solar cells, the equilibrium $T_3^+$ defects have their (0/+)

electronic transition levels above $E_F$ and the equilibrium $T_3^-$ defects have (-/0) levels below $E_F$. Because structural relaxation does not occur on the time scale of optical excitation, the optical levels are deeper than the thermodynamic transition levels that govern equilibrium occupation.[11] As shown in Figure 1(a), Reaction 1 produces $T_3^0$ (sp$^3$) defects that are stable against emitting an electron to $E_F$ without reconfiguration, but *metastable* against emitting the electron and reconfiguring to sp$^2$. This means that their (0/+) thermodynamic transition levels are actually *filled* but *above* $E_F$. Similarly, Reaction 2 produces $T_3^0$ (sp$^3$) defects that are stable against emitting a hole to $E_F$ without reconfiguration, but metastable against emitting the hole and reconfiguring to s$^2$p$^3$.

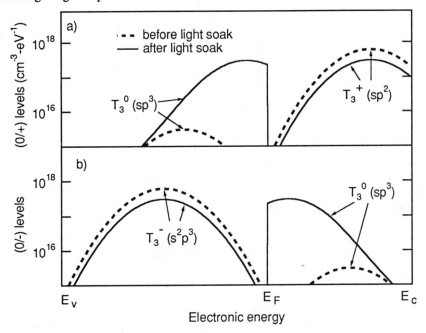

Fig. 1. Schematic diagram of the density of states for dangling-bond optical transitions before (- - -) and after (——) light soaking according to (a) Reaction 1 and (b) Reaction 2 of the Adler model. For simplicity, the predicted shifts of $E_F$ are not shown.

A signature of the Adler metastability is charge-trapping at the defect during creation and emission of charge during annealing. In order to metastably trap a carrier, the defect must surmount the barrier to reconfiguration before the trapped carrier escapes. This results in the low observed efficiency of metastable defect production. Both the energy released by the photoexcited carrier upon trapping and thermal phonons may contribute energy needed to surmount the barrier.

To anneal, the defect emits the trapped carrier to an unoccupied extended state and surmounts the barrier to reconfiguration. To anneal $T_3^0$ back to $T_3^+$, the electron emitted to the conduction band comes from below $E_F$ as is clear from Figure 1(a). To anneal $T_3^0$ back to $T_3^-$, the hole is emitted to the valence band from above $E_F$. The carrier excitation energies are lower bounds to the annealing energies; the barrier energies are likely larger and dominant. If the two processes can occur sequentially, the annealing activation energy will be the *greater* of the barrier energy and the carrier excitation energy. If the processes must be simultaneous because of carrier retrapping, the annealing energy will be the *sum* of the barrier and carrier excitation energies.

There is a shift of Fermi level associated with each of the Adler metastability reactions. In Reaction (1), an electron is removed from communication with the Fermi sea when it is captured by the dangling-bond defect. Consequently, $E_F$ is lower after Reaction (1). Reaction (2) traps and immobilizes a hole, and this raises $E_F$. When Reactions (1) and (2) occur at equal rates, $E_F$ is unchanged.

## EVIDENCE FOR CHARGE-TRAPPING

The Adler model suggests that one-carrier injection, in the absence of recombination, will cause the Staebler-Wronski metastability by carrier trapping. It is well established[12] that two-carrier injection in the absence of illumination will cause metastable degradation in solar cells. Single-carrier degradation is much more difficult to observe because of space-charge limitations. In a one-carrier device, space charge limits the density of injected charge to roughly CV, where C is the device capacitance and V is the applied voltage. Application of 1 V to a 1 µm device (field strength of $10^4$ V-cm$^{-1}$) yields charge densities less than $10^{15}$ cm$^{-3}$. Degradation by illumination or double injection can easily yield three orders of magnitude greater carrier density.

Despite the problem of space-charge limitation, single-carrier-defect creation has been observed by many workers. Nakamura *et al.*[12] and other groups[13, 14] observe degradation in the i layer of p-i-p devices due to single-carrier injection of holes. van Berkel and Powell[15] create states near $E_F$ in the undoped layer of a-Si:H thin-film transistors by electron injection. Yamagishi *et al.*[16] degrade p-i-n solar cells by blue light soaking from either the p- or n-side and conclude that single-injection of either holes or electrons can produce the Staebler-Wronski effect.

The observed change in $E_F$ during light soaking[1] strongly suggests there are defect reactions that trap and stabilize charge. One carrier is removed from the Fermi sea by the charge-trapping and reconfiguration, changing $E_F$. In the Adler model, Reaction (1) lowers the electron Fermi level and Reaction (2) raises it, as described above. Of course, charge-trapping in a-Si:H could be due to defects different than those proposed by Adler.

Charge-sensitive measurements supply more direct evidence for charge-trapping metastability in a-Si:H. In a diode structure, a change in the charge state of a defect is reflected in the depletion width and Fermi level and can be measured as a change of ac capacitance. This technique has been widely used in the study of metastable charge-trapping defects in crystalline semiconductors.[4] It was first used in a-Si:H by Crandall[17] to detect carrier-induced metastability in the i layers of p-i-n and Schottky-barrier solar cells.

Charge-induced metastability in doped a-Si:H was first observed at Bell Laboratories in the n-layer of a Schottky-barrier structure.[18] Reverse biasing the capacitor at 200°C followed by cooling in reverse bias increased the room temperature capacitance by 50% to 300%. This metastable capacitance change could be reversed by an anneal with 0 V bias applied. Evidently, after emitting an electron in reverse bias, some defect or defects reconfigure and do not readily retrap the electron when the bias is removed at room temperature. The metastable excess positive charge on the defect raises $E_F$ and reduces the depletion width to increase the capacitance.

A similar effect in p-type a-Si:H is reported elsewhere in this volume.[19] Annealing under reverse bias increases the capacitance by a factor of three or more, and this effect is fully reversible by a zero-bias anneal. The hole emission producing the metastable state has an activation energy of about 1.3 eV and the hole recapture is activated with about 0.9 eV. Similar kinetics were observed for metastable carrier trapping in undoped a-Si:H (see below). This may be a hole-controlled variant of Reaction (1) in which $T_3^+$ defects emit holes and reconfigure during the reverse bias anneal treatment, according to $T_3^+(sp^2)$ ---> $T_3^0(sp^3)$+ $h^+$. Alternatively,[4] the B-dopant atoms may participate in p-type charge-trapping metastability, and the P-dopant atoms may participate in n-type metastability.

Capacitance changes due to charge-trapping metastability in undoped a-Si:H were studied by Crandall.[17, 20] He used Pt/a-Si:H Schottky barrier, p-i-n and n-i-p structures and observed both hole-trapping and electron-trapping metastability. These metastabilities can be introduced by forward-bias carrier injection or by illumination. Jackson, *et al.*[21] observed similar metastable changes in capacitance after trapping of either electrons or holes in a metal/insulator/a-Si:H structure.

The emission of the metastably trapped electron (annealing) occurs at about 440K during capacitance deep level transient spectroscopy (DLTS) and the measured activation energy[20] for the emission is nearly always between 0.8 and 1.1 eV. Because the electron emission activation energy is greater than or equal to half the mobility gap, this is likely emission over a reconfiguration barrier, not emission from a simple trap. Crandall found that this electron-trapping defect controls the light-induced photocurrent degradation in solar cells. Emission of the metastably trapped hole is activated with 1.4 to 1.8 eV, clearly emission over a barrier. In the Adler model, the electron emission process is identified with annealing of Reaction (1), while hole emission is identified with annealing of Reaction (2).

Creation experiments above about 380K showed thermally activated carrier-trapping kinetics with characteristic energies slightly larger than the emission energy of the same defect.[17, 20] Creation experiments between 320 and 380K show a weakly temperature-activated process.[22] This suggests that energy released in carrier-trapping alone drives the low-temperature reconfiguration, but thermal energy is involved at high temperatures.

In recent transient photoconductivity decay experiments, McMahon and Crandall[23] observe a deep ('safe') hole trap that disappears as light soaking proceeds. They conclude that the hole trap is the precursor to the Staebler-Wronski effect defect. Branz and Silver[8] suggest that the safe hole trap is an intermediate step of Reaction (2) in which the hole is trapped without defect reconfiguration, according to

$$T_3^- (s^2p^3) + h^+ ---> T_3^0(s^2p^3). \qquad (3)$$

Usually, the hole is reemited, but the $T_3^0(s^2p^3)$ defect sometimes reconfigures into the metastable $T_3^0 (sp^3)$ defect to complete Reaction (2).

## PROBLEMS WITH STAEBLER-WRONSKI EFFECT MODELS

In this section, we briefly describe a few of the difficulties of the Adler model and several other widely-discussed models of the Staebler-Wronski effect. There is a wealth of complex and sometimes conflicting data on metastability in a-Si:H. No theory of light-induced degradation accounts for all the data, though all are consistent with some of it.

The greatest difficulty of the Adler model of metastability is the lack of unquestioned evidence for $T_3^+$ and $T_3^-$ defects in undoped a-Si:H. However, charged dangling-bond defects would be spinless and difficult to observe, especially if their optical transition levels are hidden under the bandtail states.[11] Without equilibrium $T_3^+$ and $T_3^-$ defects another microscopic model of the charge-trapping defects must be sought. This might suggest an impurity-related defect, similar to the charge-trapping defects found in crystalline Si. For example, Redfield and Bube[24] propose that in the Staebler-Wronski effect a column V impurity such as P reconfigures from four-fold to three-fold coordination, and this produces a $T_3^0$ defect. We suggest it is most reasonable that reconfiguration of the P atom from $P_4^+$ to $P_3^0$ would be driven by an electron capture into the $P_4^+$ defect.[4]

One might also expect that partial reconfiguration of the light-induced $T_3^0$ defect would sometimes occur, leaving a different electron spin resonance (ESR) signal. That this is not observed most likely reflects the variety of configurations in which the equilibrium $T_3^0$ are found.

Many alternative models of the Staebler-Wronski effect assume that a 'weak Si-Si bond' is broken by the energy of carrier recombination, leading to two $T_3^0$ defects. To explain why exchange narrowing of the characteristic $T_3^0$ ESR signal is not observed, most theories assume these defects are separated by a hydrogen hopping process. The greatest difficulty is that these models do not account for the experimental data (summarized above), which shows that charge trapping is an integral part of the metastable degradation processes. In addition, no effect on the annealing kinetics of the Staebler-Wronski effect is observed[25] when deuterium is substituted for hydrogen in glow-discharge-deposited amorphous silicon.

Stutzmann[26] examined the energetics of the bond-breaking process in detail using observed defect electronic energies levels in the a-Si:H energy gap and concluded that hole trapping in a valence band tail state can break the bond. However, he identifies the valence band tail states with the 'weak bonds' by assuming that the total energy of a defect is equal to the energy level of its most easily ionized electron. This equality is not expected[27] especially in a material such as a-Si:H with strong electron-phonon coupling at the defects.[11, 28]

Other models suggest that impurity atoms may be involved in the Staebler-Wronski effect.[24, 29] Certainly, impurity models are plausible in light of studies[17, 29-32] that correlate the concentrations of C, O, and N with light-induced metastability. However, in all these studies, contamination levels were above $10^{18}$ cm$^{-3}$ and would introduce short-range potential fluctuations[8] and inhomogeneity, which complicate interpretation of the impurity-induced metastability.

Kinetic data is unlikely to distinguish among the various models of the Staebler-Wronski effect. The observed saturation behavior can result from from annealing, light-induced back reactions[33] or a fixed number of defect sites susceptible to metastability. Any ensemble of metastable defects with a distribution of anneal barriers will produce both the stretched exponential time dependence and Meyer-Neldel behavior characteristic of metastability in many material systems.[34] The distribution of annealing time constants due to disorder makes stringent tests of any kinetic theory difficult. ESR and charge-sensitive spectroscopies are, therefore, important experiments if we are to better understand the Staebler-Wronski effect.

## CONCLUSIONS

The evidence for charge-trapping metastable defect reactions in a-Si:H analogous to those found in crystalline semiconductors is strong. It is possible that several metastability mechanisms are active in a material as complex and as far from its equilibrium structure as a-Si:H. More charge-sensitive measurements are needed to study the charge-trapping metastabilities and determine whether they dominate the Staebler-Wronski effect.

The only well-developed model of charge-trapping metastability in undoped a-Si:H is the Adler model, which suggests that charged dangling-bond defects trap carriers and reconfigure into metastable $T_3^0$ defects. Inhomogeneity in a-Si:H films and proximity to the doped layers of devices cause the equilibrium $T_3^+$ and $T_3^-$ defects that are the source of the metastability. Further study of charged defects in undoped a-Si:H and of charge-trapping metastability are needed.

Alternative Staebler-Wronski effect models explicitly including the charge trapping should be developed. ESR in conjunction with charge-sensitive metastability experiments could supply the detailed information about the initial, light-induced and final metastable defect configurations which are needed for improved microscopic modelling of the charge-trapping metastability.

Efforts to reduce the impact of the Staebler-Wronski effect in a-Si:H should focus on reducing the densities of charged dangling-bond defects by improving material homogeneity. In solar cells, charged impurity-related defects could be introduced near the interfaces to reduce the harmful effects of the electric field collapse that results from metastable neutralization of charged dangling bonds.[9]

## ACKNOWLEDGEMENTS

We thank Thomas McMahon and Alex Zunger for helpful discussions. This work was supported by the U.S. Department of Energy under Contract No. DE-AC02-83CH10093. One of us (M.S.) acknowledges the Solar Energy Research Institute for financial support through a subcontract.

## REFERENCES

1. D. L. Staebler and C. R. Wronski, Appl. Phys. Lett. **31**, 292 (1977).
2. D. Adler, J. de Phys. **42-C4**, 3 (1981).
3. D. Adler, Solar Cells **9**, 133 (1983).
4. H. M. Branz, Phys. Rev. B **38**, 7474 (1988) includes a list of references to the extensive literature on this subject.
5. M. Levinson, M. Stavola, P. Besomi, and W. A. Bonner, Phys. Rev. B **30**, 5817 (1984).
6. A. Chantre, Phys. Rev. B **32**, 3687 (1985).
7. Y. Bar-Yam, D. Adler, and J. D. Joannopoulos, Phys. Rev. Lett. **57**, 467 (1986).
8. H. M. Branz and M. Silver, Phys. Rev. B **42**, 7420 (1990).
9. H. M. Branz and R. S. Crandall, Solar Cells **27**, 159 (1989).
10. H. R. Park, J. Z. Liu, and S. Wagner, Appl. Phys. Lett. **55**, 2658 (1989).

11. H. M. Branz, Phys. Rev. B **39**, 5107 (1989).
12. N. Nakamura, K. Watanabe, M.Nishikuni, Y. Hishikawa, S. Tsuda, H. Nishiwaki, M. Ohnishi, and Y. Kuwano, J. Non-Cryst. Solids **59/60**, 1139 (1983).
13. W. den Boer, M. J. Geerts, M. Ondris, and H. M. Wentinck, J. Non-Cryst. Solids **66**, 363 (1984).
14. W. Kruhler, H. Pfleiderer, R. Plattner, and W. Stetter, in *Optical Effects in Amorphous Semiconductors*, edited by P. C. Taylor and S. G. Bishop (American Institute of Physics, New York, 1984), p. 311.
15. C. van Berkel and M. J. Powell, Appl. Phys. Lett. **51**, 1094 (1987).
16. H. Yamagishi, H. Kida, T. Kamada, H. Okamoto, and Y. Hamakawa, Appl. Phys. Lett. **47**, 860 (1985).
17. R. S. Crandall, Phys. Rev. B **24**, 7457 (1981).
18. D. V. Lang, J. D. Cohen, and J. P. Harbison, Phys. Rev. Lett. **48**, 421 (1982).
19. R. S. Crandall, K. Sadlon, S. J. Salamon, and H. M. Branz, this volume.
20. R. S. Crandall, Phys. Rev. B **36**, 2645 (1987).
21. W. B. Jackson, J. M. Marshall, and M. D. Moyer, Phys. Rev. B **39**, 1164 (1989).
22. M. Stutzmann, W. B. Jackson, and C. C. Tsai, Phys. Rev. B **32**, 23 (1985).
23. T. J. McMahon and R. S. Crandall, Phil. Mag. B **61**, 425 (1990).
24. D. Redfield and R. H. Bube, Phys. Rev. Lett. **65**, 464 (1990).
25. M. Stutzmann, W. B. Jackson, A. J. Smith, and R. Thompson, Appl. Phys. Lett. **48**, 62 (1986).
26. M. Stutzmann, Phil.Mag. B **56**, 63 (1987).
27. V. Heine, in *Solid State Physics*, edited by H. Ehrenreich, F. Seitz, and D. Turnbull (Academic Press, New York, NY, 1980), p. 92.
28. Y. Bar-Yam and J. D. Joannopoulos, Phys. Rev. Lett. **56**, 2203 (1986).
29. S. Tsuda, N. Nakamura, M. Nishikuni, K.Watanabe, T. Takahama, Y. Hishikawa, M. Ohnishi, Y. Kishi, S. Nakano, and Y. Kuwano, J. Non-Cryst. Solids **77/78**, 1469 (1985).
30. R. S. Crandall, D. E. Carlson, A. Catalano, and H. A. Weakliem, Appl. Phys. Lett. **44**, 200 (1984).
31. D. E. Carlson, A. Catalano, R. V. D'Aiello, C. R. Dickson, and R. S. Oswald, in *Optical Effects in Amorphous Semiconductors*, edited by P. C. Taylor and S. G. Bishop (American Institute of Physics, New York, 1984), p. 234.
32. T. Unold and J. D. Cohen, in *Amorphous Silicon Technology-1990*, edited by P. C. Taylor, M. J. Thompson, P. G. LeComber, Y. Hamakawa, and A. Madan (Materials Research Society, Pittsburgh, 1990), p. 719.
33. D. Redfield, Appl. Phys. Lett. **49**, 1517 (1986).
34. R. S. Crandall, Phys. Rev. B, Feb. 15, in press (1991).

# THEORETICAL AND EXPERIMENTAL INVESTIGATION OF HYDROGEN BONDING CONFIGURATIONS IN Si

W. B. Jackson, S. B. Zhang, C. C. Tsai, and C. Doland
*Xerox Palo Alto Research Center*
*3333 Coyote Hill Road, Palo Alto, California 94304*

## Abstract

In this work, the local density total energy calculations of various bonding configurations for H in crystalline Si are used to develop a density of states for H trapping in amorphous silicon (a-Si:H). This density of trapping states is compared with various experimental results and is used to interpret H transport measurements.

## Introduction

Hydrogen transport and trapping in Si is of fundamental importance in the growth, metastability and mesoscopic structure of hydrogenated amorphous Si. Since H is known to move at temperatures near 150 C or higher, equilibration and transport between various H bonding configurations determines many aspects of the material. Recently, significant progress has been achieved in theoretically determining the hydrogen bonding energies for various configurations in c-Si.[1-5] Prominent among these findings is that the lowest energy configurations are clusters: pairs and larger, 2-D platelet structures.[2,3] In the work presented in this paper, the various isolated and clustered H bonding configuration energies derived from calculations are used to develop a hydrogen trapping density of states which is tested against a variety of experiments.

Previous work has been involved in understanding H bonding in amorphous Si. Early measurements of hydrogen diffusion in Si established that the activation energy of H diffusion is about 1.4-1.55 eV.[6-8] and diffusion as a monatomic species.[9] The diffusion rate also depends on doping level and deposition conditions.[7,8] H evolution experiments have shown that the majority of hydrogen evolved below about 525 C results in few dangling bonds.[10,11] Hydrogen evolution above 525C causes significant increases in dangling bond densities. These results have been interpreted in terms of two types of H binding: weakly bound H and H strongly bound in Si dangling bond states.[10]

Another important development has been the introduction of the idea of the negative U for H.[12,13] Because the rupture of a Si-Si bond by H creates two dangling bonds, it seems likely that H would tend to remove both the dangling bonds. Consequently, a given Si-Si bond tends to be occupied by two H atoms or zero; single occupancy is much higher in energy. This concept explains the absence of dangling bonds and isolated H configurations and the observed temperature dependence of the dangling bond concentration as well as the passivation of the dangling bonds.

Finally, weak Si-Si bonds have an effect on H bonding and transport.[13] Since H breaks Si-Si bonds to form Si-H bonds, weak Si-Si bonds are easier to break than strong bonds. Weak bonds therefore act as traps for H (lower energy sites than normal Si-Si bonds).[13,14] The interaction between weak Si-Si bonds and H, according to the weak bond model, gives rise to metastability and equilibrium defect densities. Hydrogen removes weak bonds lowering the Si network disorder.

While these ideas have been developed more or less independently, they are not quantitative nor has the microscopic structure been elucidated in significant detail. Recent calculations of H bonding configurations have been performed on H in c-Si. These results are used to construct a density of states for hydrogen trapping. This density of states is quantitatively based on total energy calculations of H bonding in c-Si and provides details regarding the microscopic identification of various types H traps. In this paper, the H trapping densities of states is compared with experiment particularly structural and transport measurements of H bonding configurations.

## H Bonding Energies in Si

A summary of recent calculated average total energies per H atom for various H configurations in Si is presented in Fig. 1 relative to atomic H in the vacuum.[1-5] These calculations, performed by local density calculations, have a relative accuracy of about 0.2 eV or less. The disorder of the amorphous material is expected to broaden the energy levels in bands. Because the energy depends predominately on the local structure of the Si-Si bond, the average energy of the configurations should be similar between amorphous and crystalline Si. The configurations are classified according to the number H atoms involved in the configuration: 1H, 2H (pairs), and $2n$H(large platelet structures).

The 1H states have a number of distinguishing properties. The bond centered (BC) and tetrahedral interstitial sites ($T_d$) are high in energy, characterized by an unpaired electron in a gap state,[1,2] and have a low energy for migration $<0.48$ eV according to experiment[15] and molecular dynamic calculations.[4,5] The diffusion or percolation threshold for the bulk is therefore estimated to be about 0.5 eV below the vacuum energy. H in the BC site of a strained bond will be lower by an amount which depends on the bond strain but is on the order of 0.3-0.5 eV.[16] The strained bonds act as very shallow traps for the interstitial H as it diffuses through the material. Because of the gap states associated with the mobile H, the energy of the interstitial, the percolation threshold, and therefore H diffusion depend on the Fermi level position. The lowest energy position for H in an Si-H bond, ie., occupation of an isolated dangling bond, is nearly 2.5 eV below the percolation threshold.[3,10] Distinguishing properties for trapping in Si dangling bonds are: (1) changes in occupancy of an isolated Si-H bond result in changes in the Si dangling bond density, (2) observations of Si-H vibrations and (3) a narrow line in H nuclear magnetic resonance (NMR).

The 2H or H pair configurations have recently received considerable attention.[1-3] The $H_2$ molecule in the $T_d$ site is a low energy configuration. This configuration is characterized by the absence of an Si-H bond and a doublet in the H NMR signal.[17] Alternatively, one H atom can break a Si-Si bond occupying the BC site while another H in the $T_d$ site passivates the second dangling bond thereby forming the $H_2^*$ configuration. This configuration is only about 0.15 eV higher in energy, has no states in the

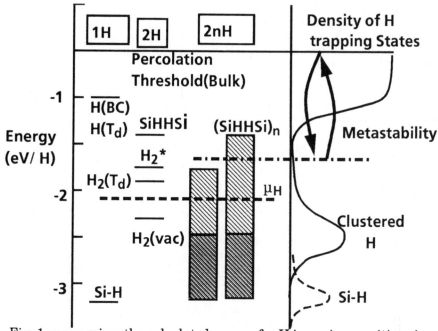

Fig. 1 summarizes the calculated energy for H in various positions in c-Si using LDA in eV per H atom for configurations of the number of H atoms. The light shading indicates the range of energies for platelets forming in bulk Si. The dark shading is the range of energies for 2-D H configurations forming in highly strained regions or as a result of hydrogenation of void surfaces. The dashed line indicates the position of the chemical potential. On the right is a depiction of the approximate density of trapping states. The majority of H is in the clustered phases but a small fraction is in the Si dangling bond states (Si-H) which is deeper than the other positions. Metastability involving bulk transport must occur from levels within 1 eV of the transport level (arrows).

gap, and exhibits Si-H vibrations.[3]   Because the energy of this configuration is significantly less than two isolated H atoms, this center is a microscopic model for the hydrogen negative U.[12] Two H atoms occupy (break) an Si-Si through the formation of an $H_2^*$. The difference between this model and the negative U model[12] is that in this model, spins are due to single H configurations rather than removal of H from Si-H bonds to form pairs. Insertion of both H atoms in between the Si atoms (SiHHSi) is energetically costly and therefore is less likely.

The multi-H platelets have been observed experimentally in c-Si[18] and their energies have been recently calculated.[3] The strain in the backbonds of the $H_2^*$ drives the clustering of such pairs into large {111} 2-D platelets $(H_2^*)_n$ structures. The clustering lowers the energy per H atom to roughly 0.05 eV per H atom below the interstitial $H_2$ molecule. If a sufficient number of H pairs cluster, the resulting structure dilates further lowering the total energy of the structure. The final energy depends on the number of H atoms in the structure indicated by the shaded energy regions. The energy can be as low as -2.4 eV per H atom. These platelets are "closed" void structures in that if the H is removed, a normal Si-Si network reforms with full strength Si-Si bonds. If the cluster is larger allowing increased expansion of the platelet, the (Si-HH-Si) platelet becomes comparable in energy (further details in Refs. 3 and 19). These platelets represent the lowest energy configurations for H in bulk Si. They are characterized by Si-H vibrations, a broad NMR component, and upon H evolution, the number of dangling and weak Si bonds does not change appreciably. If there are void surfaces in the material, the H can bond on these surfaces as well. Because a full strength Si-Si bond does not result following removal of the H, the energy of these "open" void configurations range from the lowest $(H_2^*)_n$ configurations to the isolated dangling bond (dark shaded regions). Characteristics of these configurations are (1) Si-H vibrations, (2) the presence of voids and density deficits, (3) H removal results in the formation of weak Si-Si bonds, band tail states and voids. The hydrogenated void surfaces may also give rise to a broad NMR component.

Because of the variation in energy due to the size of the clusters and disorder, the various configurations overlap forming a continuous density of sites for H. A possible density of H trapping sites is depicted on the right hand site in Fig. 1. The H diffusion is dominated by the trapping and release from the clustered H phases. The spread in energy of the clustered phases and of the mobility barrier give rise to dispersive H diffusion.[8,20] The unoccupied dangling bonds form deep traps which are somewhat deeper than the clustered phase sites.

## Comparison with c-Si Experiments

The energies presented above compare well with experiment. A summary of c-Si experimental results supporting the energy position calculated for c-Si are shown in Fig. 2. The relative placement of the

**Energy eV/H**

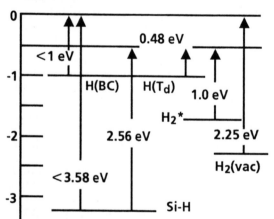

Fig. 2 The energies for various transitions of H in c-Si obtained from experiment. The energies indicated for the transitions from left to right are from Refs. 21-24, respectively. The placement and scale on the left correspond to the calculated positions from Fig. 1. The observed energies are in good agreement with calculated values.

interstitial relative to the vacuum level is obtained from H solubility measurements.[21] The dissociation of H from a Si surface provides an estimate of the Si-H trapping energy.[22] The release of H from Si dangling bonds at the Si/SiO$_2$ interface estimate of the barrier to release into bulk Si of 2.56 eV.[23] The low concentration and high temperature diffusion measurements suggest that the motion barrier for interstitials is about 0.48 eV.[15] The energy for H$_2$* dissociation are approximately 1.0 eV.[24] The energy difference between µH and the percolation threshold in a-Si:H is obtained directly from the 1.4-1.55 eV activation energy of diffusion in a-Si:H.[6-10] The formation energy of H$_2$ in the vacuum comes from chemical data. Comparing Fig. 1 and 2, we see that the computed energies for c-Si agree well with the experimentally observed values rather well demonstrating that these calculations are quantitatively correct. Further discussion has been given in Ref. 19. The rest of this paper will be devoted to testing these results for interpreting posthydrogenation of a-Si:H given that the results seem correct for c-Si.

## Posthydrogenation

The anneal-stable, posthydrogenated samples were prepared by annealing a-Si:H samples deposited at 350C at temperature of 550C for up to 9 h to drive the hydrogen from the sample. The resulting material has no measurable Si-H vibrations, a spin density of 2X10$^{19}$ cm$^{-3}$ and an H concentration of 8X10$^{19}$ cm$^{-3}$. After a rinse in a dilute HF solution to remove any oxide, these samples were exposed to an optically baffled microwave H or D plasma at temperatures ranging from 320 C to 450C.[25] The samples were then examined using Raman, electron spin resonance (ESR), photothermal deflection spectroscopy (PDS), and secondary ion mass spectroscopy (SIMS).

Si-H vibrations in the Raman spectra indicate that most H (or D) bonds to Si in Si-H bonds and is not in the molecular forms of H. A typical SIMS profile is shown in Fig. 3 by the points. The D concentration reaches as high as 7 atomic % in the near surface region. The high concentration region is reasonably accurately described by a complementary error function profile. The activation energy, 1.3-1.5 eV, and the magnitude of the diffusion coefficient is similar to that found for a-Si:H[6-8] Therefore, the µH is roughly 1.3-1.5 eV below the percolation threshold as indicated in Fig. 1. For concentrations below 8X10$^{19}$ cm$^{-3}$, the profile is given by an exponential whose slope is probably limited by the SIMS resolution. Such behavior is characteristic of trap dominated diffusion where D falls into deep traps separated from the states controlling normal diffusion.[26] The concentration where the profile deviates from the complementary error function is an estimate of the deep trap density. In these samples, a trap density of 8X10$^{19}$ cm$^{-3}$ is estimated. The spin density of the sample was measured during a series of controlled etches to profile the spin density shown in Fig. 3 (details are given in Ref. 27). The spins are greatly reduced in regions where the D concentration is highest. The expected unoccupied Si dangling bond profile and D profile calculated for the combined diffusion and trapping equations are shown by the solid lines.[26] From these data and modeling, we can conclude that to within a factor of 3-4, the deep traps are associated with a dangling Si bond before trapping. The difference by a factor of 3-4 may be within the combined error of the SIMS and ESR densities or may reflect the likely possibility that not all deep H(D) traps have a spin. Because of the sharp break between the complementary error function and the exponential, the deep traps have a significantly slower

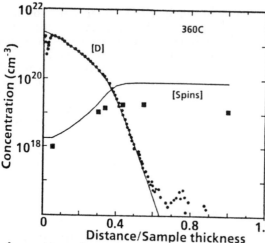

Fig. 3 SIMS Deuterium profile (circles) and Si dangling bond spin density (squares) versus normalized sample thickness (320 nm) of a post-deuterated films. The solid lines represent solutions to the the combined trapping diffusion equations. The complementary error function changes shift to an exponential at concentrations below $8 \times 10^{20}$ cm$^{-3}$.

release time than those responsible for the high concentration diffusion. The Si dangling bonds are therefore deep traps for H and the traps are separate in energy from those traps controlling the diffusion in the high concentration region of the sample. This result is in full agreement with previous results showing that the last H to be released causes the largest increase in dangling bonds.[10,11] Raman spectra of the sample did not indicate any significant change in the transverse optical phonon peak despite the incorporation of almost 7 at. % H. Nor did the PDS show any change in the Urbach edge even if the D diffused was diffused completely through the sample. This result is similar to observations that the Urbach edge remains constant as hydrogen is evolved.[28] The independence of Urbach edge and the TO peak from H content[29] indicates that for posthydrogenation or H evolution, there is not a significant change in the number of band tail states or weak bonds.

## Discussion

The results of posthydrogenation provide support for the density of hydrogen trapping states for amorphous silicon derived from c-Si calculations. In this density of states, the first H eliminates the dangling bond states and therefore the spin associated with them. These states are the deepest H trapping states. Once these states are filled by H concentrations exceeding $8 \times 10^{19}$ cm$^{-3}$, the H is forced to occupy higher energy sites. Some of these sites are probably surfaces of hydrogenated voids and weak bonds. However, the amount of H easily exceeds the number of weak bonds as estimated by tail states and furthermore does not seem to alter the disorder of the Si network.[26,29] Hence, these results support the idea that most of the H exceeding $8 \times 10^{19}$ cm$^{-3}$ tends to bond in "closed" hydrogen platelets rather than on the surfaces of voids. Only a small fraction ($<10\%$) of the H bonds into $H_2$ molecules. The interstitial molecule is unstable to platelet formation and $H_2$ in the vacuum of a void requires the presence of a void which costs significant energy to form. H diffusion is dominated by H platelets.

These results are supported by other studies of a-Si:H. A significant fraction of the H is present in the clustered NMR phase.[17] The 25kHz linewidth is consistent with the "closed" void platelets. H evolution will tend to occur from the less tightly bound clustered H, leaving the isolated H responsible for the narrow NMR line. The fact that films shrink markedly following H evolution is also evidence that much of the H is bonded in "closed" voids whose sides interact with each. Evolution of H from "open" hydrogenated voids would not be expected to result in significant film shrinkage. The IR Si-H vibration spectra are also consistent with this picture. H on Si surfaces[22,30] and in poly-Si[31] is known to give rise to vibrations at 2100 cm-1 region. Hence hydrogenated "open" voids observed in low density, high H concentration material deposited near room temperature exhibit 2100 cm-1 vibrations but the high quality films exhibit 2000 cm-1 vibrations characteristic of "closed" voids which interact with the Si-H bond changing the vibration frequency. The Si(111) 7X7 surface has some interior dangling bond sites which when hydrogenated exhibit IR vibrations below 2000 cm-1.[32] Lower stretching frequencies are characteristic of vibrations of Si-H bonds interacting with a nearby Si network such as occur in "closed" platelet structures.

The density of H trapping states in Fig. 1 has implications for metastability in a-Si:H. If metastability is caused by H motion over the bulk percolation barrier defined by squeezing through two bonded Si atoms, then the states responsible for metastability with a 1 eV activation energy barrier must be in the $H_2^*$ region of H trapping energies. This implies that isolated $H_2^*$ pairs or H near the edges of the clustered platelets are responsible for metastability as shown by calculations.[33,34] Carriers efficiently dissociate $H_2^*$ pairs with barriers consistent with experiment. Because simultaneous trapping of e's and h's on an $H_2^*$ lowers the barrier, defect generation depends on e-h product. Reduction of metastability can be achieved by lowering the H concentration in the film decreasing the concentration of the weakly bonded H while maintaining a sufficient H concentration to passivate the isolated dangling bonds. The more stable films tend to exhibit a slower H diffusion rate.

Thus, hydrogen bonding in a-Si:H and c-Si are fairly similar. The density of trapping sites calculated for c-Si appear to provide a reasonable explanation for the behavior of H in Si. The deepest traps are Si dangling bonds while the rest of the H is bonded primarily in 2-D clusters of $H_2^*$ platelets with perhaps a smaller fraction existing on hydrogenated void surfaces. These platelet structures are characterized by strong interaction between the opposite sides of the structure and strong Si-Si bonds result upon H elimination. Changes in H content do not seem to significantly alter the density of weak or strained bonds. Metastability caused by H moving through the bulk must involve configurations whose energy and properties match $H_2^*$ complexes.

We would like to thank R. Street, N. Johnson, J. Northrup, C. Herring and P. Santos for many stimulating and useful discussions. This work was supported in part by the Solar Energy Research Institute.

## References

1. C. G. Van de Walle, P. J. H. Denteneer, Y. Bar-Yam, and S. T. Pantelides, Phys. Rev. B, **39**, 10791, (1989).
2. K. J. Chang and D. J. Chadi, Phys. Rev. B. **40**, 11644 (1989).
3. S. B. Zhang and W. B. Jackson (to be published).

4. F. Buda, G. L. Chiarotti, R. Car, and M. Parinello, Phys. Rev. Lett. **63**, 294 (1989).
5. P. E. Blöchl, C. G. Van de Walle, and S. T. Pantelides, Phys. Rev. Lett. **64**, 1401 (1990).
6. D. E. Carlson and C. W. Magee, Appl. Phys. Lett., **33**, 81 (1978).
7. W. Beyer, J. Herion and H. Wagner, Jour. of Non-Crystalline Solids **114**, 217 (1989).
8. R. A. Street, C. C. Tsai, J. Kakalios, and W.B. Jackson, Phil. Mag. **56** (1987) 305.
9. W. Beyer and H. Wagner, Jour. of Non-Crystalline Solids, **59&60**, 161 (1983).
10. K. Zellama, P. Germain, S. Squelard, B. Bourdon, J. Fontenille, and R. Danielou, Phys. Rev. B, **23**, 6648 (1981).
11. D. K. Biegelsen, R. A. Street, C. C. Tsai, and J. C. Knights, Phys. Rev. B, **20**, 4839 (1979).
12. S. Zafar and E. A. Schiff, Phys. Rev. B **40**, 5235 (1989).
13. R. A. Street, M. Hack, and W. B. Jackson, Phys. Rev. B **37**, 4209 (1988).
14. R. A. Street and K. Winer, Phys. Rev. B **40**, 6236 (1989).
15. C. H. Seager, R. A. Anderson, and J. K. G. Panitz, J. Mater. Res. **2**, 96 (1987);A. Van Wieringen and N. Warmholtz, Physics **22**, 849 (1956).
16. E. Tarnow (unpublished)
17. P. C. Taylor, in *Semiconductors and Semimetals, Vol. 21C: Hydrogenated Amorphous Silicon*, J. I. Pankove, ed., (Academic Press, Orlando, 1984), p. 99 and references therein.
18. N. M. Johnson, F. A. Ponce, R. A. Street, and R. J. Nemanich, Phys. Rev. B, **35**, 4166 (1987).
19. More detailed descriptions are found in W. B. Jackson and S. B. Zhang in *Advances in Disordered Semiconductors, Vol 3* edited by H. Fritzsche (World Scientific, Singapore) (to be published).
20. J. Shinar, R. Shinar, S. Mitra, and J.-Y. Kim, Phys. Rev. Lett. **62**, 2001 (1989).
21. C. Herring and N. M. Johnson in *Hydrogen in Silicon:Semiconductors and Semimetals, Vol. 54.* edited by J. I. Pankove and N. M. Johnson, (Academic press,New York, 1991), Chapt. 10.
22. P. Gupta, V. L. Colvin, and S. M. George, Phys. Rev. B **37**, 8234 (1988).
23. K. L. Brower and S. M. Myers, Appl. Phys. Lett. **57**, 162 (1990).
24. N Johnson (unpublished).
25. N. M. Johnson and M. D. Moyer, Appl. Phys. Lett. **46**, 787 (1985).
26. W.B. Jackson, C. C. Tsai, and C. Doland, (unpublished).
27. W. B. Jackson, C. C. Tsai, and R. Thompson, Phys. Rev. Lett. **64**, 56 (1989).
28. S. Yamasaki, Phil. Mag. B. **56**, 79 (1987).
29. D. Beeman, R. Tsu, and M. F. Thorpe, Phys. Rev. B, **32**, 874 (1985).
30. G. S. Higashi, Y. J. Chabal, G. W. Trucks, and K. Raghavachari, Appl. Phys. Lett. **56**, 656 (1990).
31. H. Richter, J. Trodahl, and M. Cardona, Jour. of Non-Crystalline Solids **59&60**, 181 (1983).
32. Y. J. Chabal, G. S. Higashi and S. B. Christman, Phys. Rev. B **28** 4472 (1983).
33. W.B. Jackson, Phys. Rev. B. **41**, 10257 (1990)
34. S. B. Zhang, W. B. Jackson, and D. J. Chadi, Phys. Rev. Lett. **65**, 2575 (1990).

# NEW INTERPRETATIONS OF THE STAEBLER-WRONSKI EFFECT IN A-SI:H WITH MOLECULAR DYNAMICS SIMULATIONS

R. Biswas
Microelectronics Research Center, Iowa State University
Ames, IA 50011

I. Kwon
Department of Physics, Iowa State University and
Ames Laboratory- U.S. Department of Energy
Ames, IA 50011

## ABSTRACT

We discuss a new molecular dynamics approach for investigating the light-induced degradation in a-Si:H. In this approach Si-Si and newly developed Si-H interatomic potentials have been utilized to describe a 60-atom a-Si:H model containing 10% H, similar to device-quality material. Molecular dynamics schemes for investigating hydrogen-induced defects and bond-breaking models of the Staebler-Wronski effect are discussed.

## INTRODUCTION

The Staebler-Wronski effect[1] or the light-induced degradation in a-Si:H is among the most important problems facing improvements in the efficiency and quality of solar-cells. Although the light-induced degradation effect and the resulting metastable defect species have been extensively studied experimentally,[2-4] there is no consensus on the microscopic atomic mechanisms underlying the degradation phenomena. Although the saturation of the light-induced defects in a-Si:H has been observed[3] and the saturated defect density is found to correlate with the hydrogen content and the energy gap[4], there is no satisfactory theoretical model to account for these observations.

Various mechanisms have been proposed to account for the Staebler-Wronski effect that include the weak Si-Si bond-breaking induced by non-radiative recombination of photo-excited carriers,[2-5] charge-trapping models involving the capture of carriers at charged dangling bond sites and formation of neutral dangling bonds,[6,7] generation of pairs of floating and dangling bond defects,[8] metastable defect states of impurity or dopant atoms,[9] and the role of H-induced defects in a-Si:H.[10-12]

A drawback of many of these previous studies is the lack of a realistic atomic model of a-Si:H, and hence analogies with properties of H in c-Si had to be used. We overcome this difficulty by describing, in this paper, a new molecular dynamics approach that can directly deal with realistic atomic models of a-Si:H. We emphasize that the molecular dynamics approach offers great promise in investigating the mechanisms of light-induced degradation and the various models of the proposed light-induced defects.

The crucial ingredient in the molecular dynamics approach is the description of interatomic interactions. We have used a classical model of

Si-Si interactions that has had much success in describing the structure of a-Si[13] (without H). In this paper we develop a new Si-H potential that can be used for modeling the hydrogenated amorphous silicon structure. These interatomic interactions have been used to describe a realistic structural model of a-Si:H that contains 10% H, similar to device quality films. We perform molecular dynamics simulations of these a-Si:H models at any desired temperature.

## THEORETICAL MODELS

We use a model of two- and three-body interatomic potentials that can represent the energy of any arbitrary configuration of Si and H atoms. The Si-Si interactions have been previously developed and used very successfully for structural and vibrational properties of a-Si (without H). For Si-H, the two-body potential was developed by fitting to abinitio quantum-chemical calculations for H on a Si(100) surface.[14] The three-body Si-H potential, which describes Si-Si-H, H-Si-H, and even Si-H-Si interactions was developed by adjusting its strength to give satisfactory energies for silane and the amorphous phase. Details of the potential development are presented elsewhere.[15] A weakly repulsive H-H interaction, similar to that used by other workers[16], was used to prevent unphysical clustering of the H.

Using the two- and three-body interatomic potential model we can analytically find the force on each atom in the system due to its atomic environment. In the molecular dynamics technique, we numerically integrate Newton's equations of motion during each time step, for the positions and velocities of all the atoms in the system. The system can be equilibrated at any desired temperature. Alternatively, in studying the energy of static structures, we perform a steepest descent calculation, where we relax the atomic positions to their nearest potential minimum, in the energy surface.

The model of a-Si:H that we employ was that proposed by Guttman and Fong,[17] of 54 Si and 6 H atoms i.e.. 10% hydrogen, with periodic boundary conditions. All the H was present in monohydride bonding configurations in the amorphous network. This structural model was derived[17] by starting with an a-Si network without H. Pairs of H were then introduced so that Si atoms with Si-H bonds were at least second nearest neighbors or further apart.

We have relaxed the above a-Si:H model to a new minimum of the potential energy with our classical model. The H remains in monohydride form, and there are no dangling or floating bonds if we allow the density of the system to also fully relax (Fig. 1). The most relaxed structure has a mass density about 9% smaller than c-Si (Fig. 1). The root mean square bond angle deviation is 10.5°, and the rms bond-length deviation is 0.10Å for Si-Si bonds, and 0.01 Å for Si-H bonds. The presence of H relaxes the structure substantially in comparison to the model without H. We verified that this model was stable to thermal annealing up to 700K. Annealing followed by steepest descent quenches returned the network to its original configuration. The a-Si:H model was then a very convenient starting point for further investigations of the stability problem.

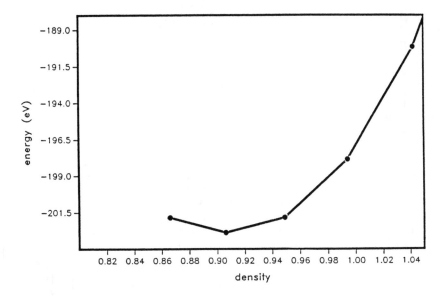

Fig. 1    Energy of the 60 atom a-Si:H model containing 10 % hydrogen, as a function of the mass density. The mass density is in units of the density of c-Si.

We note that alternative computer-generated models of a-Si:H include i) the N-atom models generated by Mousseau and Lewis[16] from an a-Si structure with 4-16% H and N > 216 atoms, and ii) the 105-atom model of Winer and Wooten[18] that has 23% H. The Mousseau-Lewis[16] model, although larger, does have a finite number of coordination defects. The Winer-Wooten model[18] has one dihydride and 22 monohydride species, but uses a non-orthogonal unit cell.

## APPROACHES TO MODELING THE STAEBLER-WRONSKI EFFECT

We first survey how the molecular dynamics approach can address the bond-breaking models of the Staebler-Wronski effect.   It is commonly believed that non-radiative recombination of photo-excited carriers is responsible for the metastable light-induced defects created by breaking weak bonds.[2,5] This transfer of recombination energy to the lattice is a difficult electron-phonon problem. A simple and computationally feasible way to model the energy   transfer is to assume that the recombination creates a "hot spot" or localized region of a few excited atoms in the network.[19] This process is analogous to exciting an electron-hole pair on a strained bond and transferring the recombination energy of the electron-hole pair to the strained bond. Generally we expect this non-radiative decay channel to be much weaker than the dominant radiative decay mode.

Such local excitations have already been tested on models of a-Si (without H).[19] The extension of these techniques to study the stability of a-Si:H including the stability of both Si-Si and Si-H bonds is under study.[15] Structural distortions of the network in this local excitation process can be identified by the displacements of atoms from their initial positions, and analysing these displacements as a function of the distance from the hot spot site. The effects of local excitations will also be tested on a-Si:H models[17] with dihydride groups. The effects of the $SiH_2$ groups in destabilizing the structure will be tested.

An alternative approach is to study the role of H-induced defects. In this case we note that a drawback of finite size models of upto a few hundred atoms in size is that the Staebler-Wronski defect would not be expected to be formed during the growth process of the computer simulation. This is because for saturated defect densities of the order of $10^{16}$ to $10^{18}$ cm$^{-3}$, the probability of encountering a light-induced defect is $10^{-6}$ to $10^{-4}$ per atom. We can however overcome this deficiency of finite-size networks by introducing relevant defects in our amorphous network and studying the stability of these defects. This is the approach we adopt in this study.

A particularly relevant H-induced defect is formed by the addition of an extra H atom to the otherwise stable network. This H atom has the lowest energy when bonded to two Si atoms in a bridge-bonded configuration. The Si-H-Si bond angle varies between 140° and 150°, so that the configuration is not colinear. This is the analogue of the bond-centered H in c-Si. At the bridge-bonded H defect the H may be displaced towards one or the other Si atoms resulting in a strong Si-H bond and a dangling bond on the other Si atom (Fig. 2). Although the dangling bond may be created on either of the two Si-atoms the energies of the two configurations (Fig. 2) are different. Network relaxation is essential in accommodating the dangling bond. Our calculations of the energetics[15] indicate that this structural change is a very viable model of the Staebler-Wronski effect, with the annealed state represented by the Si-H-Si interstitial defect and the light-soaked state by the dangling bond configuration. This structural change can qualitatively account for the metastability properties, stretched exponential kinetics and the finite density of saturation defects. The short-range motion of H is necessary for both the formation and annealing of these light-induced defects.

A definite outcome of this model for the Staebler-Wronski effect, is the presence of a H-atom in the vicinity of the metastable dangling bond defects. In principle a hyperfine interaction should exist between the H (or D) nuclear moment and the dangling bond spin. However there should be a distribution of distances between the H and the Si-dangling bond arising from the range of amorphous environments, and this may lead to a broadening of the electron spin resonance (ESR) line rather than a sharp hyperfine feature. Further experimental studies and electronic structure calculations are needed to study this aspect further.

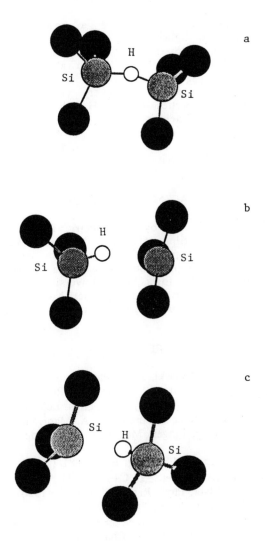

Fig. 2   The structural configuration for the bridge-bonded H (unshaded) interstitial defect where the H is weakly bonded to two Si atoms (shaded) in (a).   In the metastable higher energy configuration (b) and (c) the H forms a strong Si-H bond with one Si atom leaving a dangling bond on the other Si.   The back-bonded Si atoms are the dark circles.

In summary, we have described a new molecular dynamics approach for studying the light-induced degradation in a-Si:H. We have developed a new model of Si-H interactions and utilized it to study a structural model of a-Si:H with 10% H with all the H incorporated in monohydride bonded groups. This a-Si:H model has features in common with device-quality material and is a very viable model to study the stability and structural properties of a-Si:H. The investigation of bond-breaking models of the Staebler-Wronski effect have been surveyed, employing the molecular dynamics approach. We have also surveyed the role of H-induced defects and find that the bridge-bonded H-interstitial is a viable defect configuration that can account for several features of the Staebler-Wronski effect.

## ACKNOWLEDGEMENTS

This work was supported by the Electric Power Research Institute under the amorphous thin film solar cell program. We acknowledge a National Science Foundation grant of supercomputer time at the National Center for Supercomputer Applications, Champaign, Illinois. Work at the Ames Laboratory, operated by Iowa State University for the United States Department of Energy (USDOE) under contract No. W-7405-ENG-82, was supported by the Director for Energy Research, Division of Material Sciences, USDOE.

## REFERENCES

1.  D. L. Staebler and C. R. Wronski, Appl. Phys. Lett. 31, 292 (1977).
2.  M. Stutzmann, W. Jackson, and C. C. Tsai, Phys. Rev. B 32, 33 (1985).
3.  H. R. Park, J. Z. Liu and S. Wagner, Appl. Phys. Lett. 55, 2658 (1989).
4.  H. R. Park, J. Z. Liu, P. Roca i Cabarrocas, A. Maruyama, M. Isomura, S. Wagner, J. Abelson and F. Finger, Mat. Res. Soc. Symp. Proc. 192, 751 (1990).
5.  R. Jones and G. M. S. Lister, Phil Mag. B 61, 881 (1990).
6.  D. Adler, J. Phys. (Paris) Colloq. 42, C4-3 (1981).
7.  H. M. Branz and M. Silver, Phys. Rev. B 42, 7420 (1990).
8.  S. T. Pantelides, Phys. Rev. B 36, 3479 (1987).
9.  D. Redfield and R. H. Bube, Phys. Rev. Lett. 65, 464 (1990).
10. S. B. Zhang, W. B. Jackson, and D. J. Chadi, Phys. Rev. Lett. 65, 2575 (1990).
11. W. Jackson, Phys. Rev. B 41, 10257 (1990).
12. R. Street, Solar Cells to be published.
13. R. Biswas, G. S. Grest, and C. M. Soukoulis, Phys. Rev. B 36, 7437 (1987).
14. A. Selmani, D. R. Salahub, and A. Yelon, Sur. Sci. 202, 269 (1988).
15. R. Biswas, I. Kwon, and C. M. Soukoulis, to be published.
16. N. Mosseau and L. J. Lewis, Phys. Rev. B 41, 3702 (1990).
17. L. Guttman and C. Y. Fong, Phys. Rev. B 26, 6756 (1982).
18. K. Winer and F. Wooten, Phys. Stat. Solidi 124, 473 (1984).
19. I. Kwon, R. Biswas, and C. M. Soukoulis, Phys. Rev. B 43, 1859 (1991). (Rapid Communications).
20. C. G. VandeWalle, P.J.H. Denteneer, Y. Bar-Yam and S. T. Pantelides, Phys. Rev. B 39, 10791 (1989).

# SATURATION OF LIGHT-INDUCED DEFECTS IN a-Si:H

P.V. Santos, W.B. Jackson, and R.A. Street
Xerox Palo Alto Research Center
3333 Coyote Hill Road, Palo Alto, California 94304

## Abstract

The steady-state defect density in hydrogenated amorphous silicon (a-Si:H) under illumination was investigated for a wide range of illumination intensities and temperatures. The saturation defect density under illumination is both temperature and light intensity dependent. A chemical equilibrium model for light-induced defect generation is proposed. According to the model, defect generation is enhanced under illumination due to the reduction of the defect formation energy when the bands are populated by photogenerated carriers. Defect generation is a self limiting process and the defect density reaches a saturation value at long illumination time despite the existence of an extended distribution of defect formation sites.

## INTRODUCTION

Metastable defects are formed in hydrogenated amorphous silicon (a-Si:H) when the material is subjected to external perturbations such as strong illumination, [1, 2] doping,[3] carrier accumulation,[4] and rapid thermal quenching from high temperatures.[5] Metastable defect formation is one of the principal drawbacks for the wide spread use of a-Si:H as the active layer in devices such as solar cells, light detectors and thin film transistors, and considerable effort has been devoted to understanding the origin and the mechanisms responsible for defect formation. An important question regarding metastability is whether the observed saturation in the density of metastable defects is due to the total exaustion of defect creation sites, or to a balance between defect generation and annealing. In the first case the saturation densities should be independent of temperature and of the intensity of the perturbation, whereas in the second case saturation occur despite the existence of an extended distribution of precursor sites for defect formation.

In this paper, we investigate the saturation behavior of the density of light-induced defects in a-Si:H over a wide range of temperatures and light intensities. The saturation density of light-induced defects is both temperature and light intensity dependent, indicating that saturation is not caused by total exaustion of defect formation sites, as proposed in some models.[6] The saturation density, and its temperature and light intensity dependence, can be understood in the framework of a chemical equilibrium process involving defects and weak Si-Si bonds, similar to the used to describe defect creation by doping[7] and by thermal quenching.[8] The model takes into account the existence of an exponential distribution of weak bonds and provides a connection between the Stabler-Wronski effect and chemical equilibration in a-Si:H.

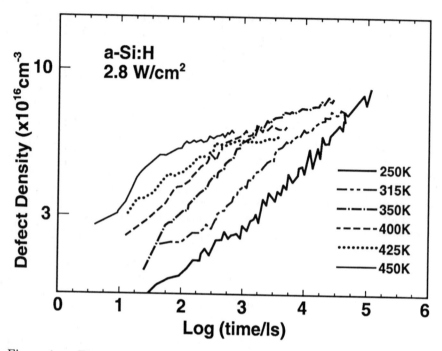

Figure 1:    Time evolution of the light-induced defect density in a-Si:H measured at different temperatures. The sample was irradiated with 2.8 $W/cm^2$ of monochromatic light (1.92 eV) from a $Kr^+$-laser.

## RESULTS

The samples used in this study are undoped a-Si:H grown by the glow discharge decomposition of pure silane at 230°C. In the light soaking experiments, the light source was either a tungsten-halogen lamp with filters to cut wavelengths outside the range from 900 to 630 nm, or, for high illumination intensities, the 6471 Å line of a $Kr^+$. Prior to each soaking experiment, the samples were annealed in dark at 500 K for at least 2 hr and cooled down to the soaking temperature at a rate of $\sim 2K/min$. The defect densities were determined by the constant photocurrent method (CPM). These measurements were performed in the coplanar configuration with ohmic contacts consisting of a 600Å chromium layer over a 200 Å thick 1% phosphorus doped a-Si:H layer.

The time dependence of the defect density during soaking was followed by a variation of the CPM method which consists in interrupting the illumination and determining the photon flux ratio $r_{CPM} = F(1.15eV)/F(2.0eV)$ that yield the same AC photo-current at the energies of 1.15 eV and 2.0 eV, respectively. The defect density was then taken to be proportional to $r_{CPM}$. The energy of 1.15 eV was chosen because it lies deep in the region of defect absorption and sufficiently far away from the band tail region. The beam of 2 eV is completely absorbed by the sample and the the CPM signal is weakly energy dependent in this region.

This procedure allows the determination of the defect density in a relatively short time (30 to 40 s) and a full justification for its use to determine the defect density will be given elsewhere.[9] Absolute values for the defect density were obtained by recording full CPM spectra before and after each soaking experiment.

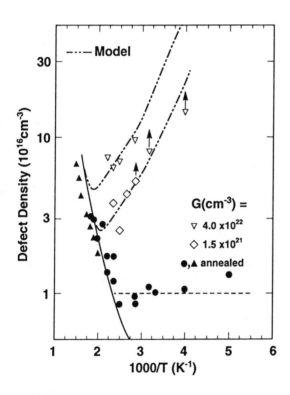

Figure 2: Defect density in the annealed state (solid dots) and after irradiation with 2.8 $W/cm^2$ (inverted triangles, corresponding to a carrier generation rate $G = 4.0 \times 10^{22} cm^{-3}$) and 0.065 $W/cm^2$ (diamonds, $G = 1.5 \times 10^{21} cm^{-3}$). The horizontal dashed line is the frozen-in defect density. The solid triangles represent the ESR-spin density in undoped a-Si:H films quenched from high temperatures (from Ref. [8] ). The solid line is the calculated saturation defect density in the dark. The upper and lower dotted curves are for generation rates of $4.0 \times 10^{22}$ and $1.5 \times 10^{21} cm^{-3}$, respectively.

Figure 1 shows the time dependence of the light-induced defect density for different temperatures, for a fixed light flux of 2800 $mW/cm^2$, corresponding to a generation rate $G = 4 \times 10^{22} cm^{-3} s^{-1}$. Above room temperature the defect density saturates around $\sim 10^{17} cm^{-3}$ for the highest illumination intensity used. The saturation values decrease slightly with increasing temperatures. By observing the decay of the defect density after the light is shut off, we verified that the saturation is not due to partial annealing of light-induced defects when the illumination is interrupted to measure the defect density. At room temperature and below, a clear saturation is not observed, even after $\sim 10^5$s of strong illumination. Except for the highest temperatures the curves are approximately parallel in the mid range of defect densities, following a power law time dependence with an exponent $\eta$ varying from 0.22 to 0.30. A similar time dependence has been measured by Lee and co-workers [12] and is slighty weaker then the dependence $N_s \sim t^{-\frac{1}{3}}$ reported by Stutzmann *et al.*[2]

Figure 2 shows the temperature dependence of the defect density in the annealed (dots) and in the light-soaked state (open symbols). The defect density in the annealed state is constant below 390K. Above this temperature, the defect density increases with an activation energy of 0.20 - 0.22 eV. We attribute the increase in the defect density to thermal equilibration of defects in the a-Si:H network.[13] For comparison, the solid triangles in the figure display the equilibrium spin density determined by ESR in undoped a-Si:H films quenched from high temperatures.[8]

The additional symbols in Fig. 2 correspond to the saturated defect density under different illumination intensities. The saturation density decreases with increasing temperature, and the effect is stronger for low illumination intensities. Where a clear saturation was not achieved, an arrow was added to the data point to indicate a lower limit for the saturation density. A remarkable point is that the saturation density depends very weakly on the illumination intensity. An increase in the illumination intensity by a factor of $\sim$ 30 at 400 K leads to a doubling of the saturation density. If a power law dependence of the type $N_{s,\infty} \sim G^\epsilon$ is assumed, the exponent $\epsilon \sim 0.2 - 0.3$ at 400 K and decreases at lower temperatures. These results are in contrast with those reported recently by Park et al,[10] who reported a saturation density at room temperature independent of the light-induced carrier generation rate. On the other hand, the light intensity dependence of the saturation defect density is much weaker than the predict by the model for light-induced defect formation proposed by Stutzmann et al.[2]

A clear evidence that saturation depends on light soaking intensity is provided by Fig. 3. The first curve in this figure shows the increase in defect density when the sample is soaked with 580 $mW/cm^2$ at 350K. The illumination intensity was then decreased by a factor of 10 using neutral density filters. The defect density decreases, indicating a reduced saturation level for the weaker illumination. By re-applying the original illumination intensity , the defect density increases again, as is illustrated in the third curve. Finally, the last curve displays defect annealing in the dark. Note that neglegible annealing occurs in the time scale of $\sim$ 1 min, when the illumination is interrupted to measure the defect density during soaking.

## DISCUSSION

The dependence of the saturation density $N_{s,\infty}$ on the illumination intensity and on temperature suggests that saturation is due not to depletion of precursor sites for defect formation, but rather to a balance between defect generation and annealing. Recently, Winer and Street[8, 7] examined thermodynamic models for the equilibrium defect density in undoped a-Si:H based on a chemical equilibrium between weak Si-Si-bonds $(SiSi)$ and neutral dangling bonds $(Si-)$ in the amorphous network. The defect density in the dark is well explained by a chemical equilibrium reaction expressed by: [8]

$$SiH + (SiSi) \rightleftharpoons (Si-) + (SiHSi-) \qquad (1)$$

Chemical equilibration at low temperatures takes place through the motion of hydrogen atoms from bond terminating sites $(SiH)$ to a weak-bond sites $(SiSi)$ that make the valence-band tail of localized states, leading to the formation of two defects: (i) an isolated dangling bond $Si-$ in the original hydrogen site $SiH$ and (ii) a complex consisting of a dangling bond adjacent to a SiH bond, which can not be separated from each other. Both defects have the same density $N_D$ so

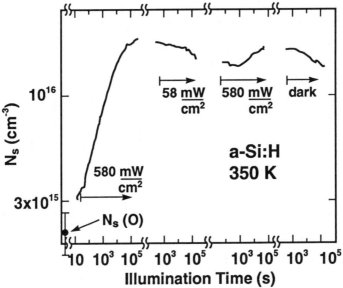

Figure 3: Time evolution of the defect density for different soaking intensities

that the total defect density is $N_s = 2N_D$. If all but the electronic contribution to the energy difference between reactants and products are neglected, the change in enthalpy $2U_{f0}$ for the forward reaction in Eq. 1 corresponds to the energy necessary to promote two electrons from the initial weak bond state of energy $E_{VB}$ in the valence bond tail to the defect state of energy $E_D$ in the gap, i.e., $2U_{f0} = 2(E_D - E_{VB})$. Assuming an exponential energy distribution of the weak bonds given by $N_v(E_{VB}) = N_{V0}exp(-E_{VB}/kT_v)$, where $T_v$ is the exponential slope of the valence-band tail,[11] and taking the energy reference to be the valence band mobility edge, the equilibrium density $N_D$ of converted bonds is the solution of the following integral equation:[8, 13]

$$N_D = \int_0^\infty \frac{N_v(E)dE}{1 + exp\left(\frac{2(\mu_0 - E)}{kT}\right)} \qquad \text{with} \qquad \mu_0 = E_D + \frac{kT}{2}ln\frac{N_D}{N_H} \qquad (2)$$

Here, the integration is over the distribution of valence band tail states and $\mu_0$ is an effective defect chemical potential describing the equilibrium between defects and weak bonds. It is easy to see from Eq. 2 that at a given temperature, most of the weak bonds with energy above $\mu_0$ are converted to defects. Equation 2 satisfactory explains the temperature dependence of the equilibrium defect density in a-Si:H.[8]

A strict thermodynamic analysis of the chemical reactions cannot be made under illumination, since the material is not in equilibrium. In order to extend the chemical equilibrium model to calculate the steady-state defect density under illumination, we assume that the sole effect of illumination is to shift the electron (hole) quasi-Fermi, $E_{Fn}$ ($E_{Fp}$), level towards the conduction (valence) band. The defect formation energy is reduced if the newly formed defects capture photo-generated carriers from the bands, or, alternatively, if an electron-hole pair is involved in the defect formation reaction, so that more defects can be thermally generated at the same temperature. Different defect formation reactions involving photo-generated carriers can be envisaged.[9] We will only consider here the simple process involving an electron-hole pair $(e - h)$ described by the reaction:

$$SiH + (SiSi) + e + h \rightleftharpoons (Si-) + (SiHSi-) \tag{3}$$

The enthalpy change for this reaction is smaller than that for Eq. 1 by the energy difference $E_{Fn} - E_{Fp}$ between the electron and hole quasi-Fermi levels. The defect chemical potential under illumination $\mu$ is expressed by:

$$\mu = 2E_D - (E_{Fn} - E_{Fp}) + \frac{kT}{2} ln \frac{N_D}{N_H} \tag{4}$$

$\mu$ depends on the quasi-Fermi levels and therefore on the generation rate $G$. With increasing illumination levels, $\mu$ moves closer to the valence band edge, and more weak bonds are transformed into defects. The defect creation process is self-limiting since the newly created recombination centers reduce $E_{Fn}$ and $E_{Fp}$, inhibiting further defect formation.

In order to obtain the defect chemical potential from Eq. 4 it is necessary to relate the quasi-Fermi levels $E_{Fn}$ and $E_{Fp}$ to the defect density and to the carrier generation rates $G$. We will assume the following simple approximation for $E_{Fn}$:

$$E_c - E_{Fn} = -kT ln \left( \frac{G^{\gamma_n}}{A_n N_c N_s^{\delta_n}} + e^{-\frac{E_a}{kT}} \right) \tag{5}$$

with $\gamma_n \sim 1$ and $0.6 < \delta_n < 1.2$.[14, 15] Here, $E_c$ is the mobility gap, $N_c = 2.9 \times 10^{19} \left( \frac{T}{300K} \right)^{3/2} cm^{-3}$[16] is the effective density of states at the mobility edge, $E_a$ is the activation energy for the dark conductivity, and $A_n$ is the effective electron capture probability of the recombination centers. A similar expression was used to calculate the hole quasi-Fermi level.

Substitution of Eqs. 4 and 6 in Eq. 2 yields the following analytical approximation for the saturation defect density $N_{s,\infty}$ in the low temperature range $(T << T_v)$:

$$N_s^{\infty} \approx 2 \left[ k \left( \frac{2T_v^2}{2T_v - T} \right) N_{v0} \right]^{\frac{2}{\zeta}} exp \left[ -\frac{2E_D - E_c}{\zeta kT_v} \right] \left( \frac{N_H G^{(\gamma_n + \gamma_p)}}{A} \right)^{\frac{\beta}{\zeta}} \tag{6}$$

with $\beta = T/T_v$, $A = 2^{(\delta_n + \delta_p)} A_n N_c A_p N_v$, and $\zeta = 2 + \beta(1 + \delta_n + \delta_p)$

Equation 6 predicts a power-law dependence of the saturation defect density on the generation rate G with a temperature dependent exponent $\epsilon = (\gamma_n + \gamma_p)\beta/[2 + \beta(1 + \delta_e + \delta_p)]$. Assuming a characteristic tail temperature $T_v = 550K$, and $\delta_{n,p} = \gamma_{n,p} = 1$, we obtain an exponent $\epsilon$ that varies between $\epsilon = 0.27$ at 250K and $\epsilon = 0.35$ at 400K, which agrees with the experimental value in this temperature range. Lower $\epsilon$'s are obtained if the $\gamma_{n,p}$ are smaller.

In order to test the chemical equilibrium model described by Eq. 3, the steady-state defect density under illumination was calculated from Eqs. 2 and 5 (without using the suppositions that led to Eq. 6), for the experimental conditions of Fig. 2. The results are the lines superposed on the experimental points in Fig. 2. The

| Parameter | Description | Value* |
|-----------|-------------|--------|
| $E_G$ | Band-gap | 1.6eV |
| $E_D$ | Defect level | 0.6 eV |
| $T_v$ | VB-Temperature | 550 K |
| $N_{v0}$ | Density of VB tail states | $2 \times 10^{20}\ cm^{-3} eV^{-1}$ |
| $A_n^\dagger$ | electron capture probability | $3 \times 10^{-9} cm^3/s$ |
| $A_p^\dagger$ | hole capture probability | $0.8 \times 10^{-9} cm^3/s$ |
| $N_H$ | Hydrogen density | $5.0 \times 10^{21} cm^{-3}$ |

\* Energy values measured with respect to the valence-band-mobility edge.
† From Ref. [2]

Table I: Parameters used in the calculations of the saturation densities

parameters used in the calculations are summarized in Table I, and are similar to those used by Street and Winer [8] to describe the temperature dependence of the equilibrium defect density in the dark. We restrict ourselves again to the simple case where $\delta_{n,p} = \gamma_{n,p} = 1$ and we considered the equilibrium Fermi-level to coincide with the defect level $E_D$, i.e., $E_a = E_c - E_D$. Finally, in order to shift the curves vertically and obtain better agreement with the experimental data, we scaled the capture probabilities $A_n$ and $A_p$ by a constant factor $k_s = 2.5 \times 10^4$. It should be noted, however, that similar results are obtained if other parameters such as $E_D$, $\gamma_{n,p}$, or $N_{v0}$ are scaled instead of $A_n$ and $A_p$, and the temperature and light intensity dependence of the defect density is not substantially affected.

The solid line in Fig. 2 is the equilibrium defect density in the dark. It reproduces well the experimental data in the annealed state for temperatures above the freeze-in temperature of $\sim 400$ K. For each model, the defect density under illumination was calculated for carrier generation rates of $4 \times 10^{22} cm^{-3}$ (upper curve) and $1.5 \times 10^{21} cm^{-3}$ (lower curve). The model predicts a weak dependence of the saturation density on temperature. In the low temperature region, the saturation defect density decreases with an effective activation energy of 0.10 - 0.15 eV. The model gives a reasonable agreement to the experimental data in the high temperature region. Unfortunately, precise comparison between these two models and experimental data in the low temperature region is difficult since a clear saturation in the density of light-induced defect in not reached even after long illumination times (see Fig. 1). This is indicated by the arrows on the experimental points in Fig. 2, which in this case correspond to a lower limit for the saturation density.

## CONCLUSIONS

We presented new experimental data on the saturation value for the light-induced defect density showing that it depends both on temperature and on illumination intensity. The saturation defect density is well explained by a chemical equilibrium model for the defect density under illumination, similar to the one used to account for the equilibrium defect density in the dark. According to the model, the increased defect density under illumination is due to a decrease in the defect formation energy, when the electron and hole quasi-Fermi energy is moved towards the conduction and valence band, respectively.

This research was supported by the Solar Energy Research Institute (Golden, CO).

# References

[1] D.L. Stabler and C.R. Wronski, *Appl. Phys. Lett.* **31**, 292 (1977).

[2] M. Stutzmann, W.B. Jackson, and C.C. Tsai, *Phys. Rev. B* **32**, 23 (1985).

[3] R.A. Street, M. Hack, and W.B. Jackson, *Phys. Rev. B* **37**, 4209 (1988).

[4] W.B. Jackson and J. Kakalios, in *Amorphous Silicon and Related Materials*, Eds.: Hellmut Fritzsche ( World Scientific, Singapore, 1988 ), p. 247.

[5] R.A. Street, J. Kakalios, C.C. Tsai, and T.M. Hayes, *Phys. Rev. B* **35**, 1316 (1987).

[6] D. Redfield, *Appl. Phys. Lett.* **48**, 846 (1986).

[7] K. Winer, *Phys. Rev. B* **41**, 12150 (1990).

[8] R.A. Street and K. Winer, *Phys. Rev. B* **40**, 6236 (1989).

[9] P.V. Santos, to be published.

[10] H.R. Park, J.Z. Liu, and S. Wagner, *Appl. Phys. Lett.* **55**, 2658 (1989).

[11] Z E. Smith and S. Wagner, *Phys. Rev. B* **59**, 688 (1987).

[12] C.Lee, W.D. Ohlsen, and P.C. Taylor, in *Proc. of the 18th International Conference on the Physics of Semiconductors*, Eds.: O. Engström ( World Scientific, Singapore, 1986 ), p. 1085.

[13] K. Winer, *Phys. Rev. Lett.* **63**, 1487 (1989).

[14] W.B. Jackson, *Phil. Mag. Lett.* **59**, 103 (1989).

[15] C.R. Wronski, Z.E. Smith, S. Aljishi, V. Chu, K. Shephard, D.-S. Shen, R. Schwartz, D. Slobodin, and S. Wagner, in *Stability of Amorphous Silicon Alloy Materials and Devices (AIP Conf. Proc. No. 157)*, Eds.: B.L. Stafford and E. Sabisky ( American Institute of Physics, New York, 1987 ), p. 172.

[16] N. Ashcroft and N.D. Mermin , *Solid State Physics* (Holt-Rinehardt and Winston, New York, 1976).

# INVESTIGATION OF THE STAEBLER-WRONSKI EFFECT IN A-SI:H BY SPIN-DEPENDENT PHOTOCONDUCTIVITY

Martin S. Brandt and Martin Stutzmann

Max-Planck-Institut für Festkörperforschung, D 7000 Stuttgart 80, Germany

## ABSTRACT

We investigate undoped amorphous hydrogenated silicon with spin-dependent photoconductivity (SDPC). In addition to recombination via single carriers in tail states and dangling bonds, a broad resonance is observed, attributed to recombination of excitonic states. A study of the influence of light-induced degradation on the different recombination processes in a:Si:H is presented. The experiments are well described by existing theories, which assume exciton-like states to be the electronic precursor for metastable defects.

## INTRODUCTION

Many different experimental methods have been used to study light induced defect creation[1] in amorphous hydrogenated silicon. Among the more common techniques are electron spin

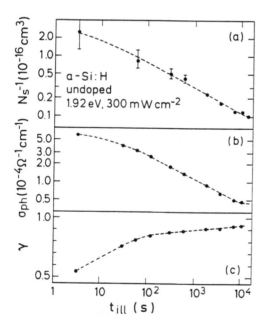

Figure 1: Changes in the inverse defect density $N_S^{-1}$, the photoconductivity $\sigma_{ph}$ and $\gamma$ of undoped a-Si:H upon illumination with homogeneously absorbed light.

resonance (ESR), photothermal deflection spectrosopy, constant photocurrent method and photoconductivity (PC). We report here results for undoped a-Si:H obtained with a combination of two of these, namely ESR and PC, called spin-dependent photoconductivity (SDPC).[2-5]

For this experiment, a thin film prepared by conventional glow discharge and equipped with interdigit Al-electrodes is placed in an X-band ESR-cavity. Under illumination from a tungsten lamp ($\approx$ 30mW/cm$^2$), the change in photoconductivity under spin resonance conditions is detected. Magnetic field modulation is used as in conventional ESR, but the spectra are numerically integrated for enhancement of broad features.

For a given recombination step, the transition from initial to final states has to be spin allowed. Initial states, whose recombination is allowed (e.g. an electron and a hole forming a spin singlet, antiparallel spins), have a shorter lifetime than initial states, whose recombination is forbidden (e.g. triplet states, parallel spins). This results in a depletion of "allowed" states relative to their thermal equilibrium occupation probability. By mixing allowed and forbidden configurations, saturation of ESR induces a net transfer of forbidden to allowed states, thereby enhancing the particular recombination step.[6] In SDPC, this is then observed as a decrease of photoconductivity at resonance.

Changes in spin density ($N_S$) and photoconductivity ($\sigma_{ph}$) of undoped a-Si:H under prolonged illumination with homogeneously absorbed light (1.91 eV, 300 mW/cm$^2$) are summarized in Fig. 1. For illumination times longer than 100 seconds, we observe $\sigma_{ph} \propto N_S^{-1}$ for defect densities $N_S > 10^{16}$cm$^{-3}$, as expected for the monomolecular case. This is corroborated by determining $\gamma$, the exponent in the power law describing the relation between $\sigma_{ph}$ and the intensity $I$ of the incident light ($\sigma_{ph} \propto I^\gamma$).[7] For long illumination times, $\gamma$ approaches the monomolecular value of 1, while we observe bimolecular recombination in the annealed sample ($\gamma \approx 0.5$). Having established these basic results, we turn to the SDPC measurements.

## SDPC EXPERIMENTS

Figure 2 shows SDPC spectra taken at 175 K after light induced degradation at room temperature for different times. Note that the magnetic field range in Fig. 2 (1000 G) is much larger than what is commonly used for spin resonance in a-Si:H. The central line,

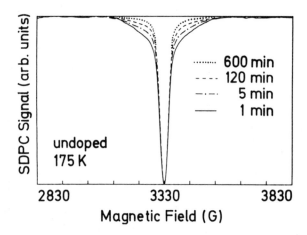

Figure 2: SDPC signal of undoped a-Si:H, normalized to the same intensity at the center resonance. The spectra are taken after light soaking at 300 K (1.91 eV, 300 mW/cm$^2$) for different durations.

which is overmodulated in this particular measurement, consists of a combined e-db line ($g = 2.0050$), due to the transition from an electron in the conduction band tail to a neutral dangling bond, and an h-line ($g = 2.01$), arising from holes hopping in the valence band tail. These are the spin-dependent steps observed for the recombination via dangling bonds and can be separated from one another by phase shift analysis.[5]

Clearly seen is a second, broader resonance ($g \approx 2$, $\Delta H_{pp} \approx 190$ G), with a gaussian lineshape. The most obvious mechanism for a gaussian broadening of this size is dipolar spin-spin-interaction,[8] where $\Delta H_{pp}$ is of the order of the dipolar energy $g^2 \mu_B^2 / r^3$. The distance $r$ between the interacting spins can thus be estimated to be approximately 5Å. A likely candidate for such a signal is a Frenkel-type exciton, consisting of an electron in the conduction band tail and a hole in the valence band tail. Excitons can either be formed in the process of bimolecular recombination of free carriers or during the thermalization of geminate pairs. The recombination of such excitons, trapped at weak bonds (Fig. 3) has been suggested as the cause for light induced degradation of a-Si:H, the Staebler-Wronski-Effect.[9] According to Fig. 2, light soaking decreases the influence of trapped excitons to the recombination in a-Si:H, as detected by SDPC. Increasing the spin density $N_S$ enhances the recombination via dangling bonds, thereby effectively quenching the relative contribution of excitons to the recombination.

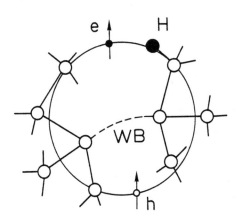

Figure 3: Schematic representation of an excitonic state, formed by an electron and a hole with a distance of approx. 5Å, localized at a weak bond. The recombination of these excitons can lead to metastable dangling bonds via breaking of the weak bond.

When observed with SDPC, the intensity of the exciton line has a distinct temperature dependence, as shown in Fig. 4. While it is hardly observable at room temperature, the exciton line attains a broad maximum at around 160 K and decreases again, when measured at lower temperatures. The behaviour at low temperatures can be readily understood from a comparison with photoluminescence under spin resonance conditions (optically detected magnetic resonance, ODMR).[10,11] The resonance can only be observed in SDPC when exciton recombination and thermal breakup occur with similar probabilities. At low temperatures ($T < 100$ K), thermal dissociation becomes unlikely, radiative recombination increases and the resonance can be observed in ODMR.[10,11]

A quantitative summary of the SDPC temperature dependence in the annealed and light soaked state of undoped a-Si:H is shown in Fig. 5. The upper part gives the temperature dependence of the dc photoconductivity, both for the annealed (left) and illuminated (right) sample. The dc photoconductivity shows the expected decrease after light soaking. In addi-

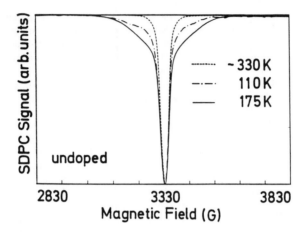

Figure 4: SDPC signal of an undoped a-Si:H sample after anneal, normalized to the same intensity at the central resonance. The spectra are taken at different measurement temperatures.

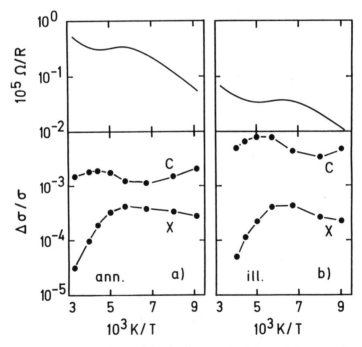

Figure 5: Temperature dependence of the dc photoconductivity and the normalized SDPC signals $\Delta\sigma_{\text{central}}/\sigma_{ph}$ for the central line (C) and $\Delta\sigma_{\text{exciton}}/\sigma_{ph}$ for the exciton line (X). Part a) shows the results for the annealed sample, b) for the sample after light soaking for 120 minutes.

tion, it exhibits thermal quenching at temperatures above 160 K, due to thermal excitation of holes trapped in the valence band tail. In the lower part of Fig. 5, the maximum change $\Delta\sigma$ of the SDPC signals at resonance is plotted, normalized to the photoconductivity $\sigma_{ph}$ at the respective measurement temperature. The normalized SDPC signal $\Delta\sigma_{central}/\sigma_{ph}$ for the central line (i.e. the recombination via dangling bonds) is essentially constant over the temperature range investigated. (The weak temperature dependence is related to the structure due to thermal quenching in the dc photoconductivity.) The SDPC signal $\Delta\sigma_{exciton}/\sigma_{ph}$ for the exciton, in contrast, exhibits the temperature dependence already discussed in Fig. 4. In the light soaked state, $\Delta\sigma_{central}/\sigma_{ph}$ has increased, while $\Delta\sigma_{exciton}/\sigma_{ph}$ is found to decrease for temperatures above 200 K and assumes the same values as in the annealed samples for $T < 200$ K.

To conclude the experimental analysis, the relative effect of light soaking on the different SDPC signals is presented in Fig. 6. in more detail. Here the ratio

$$\frac{\Delta\sigma_{exciton}}{\sigma_{ph}} \Big/ \frac{\Delta\sigma_{central}}{\sigma_{ph}} = \frac{\Delta\sigma_{exciton}}{\Delta\sigma_{central}}, \tag{1}$$

determined at a $T = 175$ K is plotted versus the photoresistance of the sample at 300 K, which we have verified to be proportional to the defect density in Fig. 1. For illumination times longer than 100 seconds, this ratio is proportional to the inverse number of dangling bonds.

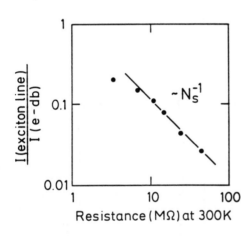

Figure 6: Ratio of $\Delta\sigma_{exciton}/\sigma_{ph}$ to $\Delta\sigma_{central}/\sigma_{ph}$, for a measurement temperature of 175 K. Due to the small magnitude of the hole contribution to the central SDPC line, we can use I(central) and I(e-db) interchangeably.

## DISCUSSION

Our experimental observations can be understood starting with two simplified rate equations[9]

$$dn/dt = G - A_X np - A_n nN, \tag{2}$$

$$dp/dt = G - A_X np - A_p pN, \tag{3}$$

where $G$ is the generation rate and $A_X$, $A_n$, $A_p$ are the transition rates for excitonic recombination, for the transition of an electron to a dangling bond and for the transition of a hole to the dangling bond, respectively. In the monomolecular case ($A_X np \ll A_n nN$, $A_p pN$), the solutions are

$$n = \frac{G}{A_n N}, \quad p = \frac{G}{A_p N}, \quad \sigma_{ph} \propto \frac{1}{N}. \tag{4}$$

The change in conductivity for the dominant contribution to the central line observed in SDPC, the transition of an electron to a dangling bond, would be

$$\Delta\sigma_{\text{central}} \propto A_n nN. \tag{5}$$

Combined with Eq. (4), this leads to a linear dependence of $\Delta\sigma_{\text{central}}/\sigma_{ph}$ on the number of dangling bonds

$$\frac{\Delta\sigma_{\text{central}}}{\sigma_{ph}} \propto \frac{\frac{1}{N}N}{\frac{1}{N}} = N. \tag{6}$$

Similarly, we have for the excitonic resonance

$$\Delta\sigma_{\text{exciton}} \propto A_X np. \tag{7}$$

As shown in Fig. 4, $\Delta\sigma_{\text{exciton}}/\sigma_{ph}$ is roughly independent of $N_S$ at temperatures below 200 K. This region coincides with the region of hole trapping. Supposing that all available hole traps which can form excitons, e.g. hole states at the weakest bonds (c.f. safe hole traps[12-14]), are filled, $\Delta\sigma_{\text{exciton}}$ would be independent of $p$ and the last equation would change to

$$\Delta\sigma_{\text{exciton}} \propto A_X nh_t, \tag{8}$$

with $h_t$ the concentration of these traps. This then leads to

$$\frac{\Delta\sigma_{\text{exciton}}}{\sigma_{ph}} \propto \frac{\frac{1}{N}}{\frac{1}{N}} = \text{const}, \tag{9}$$

as observed for $T < 200$ K. At temperature above 200 K, holes are excited from traps, so that the above approximation becomes invalid.

Combing Eqs. (6) and (9), we can now deduce

$$\frac{\Delta\sigma_{\text{exciton}}}{\Delta\sigma_{\text{central}}} \propto \frac{1}{N}, \tag{10}$$

as observed in Fig. 6. This result confirms quantitatively the main assumptions of recombination based models for the Staebler-Wronski-effect, namely that the relative importance of defect creating recombination steps (e.g. exciton recombination) decreases as recombination via dangling bonds becomes enhanced during light soaking. Since light induced degradation usually occurs under conditions of a constant generation rate, the total frequency of recombination events must also be constant. Therefore, the relative decrease of excitonic recombination seen in Figs. 2 and 6 is directly related to the slowing down of the metastable defect creation with increasing illumination time.

## CONCLUSION

We have shown that excitonic states can be observed in photoconductivity over the temperature range from 100 K to 300 K. The excitonic line as detected by SDPC is studied as a function of measurement temperature and light soaking. The recombination of such excitons localized at weak bonds is thought to be responsible for the creation of metastable states under prolonged illumination. We have shown that the behaviour of the excitonic line in SDPC is consistent with the predictions of "bond breaking" models.[9]

## REFERENCES

1. D. L. Staebler and C. R. Wronski, Appl. Phys. Lett. **31**, 292 (1977).
2. I. Solomon, D. Biegelsen and J. C. Knights, Solid State Commun. **22**, 505 (1977).
3. E. A. Schiff, AIP Conf. Proc. **73**, 233 (1981).
4. R. A. Street, Phil. Mag. B **46**, 273 (1982).
5. H. Dersch, L. Schweitzer and J. Stuke, Phys. Rev. B **28**, 4678 (1983).
6. D. Kaplan, I. Solomon and N. F. Mott, J. Phys. (Paris) **39**, L 51 (1978).
7. A. Rose, Concepts in Photoconductivity and Allied Problems (Krieger, New York, 1978).
8. J. H. van Vleck, Phys. Rev. **74**, 1168 (1948).
9. M. Stutzmann, W. B. Jackson and C. C. Tsai, Phys. Rev. B **32**, 23 (1985).
10. F. Boulitrop, Phys. Rev. B **28**, 6192 (1983).
11. K. Morigaki, J. Non-Cryst. Solids **77&78**, 583 (1985).
12. T. J. McMahon and R. S. Crandall, Phil. Mag. B **61**, 425 (1990).
13. T. J. McMahon and J. P. Xi, Phys. Rev. B **34**, 2475 (1986).
14. B. Gu, D. Han, C. Li and S. Zhao, Phil. Mag. B **53**, 321 (1986)

This work was supported by Bundesministerium für Forschung und Technologie under contract 0328962A.

# THE REHYBRIDIZED TWO-SITE (RTS) MODEL FOR DEFECTS IN a-Si:H

David Redfield and Richard H. Bube
Department of Materials Science & Engineering, Stanford University
Stanford, CA 94305-2205

## ABSTRACT

A comprehensive model for the metastable defects in a-Si:H is developed by adapting a recent theory for several kinds of defects in crystalline semiconductors, particularly the DX center in AlGaAs. This new model accounts in a unified way for all of the major observations of defects induced by light, quenching, doping, or compensation; as well as for their anneal. The stretched-exponential time dependence of defect densities with light or annealing, and saturation of the density are also explained. This model is based on foreign atoms rather than on Si-Si bond breaking, and in undoped materials it is suggested that uncontrolled impurities are the source.

## INTRODUCTION

The subject of defects and metastabilities in semiconductors is currently very active in several areas, one of which is the dangling-bond defect in hydrogenated amorphous silicon (a-Si:H). Since understandings of a-Si:H are based generally on analogous crystalline semiconductors, we have developed this microscopic model[1] for the metastable defects in a-Si:H by adaptation of a new theory that is having success in describing a number of defects and metastabilities in crystalline semiconductors. The most prominent of these currently is the DX center in AlGaAs alloys, for which the new description is given in Ref. 2. We find that this model also explains naturally the properties of defects caused by light, by doping or by quenching in a single unified picture.

There are a number of similarities in properties of the dangling-bond (DB) defect in a-Si:H and the DX center that suggest similar descriptions. Both are localized centers having a metastable state whose energy is above that of its ground state by about 0.2 eV, and with an energy barrier between the states that permits a relative stability for the metastable state at moderate temperatures. In both materials either light or temperature can induce transitions between the two states; these transitions probably all involve a large structural relaxation. The kinetics of the transitions in either direction are nonexponential, and were shown recently to be stretched exponentials: in Ref. 3 for the DX center and in Refs. 4-6 for the DB. Also, the stretch parameters of annealing kinetics in both materials are proportional to temperature. More similarities will be indicated below.

The DX center in AlGaAs is known to be associated with the presence of donors, the origin of the "D" in the name. Our "rehybridized two-site" (RTS) model is also founded on foreign atoms; as explained here, they may be donors or acceptors or any other atom having prescribed properties. After the model is described in terms of doped or compensated materials, we discuss alloys and finally, cases in which there are no added dopants.

## THE RTS MODEL

As with the Chadi and Chang model for the DX center,[2] the ground state of the RTS center in doped a-Si:H has a dopant atom at a substitutional site with 4-fold coordination. This corresponds to a normal, electrically active dopant, so a free carrier exists for each such atom. Because the dopant atom simulates a host atom in this state, there is no observable <u>localized</u> electronic property of the center (the free carrier is its only evidence). In the metastable state of the center, the dopant atom has broken one of its bonds to an adjacent Si atom and moved away on the extended axis of the ruptured bond to the neighboring interstitial site as shown pictorially in Fig. 1. In this metastable configuration, the dopant atom has 3-fold coordination, so the dopant wave functions have rehybridized from $sp^3$ in the ground state to basically $sp^2$ in the metastable state. Hence the name "rehybridized two-site" or RTS center. We hypothesize that the Si whose bond is broken retains its $sp^3$ hybrids, thus leaving a Si dangling bond (DB). In this way, a center based on a foreign atom creates a DB on a Si atom, giving the appearance of breaking an intrinsic bond of the Si itself.

A RTS center in its ground state, RTS$^g$, replaces the "weak bond" of Si-Si bond-breaking models; and a RTS center in its metastable state, RTS$^*$, is the "defect" with a dangling bond on a Si atom. Any foreign atom that can take either 3-fold or 4-fold coordination can be responsible for a RTS defect; atom types other than the usual dopants are considered toward the end.

The contribution of the Chadi and Chang calculations was to show that there is a site at which a 3-fold configuration for a donor in AlGaAs is relatively stable, with only a slightly higher energy ($\approx0.2$ eV) than that of the ground-state configuration.[2] In fact, for some alloy compositions the 3-fold configuration has a lower energy and becomes the ground state (as observed). Also, they show in other work that similar effects occur with other dopants and in other materials (see other papers cited in Ref. 1).

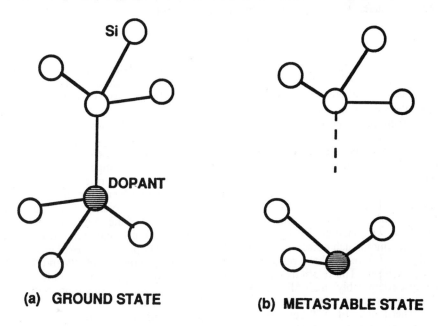

(a) **GROUND STATE**        (b) **METASTABLE STATE**

FIGURE 1. The two configurations of a RTS center. The dashed line is a dangling bond. The "dopant" may be any atom that can have either 3- or 4-fold coordination.

A model for the DB in a-Si:H similar to this RTS model was proposed[7] before the Chadi and Chang calculations and therefore lacked their key results; it seems not to have been pursued actively since. Other proposals by Wautelet et al.,[8] and Branz[9] have similarities to the RTS model, but also important differences.

A major consequence of the configuration difference is that a dopant in the 3-fold state forms a localized center having a deep level, rather than a shallow level, so its carrier is likely to be recaptured. Thus when the defect forms the conductivity of the material is reduced, as is observed in light-induced effects. Taking an acceptor (e.g., boron) in lightly doped material as an example, the $RTS^g \rightarrow RTS^*$ transition causes the center to change its charge state from negative in the ground state to neutral in the metastable state upon capture of the hole

$$a^- + h^+ \rightarrow D^o . \tag{1}$$

Here $a^-$ is the negatively charged shallow acceptor, and $D^o$ is the (deep-level) defect in its neutral state, i.e., the DB has a single electron and is thus paramagnetic. If the doping is heavy enough to lower the Fermi energy below the $D^o/D^+$ energy level of the RTS center, however, then the DB will be unoccupied and $RTS^*$ will be positive and diamagnetic. It is thus the coordinations that define the two states of the RTS center more than the charges or the spins.

For centers formed by donor atoms (e.g., phosphorus), this description is essentially the same; but rehybridization to form the metastable state involves all 5 valence electrons of the donor, so that an electron must be captured to form the metastable state. The relation between these two states is

$$d^+ + e^- \rightarrow D^o . \tag{2}$$

It can be seen that the $D^o$ are configurationally the same for both donors and acceptors; only the central-cell potential identifies the particular foreign atom involved. We note that all the normal donors and acceptors in Si tend to favor 3-fold coordination, hence easing the formation of $RTS^*$.

When the material is in equilibrium the populations of the two states have a ratio given by a Boltzmann factor $\exp[ - (\Delta E / kT)]$, with $\Delta E$ the energy difference between the two states. Because $\Delta E$ for these centers is small – only $\approx 0.2$ eV – the equilibrium population of the metastable state is appreciable at common temperatures; e.g., more than 1% at 250°C, which is the usual deposition temperature. During cooling, some of these anneal out, but because relaxation processes slow with cooling, the metastable state of some of the centers freezes in. These may appear to be another species of defect that cannot be annealed, but in reality are just a subset of the same set of RTS centers. Thus we conclude that light-induced defects are no different than those induced by quenching in such defects, or than those induced by doping. This explains why no significant differences have been found between doping-induced defects and those induced by light. The relations described in this paragraph apply to any defect model having a comparably small $\Delta E$; but the Chadi and Chang calculations support the value 0.2 eV that we have inferred for the RTS model from quenching data.

The density of RTS centers is limited by the density of the foreign atoms that form them; this limit sets the maximum for the density of observable defects in the steady state — i.e., the saturation condition. (Actually, the defect density can approach, but never equal the density of centers.) Any change in doping density, for example, generally produces a corresponding change in the density of centers (provided that the deposition conditions lead to similar bond formations). By the reasoning in the previous paragraph, the consequence of increased doping is to increase both the frozen-in density and the light-induced density of defects. This provides a simple explanation for this pair of observations that have sometimes been considered to be distinct.

## PROPERTIES EXPLAINED BY THE RTS MODEL

In Ref. 1 all of the following observations of defect properties were explained by this model in a unified way:
1. Generation of deep-level defects by doping, by either donors or acceptors.
2. Reduced density of defects in compensated materials.
3. Quench-induced defects and their relaxation.
4. Generation of defects by light, the Staebler-Wronski effect, including the conductivity reduction in doped material.
5. Dependences of the S-W effect on both doping and compensation.
6. Presence of a Si dangling bond in the metastable state and its spin.

In addition to all of these, the following observations are explained similarly:
7. Stretched-exponential kinetics of the transitions. The recent identification of transition kinetics of light-induced changes and their anneal as stretched exponentials[4-6] (SEs) is explained by analogy with the SE character of transitions of the DX center.[3] Not only are the kinetics of transitions in both types of defects described by SEs, but the stretch parameters of annealing of both are proportional to temperature. The origin of this effect seems to be a distribution of local environments of the centers caused either by nearby foreign atoms or by microstructural nonuniformities around the centers. This is discussed further below in connection with alloy effects.
8. Coordination change of P atoms from 4-fold to 3-fold caused by light.[10] Although unknown to us when Ref. 1 was written, it is observed by NMR that all 4-fold-co-ordinated P atoms (which are about 20% of the total) reconfigure to 3-fold upon exposure to light. This previously unexplained result is precisely what the RTS model calls for. Moreover, no Si-Si bond-breaking model accounts for this effect.

In several of these effects the need for carrier capture in forming a defect plays a key role. This helps explain the differences between properties of doped and compensated materials, in which carriers are available or unavailable.

## ALLOY EFFECTS IN THE RTS MODEL

There are major effects on DX centers caused by alloying GaAs with Al.[11] One effect is alteration of the energy difference between the two states, even to the extent that the 3-fold (deep-level) state can become the ground state. (High pressure on GaAs can do the same.) The Chadi and Chang theory accounts for this effect.[2]

A second effect of alloying is to change the value of the stretch parameter of the stretched exponentials that describe transitions between the two states of DX centers. In unalloyed materials (either GaAs or AlAs) this parameter is one, so that kinetics are simple exponentials, whereas in AlGaAs the stretch parameter is less than one.[3,11] This has been explained as a consequence of variations in the local environments of DX centers, due to proximity of alloy atoms to the centers.[11] It is believed that SEs arise generally from a distribution in the physical properties of such centers, e.g., a range of the barrier heights that control the transition rates.[12] There has very recently been a graphic demonstration of the way this occurs by observation of several peaks in persistent conductivity caused by DX centers; the number of peaks changes as alloy composition changes.[13] These peaks correspond to different numbers of Al atoms adjacent to DX centers, and the variation follows the composition as expected.

In a-Si:H by analogy, we expect these same effects to occur and to explain the SE kinetics. There are some significant differences in a-Si:H, however. One is the monatomic character of Si in contrast to the compounds, so there are not two different sublattices. An alloy atom in a-Si:H can thus occupy a nearest-neighbor site of the foreign atom, whereas in AlGaAs the closest that an Al atom can be to a dopant on a Ga site is the second neighbor. As a result, alloy effects in a-Si:H can be stronger than in compounds. Second, structural disorder may produce a distribution in local environments of RTS centers, apart from alloy effects; this would produce an intrinsic stretched exponential.

A third difference in a-Si:H is the presence of large amounts of H ($\approx 10\%$); this H may be regarded as an alloy material not restricted to "lattice" sites. Thus, these H atoms may occur at a variety of distances relative to RTS foreign atoms, thereby creating another influence on the details of RTS centers. This offers a way of accounting for reported effects of H on the stability properties.

## UNDOPED AMORPHOUS Si:H

The final topic to be considered in the framework of the RTS model is the existence of instability in a-Si:H that is not intentionally doped. There are two possibilities: either the host Si atoms can engage in RTS-like behavior, or there are uncontrolled foreign atoms present that form RTS centers. In Ref. 1 we gave several reasons to favor the second alternative; one of the main reasons is the saturation of the density of light-induced defects at only $\approx 10^{17}$ cm$^{-3}$ in good materials.[14] With foreign atoms as the source, this would represent approximately their density, whereas the density of Si atoms is more than 5 orders of magnitude greater.

It is well known that virtually all a-Si:H contains C, N, and O with atom densities considerably higher than $10^{17}$ cm$^{-3}$, so it is necessary to inquire about their possible roles in formation of RTS centers in undoped material. In Ref. 1 it was pointed out that N could serve as a donor in Si, and thus might form donor-type RTS centers. Carbon is also an interesting candidate for this effect for several reasons. First, the existence of both graphite and diamond shows that C easily forms either 3-fold or 4-fold coordinations, with 3 preferred. Second, the reconfiguration of such an isoelectronic atom would not involve a change in charge state, since in neither the 3-fold nor the 4-fold state would C release a free carrier. This is consistent with the absence of $10^{17}$ cm$^{-3}$ carriers in undoped material that would be expected to match that saturation density of defects, whereas any dopant-like centers would release that many free carriers. It is also consistent with the observation that light-induced degradation produces no conductivity change in intrinsic material to compare with the major changes in doped material. Third, it has been shown that light-induced degradation does rise with C content, although not linearly.[15]

Thus C in a-Si:H can play two roles: as an alloy atom and as a source of RTS centers. Although Ge is also a favored alloy material, 3-fold coordination is less common than for C. This relationship is consistent with the greater instability of a-SiC:H than of a-SiGe:H. The preparation conditions, including deposition and any further heat treatment may, however, influence the initial coordinations of these atoms.

In summary, this RTS model provides a unifying model of defects in a-Si:H that can arise from doping, quenching, or light; and explains a large number of their properties.

## ACKNOWLEDGMENT

This work was supported by the Electric Power Research Institute.

## REFERENCES

. D. Redfield and R. Bube, Phys. Rev. Lett. **65**, 464 (1990).
. D. Chadi and K. Chang, Phys. Rev. B **39**, 10063 (1989).
. A. Campbell and B. Streetman, Appl. Phys. Lett. **54**, 445 (1989).
. D. Redfield and R. Bube, Appl. Phys. Lett. **54**, 1037 (1989).
. R. Bube and D. Redfield, J. Appl. Phys. **66**, 820 (1989).
. R. Bube, L. Echeverria, and D. Redfield, Appl. Phys. Lett. **57**, 79 (1990).
. N. Ishii, M. Kumeda, and T. Shimizu, Jap. J. Appl. Phys. **24**, L244 (1985).
. M. Wautelet, L. Laude, and M. Failly-Lovato, Solid State Comm. **39**, 979 (1981).
. H. Branz, Phys. Rev. B **38**, 7474 (1988).
0. M. McCarthy and J. Reimer, Phys. Rev. B **36**, 4525 (1987).
1. E. Calleja, P. Mooney, S. Wright, and M. Heiblum, Appl. Phys. Lett. **49**, 657 (1986).
2. R. Crandall, in *Amorphous Silicon Technology—1989*, Matl. Res. Soc. Symp. Proc. **149**, 589 (1989).
3. G. Brunthaler and K. Kohler, Appl. Phys. Lett. **57**, 2225 (1990).
4. D. Redfield and R. Bube, in *Amorphous Silicon Technology—1990*, Matl. Res. Soc. Symp. Proc. **192**, 273 (1990).
5. T. Unold, and J. Cohen, Appl. Phys. Lett. **58**, (to be published).

# The Application of a Comprehensive Defect Model to the Stability of a-Si:H

N.Hata and S.Wagner
Department of Electrical Engineering, Princeton University, Princeton, NJ 08544

## ABSTRACT

Sufficient information has become available about the density and distribution of the dangling-bond defects, the defect annealing behavior, the light-induced defect generation rates $(dN_s/dt)_{LI}$, and the saturated light-induced defect densities $(N_{sat})$ in amorphous silicon (a-Si:H) to allow the development of a comprehensive defect model. The model which we have developed describes the density of defect states in function of growth conditions, of annealing and of light-soaking. We apply this model to the defect distributions which are established by various combinations of light-soaking and annealing.

## INTRODUCTION

The density and distribution of the dangling-bond defects, the defect annealing behavior, and the light-induced defect generation in amorphous silicon (a-Si:H) has been studied extensively. Enough data have become available to permit setting up a realistic numerical model of defect densities and distribution. We have introduced such a model and used it for the description of the defect profile associated with a growing surface.[1] In that study we treated the density and distribution of defects as they change from the surface into the bulk. We showed how these changes affect the total defect density and the dark conductivity as functions of film thickness. Both of these properties can be determined experimentally, and they were used to test the model.

In this paper we extend the model to light-soaking and annealing after film growth. We assume the same Gaussian envelope of the annealing energy as before. This envelope constitutes the defect pool, i.e., all defects which can possibly be created by light-soaking. Light-soaking creates this Gaussian distribution at a rate known from experiment.[2,3] We assume that saturation of light-induced defects[4] occurs by exhaustion of the possible defect sites, so that the saturated defect density $N_{sat}$ is given by the density of these sites. $N_{sat}$ in device-grade a-Si:H is experimentally found to vary from sample to sample[5] over the range between $4 \times 10^{16}$ and $2 \times 10^{17}$ cm$^{-3}$. We use the experimentally determined value of $N_{sat}$ for each sample. In our earlier work, we had used the constant value of $4 \times 10^{17}$ cm$^{-3}$, which had been determined by photoemission yield spectroscopy on one sample of a-Si:H.[6] Once $N_{sat}$ has been measured, the rate of light-induced defect creation can be computed.[7] We use this rate also in our defect model.

## THE DEFECT MODEL

We employ the defect pool $P(E_{ann})$ of Gaussian distribution $f(E_{ann})$ [3]:

$$P(E_{ann}) = N_{sat} f(E_{ann}) = N_{sat} \exp[- (E_{ann} - E_0) / 2W^2]/(2\pi W^2)^{1/2} . \quad (1)$$

$N_{sat}$, $E_{ann}$, $E_0$, and W are the saturated defect density, the activation energy for annealing, and the center and width of the annealing energy distribution, respectively. The annealing of the defect states within the defect pool is expressed by the rate equation

$$dN_s(E_{ann})/dt = - [N_s(E_{ann}) - N_{eq}(E_{ann},T)] / \tau_{ann}(E_{ann},T). \quad (2)$$

$N_s(E_{ann})$ is the density of defects with the annealing energy of $E_{ann}$. $N_{eq}(E_{ann},T)$ is the equilibrium density at the temperature T of defects with $E_{ann}$. $\tau_{ann}(E_{ann},T)$ is the annealing time constant of defects with $E_{ann}$ at T. This annealing time constant $\tau_{ann}$ is given by

$$\tau_{ann}(E_{ann},T) = v_{ann}^{-1} \exp(E_{ann} / kT) \quad , \quad (3)$$

where $v_{ann}$ is the attempt-to-anneal frequency, and k the Boltzmann constant.

Defect generation by light soaking is treated as the generation of defects of the same distribution as the defect pool. We calculate the net rate of defect buildup as the difference between light-induced generation and thermal annealing:

$$dN_s(E_{ann})/dt = (dN_s/dt)_{LI} f(E_{ann}) [1 - N_s(E_{ann}) / P(E_{ann})]$$

$$- [N_s(E_{ann}) - N_{eq}(E_{ann},T)] / \tau_{ann}(E_{ann},T) . \quad (4)$$

$(dN_s/dt)_{LI}$ is the rate of creation of light-induced defects. The factor $[1 - N_s(E_{ann}) / P(E_{ann})]$ accounts for the slowdown in light-induced defect creation as the defect density approaches its saturated value. Although there is evidence for the saturation of the light-induced defect density, the mechanism of saturation is not known. As mentioned earlier, we assume an exhaustion of possible defect sites for our model. A very plausible alternative is a balance between light-induced bond breaking and light-induced annealing. A mechanism of this kind could make Nsat depend on the conditions of the experiment without any thermal annealing, so that the defect pool would not be a precisely defined distribution. In the absence of reliable information we assume the most simple case of a fixed, well-defined, defect pool. The thermal annealing during light-soaking is assumed identical to thermal annealing in the dark.

The values of the parameters which we used in the model calculations are: $E_0 = 1.05$ eV, $W = 0.1$ eV, $\nu_{ann} = 10^{10}$ s$^{-1}$, and $N_{sat} = 5$ x $10^{16}$ to $4$ x $10^{17}$ cm$^{-3}$. For simplicity, the equilibrium defect density distribution at 250 °C is used as the starting state before light soaking. A carrier generation rate which is uniform throughout the sample is assumed. All distributions in this paper represent the $D^{0/-}$ transition.[2]

## APPLICATION OF THE DEFECT MODEL TO LIGHT SOAKING

Figure 1 shows calculated defect density distributions $N(E_{ann})$ for the as-grown state, and for various degrees of light-soaking (time, or light intensity). The addition of defects by light-soaking to the as-grown defects, all under the envelope of the defect pool, is shown in Fig. 1 (a). Fig. 1 (b) depicts the total density of defects during the early stages of light-soaking. The vertical arrow illustrates how increasing the light-soaking time or light intensity results in higher densities of light-induced defects. The initial *number* of as-grown defects is set by thermal equilibrium with strained Si-Si bonds.[8] Their *distribution* occupies the defect pool at the highest annealing energies. Light-soaking adds defect states which have the defect-pool distribution.[1-2] Thus the defect distribution after moderate light soaking exhibits two maxima which can be separated by as much as 0.2 eV, one maximum at the peak energy of the defect pool, the other in the high-energy wing of the pool. Since light-soaking increases the defect density at the lower-energy peak, the light-induced defects appear of different nature from the as-grown defects, a phenomenon observed and discussed by several researchers.[9] Our model shows that two peaks in the defect distribution may reflect light-induced and as-grown defects produced in different parts of the *same* defect pool.

Another consequence of the modelling of light-induced defects is the existence of a certain temperature below which the saturated defect density is temperature-independent and above which it decreases with temperature. Such an effect has been predicted by Bube and Redfield.[10] The threshold temperature depends on the light intensity, because it is explained by the balance between the rate of thermal annealing and the rate of light-induced defect creation. To study this effect, we first compute the temperature-dependent saturated defect

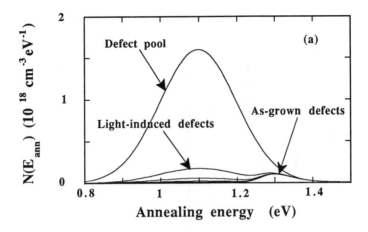

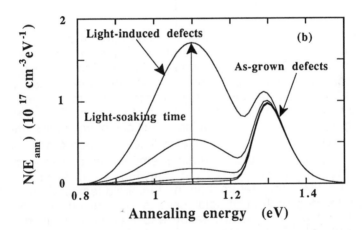

Figure 1    Distribution of as-grown and light-induced defects (a) under the envelope of the defect pool, and (b) under a magnified vertical scale without display of the defect pool.

distribution during light-soaking at the rate of creation of light-induced defects of $7 \times 10^{14}$ $cm^{-3}s^{-1}$, which is shown in Fig. 2.  At temperatures lower than the threshold temperature of about 330 K, the distribution is almost the same as the defect pool distribution, while above the threshold temperature thermal annealing during light-soaking reduces the easy-to-anneal part of the defect distribution.  By integrating the distributions over the entire range of annealing energy, the temperature dependence of saturated defect density $N_{sat}$ is obtained.  Figure 3 shows such calculated temperature dependences of $N_{sat}$ and their

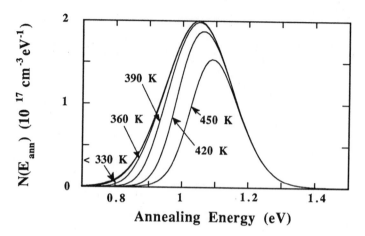

Figure 2    Saturated defect density distributions over a range of light-soaking temperatures.

agreement with two different sets of experimental data.[11-12]  The dependence of the threshold temperature on the rate of creation of light-induced defects $(dN_s/dt)_{LI}$ is shown in Fig. 4.  The increasing trend of the threshold temperature with the rate of light-induced defect creation is observed because a higher rate of thermal annealing is needed to balance to the higher rate of defect creation.  The shadowed area in Fig. 4 represents the operating conditions of solar cells: the rate of defect creation in an operating solar cell lies between $4 \times 10^{11}$ and $1 \times 10^{13}$ cm$^{-3}$s$^{-1}$, and the typical operating temperature of a solar module between 310 and 340 K.  It is clear from Fig. 4 that solar cells are operated at higher temperature than the threshold temperature, so that the long-term degradation of solar cells is determined by a balance between light-induced defect creation and thermal annealing during their operation.    The establishment of this balance has been observed.[13-14]  The annealing is enhanced by the fact that solar cells are exposed to the sun only during the day hours, while their thermal annealing proceeds at all times, although usually at lower temperature in the absence of insolation.

## DISCUSSION

Our results show that the defect pool model is consistent with experimental data on light-induced defects.  For example, the concept of "metastable defects" is not needed to explain why light-induced defects lie at energies different from "intrinsic" defects.  The defects have the same physical origin and are explained in terms of a distribution of defect annealing energies related to the defect pool

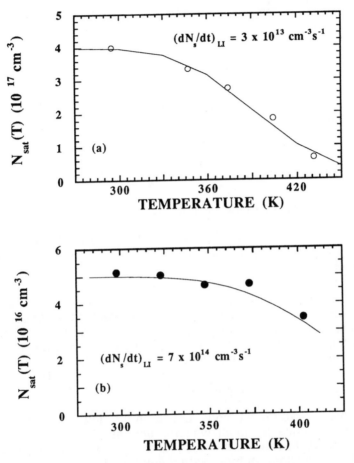

Figure 3    Calculated temperature dependences of saturated defect density for two different rates of creation of light-induced defects of (a) 3 x $10^{13}$ and (b) 7 x $10^{14}$ cm$^{-3}$s$^{-1}$, chosen to fit the experimental data of (a) Ref. 12 and (b) Ref. 13.

model.  Also, the temperature dependence of the saturated defect density is explained quantitatively by the balance between light-induced defect creation and thermal annealing.

The concept of a defect pool is common to other recent defect models[15-18] and to our quantitative model.  Questions remain about details of these models: (1) are the Gaussian distributions taken in the models correct?, and (2) do considerable concentrations of charged

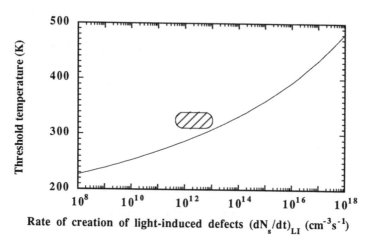

Figure 4    The dependence of the threshold temperature of the saturated light-induced defect density on the rate of creation of light-induced defects.

defects[17-18] exist in undoped a-Si:H?  Those questions need to be resolved experimentally.  Nevertheless, the defect pool concept is already accepted, and the present approach of applying the defect pool concept to the stability problem is proving quite useful and could help identify the mechanism of the light-induced defect creation.

## CONCLUSION

A comprehensive defect model based on the defect pool concept was applied to derive a model which describes changes in defect distribution upon light-soaking and thermal annealing.  This model was then used to calculate defect distributions and the results were compared with experimental data.  The distribution of light-induced defects is successfully explained by the present model without resorting to the concept of "stable" and "metastable" defects.  Questions remain about the microscopic origin of the defect pool and about the mechanism of light-induced defect creation.  Yet, the present model can be used to predict the defect distributions upon different light-soaking and thermal treatments, which is important in predicting the insolation- and temperature-dependent performance of a-Si:H solar cells.

## ACKNOWLEDGMENTS

The authors would like to acknowledge helpful discussions with M.Isomura. This work is supported by the Thin-Film Solar Cell Program of the Electric Power Research Institute.

REFERENCES

1.  N.Hata, S.Wagner, P.Roca i Cabarrocas, and M.Favre, Appl. Phys. Lett. 56, 2448, (1990).
2.  N.Hata, E.Larson, J.Z.Liu, Y.Okada, H.R.Park, and S.Wagner, Mat. Res. Soc. Symp. Proc. 192, 285 (1990).
3.  N.Hata, J.Bullock, M.Isomura, X.Xu, J.H.Yoon, and S.Wagner, in Technical Digest of the International PVSEC-5 (Kyoto, Japan, 1990) p.67.
4.  H.R.Park, J.Z.Liu, and S.Wagner, Appl. Phys. Lett. 55, 2658 (1989).
5.  H.R.Park, J.Z.Liu, P.Roca i Cabarrocas, A.Maruyama, M.Isomura, S.Wagner, J.R.Abelson, and F.Finger, Appl. Phys.Lett. 57, 1440 (1990).
6.  S.Aljishi, S.Jin, L.Ley, and S.Wagner, Phys. Rev. Lett. 65, 629 (1990).
7.  D.Redfield and R.H.Bube, Appl. Phys. Lett. 54, 1037 (1989).
8.  Z E.Smith and S.Wagner, Phys. Rev. Lett. 59, 688 (1987).
9.  D.Han and H.Fritzsche, J. Non-Cryst. Solids 59-60, 397 (1983).
10. R.H.Bube and D.Redfield, J. Appl. Phys. 66, 820 (1989).
11. H.Ohagi, J.Nakata, A.Miyanishi, S.Imao, M.Jeong, J.Shirafuji, K.Fujibayashi and Y.Inuishi, Jpn. J. Appl. Phys. 27, L2245 (1988).
12. M.Isomura and S.Wagner, in this volume.
13. K.Takigawa, H.Kobayashi, and Y.Takeda, in Technical Digest of the International PVSEC-4 (Sydney, Australia, 1989) p.777.
14. C.Jennings, C.Whitaker, and D.Summer, Solar Cells 28, 145 (1990).
15. R.A.Street and K.Weiner, Phys. Rev. B40, 6236 (1989).
16. K.Weiner, Phys. Rev. B41, 12150 (1990).
17. R.S.Crandall and H.M.Branz, in the Conf. Record of the 21st IEEE Photovoltaic Specialists Conf. (IEEE, New York, NY,1990) p.1630.
18. H.M.Branz and M.Silver, Phys. Rev. B42, 7420 (1990).

# PART III


# EXPERIMENTAL
# DATA AND
# MODEL VERIFICATION

# DEFECT EQUILIBRATION IN DEVICE QUALITY a-Si:H AND ITS RELATION TO LIGHT-INDUCED DEFECTS

T.J. McMahon
Solar Energy Research Institute, 1617 Cole Boulevard, Golden, CO 80401

## ABSTRACT

Electron spin resonance was used to characterize concentrations of thermal equilibrium defects from room temperature to 280°C in a 60-micron-thick hydrogenated amorphous silicon (a-Si:H) film. I found a defect formation energy of 0.35 eV in material with $1 \times 10^{15}$ cm$^{-3}$ spins at 190°C; an assumed two-level configuration coordinate diagram would then have a "source" density of $10^{19}$ cm$^{-3}$. Annealing of defects quenched in from 250°C yields an activation energy of 2.1 eV. When annealings of defects quenched in from high and low temperature are compared *the defects introduced at the higher temperature always anneal faster;* The metastable states with higher formation energies have smaller annealing activation energies. Additionally, the time constant for generation of defects at 205°C is 10 times longer than the corresponding annealing time constant, consistent with the very high formation barrier expected for this two-level system. The stretch parameter for defect generation is much closer to unity than for defect annealing. Easy-to-anneal, light-induced defects can be described as a very high-temperature distribution similar to that which might be quenched in as a result of $kT \cong 0.5$ eV.

## INTRODUCTION

Early observations of a structural equilibrium in a-Si:H were made on doped films in which cooling-rate-dependent conductivity changes could be readily observed.[1,2] The first observation of defect equilibration in undoped a-Si:H was made by Smith et al.[3] using excess gap-state absorption after fast cooling and slow cooling from 300°C. The model used to explain these results involves excess intrinsic-carrier recombination at higher temperatures,[4] in much the same way that Stutzmann et al.[5] used photogenerated carriers in band tails to model photodegradation. The "hydrogen glass" model was then introduced to explain the defect relaxation kinetics that occur well below the melting temperature of silicon.[6] Street et al.[2] suggested that hydrogen diffusion limits the defect relaxation rates in doped a-Si:H films.

Extensive studies of thermal defects have been made using spin density data on undoped films.[7-9] Stretched exponential time dependences are now used to describe annealing behavior.[8] Annealing times exhibit an activated behavior with an energy of 1.7 eV when spins are measured and 1.2 eV when photoconductivity is measured.[10] The lower value, determined from photoconductivity, occurs because recombination is controlled by easier-to-anneal defects nearer the dark Fermi level.[11] Frozen-in defects have been observed to anneal more slowly than light-soaking defects.[8,10] They also saturate at relatively low levels of several $10^{15}$ cm$^{-3}$ compared to light-soaking defects, which are found at levels of $10^{17}$ cm$^{-3}$.[12]

Street and Winer[8] have used hydrogen trapping into an exponential distribution of weak bonds to explain these defect kinetics; in such a model the size of the energy barrier to annealing is anticorrelated with the defect formation energy. The stretch parameter for the annealing kinetics was related to the diffusion behavior of hydrogen.

## EXPERIMENTAL

Electron spin resonance data on thermal equilibrium spin concentrations were measured from room temperature to 280°C for a 60-micron-thick, device-quality a-Si:H film produced at Glasstech Solar, Inc. The sample consisted of small flakes that peeled from the substrate after the deposition of a high-frequency (110 MHz) glow-discharge film deposited with H dilution at a rate of 2 μm/h. Of the 0.2114 g of material, which was placed in a 3-mm bore-quartz tube, 0.145 g fit within the cavity height. This resulted in spin counts of about $10^{14}$. This specimen, used in conjunction with a Wilmad high-temperature quartz dewar insert, yielded a 2% reproducibility. The as-supplied, room-temperature spin density of this film was $1.87 \times 10^{15}$ cm$^{-3}$.

Quenching the sample from the highest temperatures is achieved by placing the quartz tube in an aluminum cylinder that is inside an oven. The sample remains in the oven for a minimum of 20 min to ensure the designated temperature is reached and defect equilibration is complete. (At lower temperatures the time required to reach equilibrium during defect generation is very long.) The equilibrated tube with the sample is then dropped into cool water. Though such a method of cooling is somewhat slower than can be acheived for quenching of films on substrates,[8,10] it should be reproducible.

Spin densities, $N_s(t_A)$, are measured at room temperature between annealing intervals and plotted against the cumulative annealing time, $t_A$.

## RESULTS

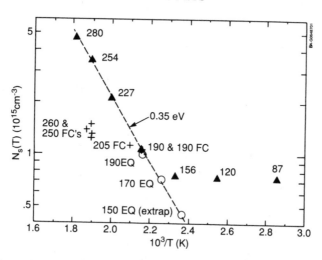

Fig. 1. Spin defect densities (▲) taken during heating at the temperatures as noted in °C. Room temperature results of fast-cooling (+) and long annealing times (○) at 190°C and 170°C are shown with an extrapolated 150°C density.

Equilibrium Spin Densities. Fig. 1 shows how this sample responded to elevated temperatures from the low-spin state. The spin density data (▲) are Curie Law corrected and measured during heat-up. They show a constant spin density of $0.72 \times 10^{15}$ cm$^{-3}$ below 170°C. The increase in spin values at and above this temperature is expected from all defect models. The four highest values indicate a defect formation energy of 0.35 eV, which is higher than the 0.30 eV reported earlier for a similar measurement[9] and much higher than the 0.20 eV reported for quenched-in defect densities.[8] The 190°C and 170°C equilibrium densities obtained after the long anneals discribed in the next paragraph were measured at room temperature. These have been plotted (O) in Fig. 1 at the appropriate value of $10^3/T$. The fact that they fall onto the fitted line comfirms the defect formation energy of 0.35 eV. Street and Winer's value of 0.20 eV[8] could be partly explained by insufficient cooling rates from higher temperatures, but it may be that samples with higher spin levels also have lower defect formation energies. The extrapolated 150°C value was obtained by extending the 0.35-eV line downward to $10^3/(273K+150°C)$ and a spin density of $0.45×10^{15}$ cm$^{-3}$.

   250°C Defect Annealing. Also shown in Fig. 1 are the results (+) of three quenches from 250°C and quenches (one each) from 260°C, 205°C, and 190°C, each to be followed by long anneals. First we see that only one third of the 250°C and 260°C defects remain. The cool-down rate is too slow to freeze in all of the 250°C defects, but it is adequate to capture most of the 205°C defects and all the 190°C defects (see Fig. 3 and discussion). The spin density data in Fig. 2 show the time dependence for a 150°C annealing of defects resulting from the first 250°C fast-cool (▲). This annealing sequence was terminated long before the extrapolated 150°C spin density was reached.

   A second 250°C quench was followed by sequential annealing at 170°C and measured as before. This set of anneal data (×) is also plotted in Fig. 2. All points, except the initial cool-down and first anneal value at 800 s, were fitted to a stretched exponential relaxation time dependence of the following form:

$$N_s(t_A) = (N_i - N_f)e^{-(t_A/\tau)^\beta} + N_f$$

where $N_i$ is the initial quenched-in density, $N_f$ is the final anneal value at $t_A = \infty$, $\tau$ is the time constant, and $\beta$ is the stretch factor. The first and second points must be deleted from the fitting process to satisfactorily fit the remaining values between 1200 and $2×10^6$ s. This fit is achieved with the following: $N_i = 1.67×10^{15}$ cm$^{-3}$, $N_f = 0.72×10^{15}$ cm$^{-3}$, $\tau(170°C) = 4.2×10^4$ s, and $\beta = 0.30$.

   The initial flatness of these annealing data at first seemed to indicate a second defect process such as that observed by Lee et al.[13] But further investigation suggests that this flatness is more likely to be an artifact of an insufficiently fast cool-down rate, i.e., the fast-to-anneal defects (approximately two thirds of the total) are lost during the cool-down from 250°C. The influence of the cool-down rate on $N_i$ is obvious. In addition, it can increase the values of $\tau$ and $\beta$ that best fit the data. The rate of cooling required to quench in defects at different equilibration temperatures can be found in the discussion of Fig. 3.

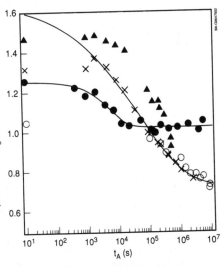

Fig. 2. Time dependence of the spin density during annealing of 250°C fast-cooling defects at 150°C (▲), 170°C (×) and 190°C (●). Lines are the stretched exponential fits discussed in the text. Annealing at 150°C of 190°C equilibrium spins is shown (O).

Even though precautions were taken to ensure fast and identical quenches, we again see a slightly smaller spin density after the third quench from 250°C. Sequential anneals were again carried out, this time at 190°C. Results are plotted (●) in Fig. 2 as is the best stretched exponential fit: $N_i = 1.24 \times 10^{15}$ cm$^{-3}$, $N_f = 1.02 \times 10^{15}$ cm$^{-3}$, $\tau(190°C) = 4900$ s, and $\beta = 0.82$. The dramatic increase in the value of $\beta$ from 0.30 to 0.82 at the higher annealing temperature is the result of having to anneal a narrower range in $t_A$ of defects. This point is discussed after the 205°C defect data are presented.

Defect Anneal and Generation Barriers. The annealing time constants are plotted in Fig. 3. $\tau(150°C) = 8 \times 10^5$ s is obtained by noting the time at which the spin density on the annealing curve is $[(N_i - N_f)/e] + N_f$, where $N_f$ is the extrapolated 150°C equilibrium spin density. They are thermally activated with a 2.1 eV activation energy between 150° and 190°C. These data give some indication of how fast this sample must be cooled in order to freeze in a significant number of defects at these higher temperatures. A simple extrapolation yields $\tau(250°C) = 10$ s. It is not surprising that only one third of the 250°C equilibrium spins is frozen in with the limited rate of cooling for a sample enclosed in an quartz tube. The defects that require less than several seconds to anneal at temperatures between 250°C and 200°C are lost from the anneal distribution during the cool-down. With such strongly activated relaxation behavior thermal defect data and its interpretation can easily be flawed.

I find the generation of spins to be extremely slow. For example, the generation time constant for this film was obtained by fitting a stretched exponential to the generation of 205°C defects where: $N_i = 0.77 \times 10^{15}$ cm$^{-3}$, $N_f = 1.05 \times 10^{15}$ cm$^{-3}$, $\tau_{gen}(205°C) = 6844$ s, and $\beta = 0.79$. The value for $\tau_{gen}(205°C)$ is shown (▲) in Fig. 3 and is 10 times longer than the time constant expected for annealing defects at 205°C, as indicated by the upward-pointing dashed line. If the attempt frequency were the same for generation and annealing, the generation barrier would be only 0.1 eV higher than the anneal barrier. Since the formation energy is much larger (0.35eV), one must invoke arguments for a difference in entropy or attempt frequency between the initial and final states.

For comparison purposes I have included data from Street and Winer[8] for a sample made at high rf power (■) that has a slope of 1.7 eV; this sample, having a higher spin density, should have more disorder and, as a consequence, smaller annealing barriers. In addition they found that the relaxation time for the 200°C annealing of spins resulting from a mild light-soaking (◆) is about 10 times faster than the 200°C annealing of a similar number of thermally generated spins.

Faster annealing for light-soaking defects was first noticed in a photoconductivity study.[10] The data from that study, along with some unpublished time constants measured in the same study, are also shown in Fig. 3. Notice that the time constants for the anneal of 250°C thermal defects (○) exhibit an anneal barrier of 1.2 eV, somewhat smaller than values determined by spin measurements (for explanation see Ref. 11). Next, the annealing relaxation time constants

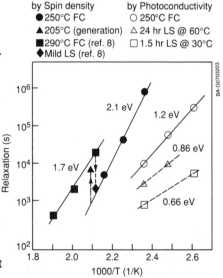

Fig. 3.   Relaxation time constants vs. $10^3/T$. Photoconductivity from ref. 10

for two light-soaking situations are shown ($\square$ and $\triangle$): the lower line with a slope of 0.66 eV is drawn through anneal time constants measured after 1.5 hr mild light-soaking periods at 30°C, which produced the same photoconductivity decrease as did the 250°C thermal defects. The anneal barrier for defects produced by light-soaking in numbers equal to thermal defects is 0.54eV smaller. Consequently they anneal faster.

Additionally, the line with a slope of 0.86 eV is drawn through the anneal time constants measured after 24-hr light-soaking periods at 60°C. These longer light-soaking defects introduced at higher temperature take longer to anneal but still not as long as the time required to anneal thermal defects.

190°C Defect Annealing. One 150°C annealing sequence was carried out from an equilibration temperature of 190°C. It started from a spin density of $1.03 \times 10^{15}$ cm$^{-3}$ after the film had been at 190°C for $10^7$ s following the 190°C anneal of 250°C defects already discussed. All of the 190°C spins can be quenched in since $\tau(190°C) = 4900$ s. These data are plotted ($\bigcirc$) in Fig. 2 and are to be compared with the 150°C anneal data for 250°C fast-cool defects ($\blacktriangle$). It is significant that the anneal data for the 250°C fast-cooling defects appear to be about to merge with or cross through the anneal data for the 190°C fast-cool as the defect level drops to $0.9 \times 10^{15}$ cm$^{-3}$. If the extrapolated 150°C anneal value is assumed for each data set, then the time required to anneal half of the 250°C defects is $3 \times 10^5$ s, and the time required for half of the 190°C defects is $3 \times 10^6$ s. This is found even though the faster-to-anneal defects are already missing from the 250°C defect distribution due to less-than-instant cool-down. It is clear that *the extra excess defects introduced at the higher equilibration temperature anneal much faster.* Explanation of this result requires a model in which metastable states with higher formation energies have smaller annealing activation energies.

205°C and 260°C Defect-Anneal Time Distributions. During the 150°C anneal experiment previously described, spin densities drop to a minimum value of $0.72 \times 10^{15}$ cm$^{-3}$. This is well above the extrapolated 150°C equilibrium value. It seems that these more permanent spins do not participate in the thermal equilibrium process and may be due to a temperature-independent surface spin density[14] as high as $2 \times 10^{12}$ cm$^{-2}$. Also the 150°C annealing data of 250°C fast-cool defects was not measured beyond the inflection point. It would be of interest to measure defect-anneal time distributions, but to do so requires that these two shortcomings be corrected.

Fig. 4 shows the *complete* 170°C annealing data measured for 205°C and 260°C fast-cooling defects along with their stretched exponential fits; the 260°C defects: $N_i = 1.42 \times 10^{15}$ cm$^{-3}$, $N_f = 0.79 \times 10^{15}$ cm$^{-3}$, $\tau(170°C) = 54000$ s, and $\beta = 0.54$; for the 205°C defects: $N_i = 1.06 \times 10^{15}$ cm$^{-3}$, $N_f = 0.80 \times 10^{15}$ cm$^{-3}$, $\tau(170°C) = 1.3 \times 10^5$ s, and $\beta = 0.69$. $\tau_{ann}(170°C)$ for 260°C defects is much shorter than for $\tau_{ann}(170°C)$ for 205°C defects. The large increase in $\beta$ for the annealing of 205°C defects will be discussed after dealing with defect-anneal time distributions.

Defect-anneal time distribution functions are derivatives, $dN_s/dln(t_A)$, of the stretched exponential fits. The distribution functions for the 170°C experimental anneal data are shown at the bottom of Fig. 4. Distribution B shows the equilibrium defect distribution as a result of fast cooling from 260°C. A is constructed from the result that the integrated density of defects is three times that of B with the extra defects on the fast-to-anneal side of distribution B, in agreement with experiment.

In Fig. 4 the annealing data merge, and as a direct result B includes C (C being the result of fast cooling from the 205°C equilibrium state). It appears as though the thermally induced defects "fill" the metastable side of the two-level system shown in Fig. 5. The barriers and formation energies shown were derived from experiments above. As the "well" is filled the additional states become easier to anneal as long as the pass-through point remains the same. Distribution B occupies states 0.040 eV

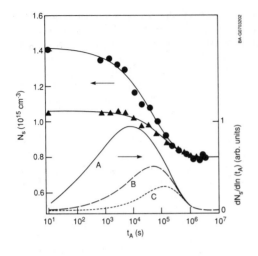
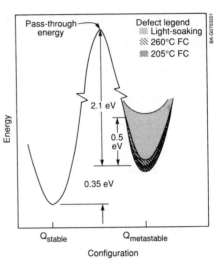

Fig. 4. Time dependence of 170°C anneals and stretched exponential fits for : (●) 260°C defects and (▲) 205°C defects. The distribution functions for each fit are shown at the bottom as curves B and C. Distribution A is discussed in text.

Fig. 5. The two-level configuration coordinate system inferred from this work; the formation energy is taken from the equilibrium data and the activation energy from the annealing time constants. Metastable populations are shown.

higher than C as depicted in Fig. 5. This corresponds to the factor of two found for the difference in the location of the peaks of B and C in Fig. 4.

Within the realm of thermal equilibrium defects, the data do support a model in which the anneal barrier is anticorrelated with the defect formation energy; Fig. 5 defines such a system. Since the spread in energy is somewhat smaller than the formation energy of 0.35 eV, a Boltzmann distribution can be used to estimate the "source" state density. Paying no attention to detailed microscopic models or entropy effects[8], a simplified two-level system has a "source" density of $10^{19}$ cm$^{-3}$.

Fig. 5 is, of course, an oversimplification of the processes involved in defect stabilization and anneal. The long time constants may not be entirely a result of the exceptionally high barriers derived from $E = kT \ln(\nu\tau_A)$, but rather the result of a defect configuration being locked in until the right combination of lattice motions occurs releasing the neighboring atoms into some alternate configuration. Such hierarchically constrained dynamics result in a stretched exponential relaxation behavior[15].

Stretched Exponential Parameter "$\beta$". The value of $\beta$ measured for different anneal temperatures or at the same anneal temperature for different quench temperatures shows that it is not simply proportional to the anneal temperature.[2] It depends more on the difference between the quench and anneal temperatures and possibly on how quickly the film is quenched. Indeed, $\beta$ varies from 0.54 to 0.69 for a fixed annealing temperature of 170°C for 260°C and 205°C quenches, respectively, simply because a narrower range of states in the anneal time distribution is available for annealing. Therefore, the dispersion is reduced and $\beta$ increases; a delta function distribution will have $\beta$=1.0. Also, such an explanation is better able to account for the large increase in $\beta$ found for the increased annealing temperatures used on 250°C defects as in Fig. 2.

It is important to notice that β for the generation of 205°C defects is 0.79 for this sample and was found to be 1.09 for a standard glow discharge sample; in each sample the β for generation is much closer to unity than for the annealing of defects. The generation of thermal spins is less dispersive than the annealing of metastable spins; this is consistent with models in which a single excitation of a fairly unique energy is required to generate a metastable spin. Anneal of a spin, on-the-other-hand, is apparently more dispersive, as though a distribution of barriers should be associated with the annealing of metastable spins. The model in which hydrogen is excited out of saturating Si-H positions into a distribution of weak Si-Si bond sites[8] is in good agreement with this new observation.

## LIGHT-SOAKING DEFECTS

Finally, differences in the anneal kinetics and saturation levels of light-soaking and thermal defects can be explained even if the same two-level system of Fig. 5 is used to discribe both. Light-soaking defects would have a distribution of formation energies with an increasing number of sites having higher formation energy, perhaps increasing exponentially as has been noted for the weak-bond model.[8] I use the experimental result that metastable states with higher formation energy have smaller anneal barriers and postulate that both the defect saturation level and their stability depend on the energy available during the defect reconfiguration. Defects created by thermal equilibrium precesses correspond to a Boltzmann distribution of the low-lying, fairly stable, metastable configurations characteristic of kT (0.04 eV). In contrast, light-induced metastable states can produce higher formation energy defects due to the higher recombination energy available from optical excitation.

Light-induced defect generation is slow (one defect is produced per $10^8$ recombination events and decreases until saturation at $10^{17}$ cm$^{-3}$ is reached[16]). Once generated, however, the 0.5 eV smaller anneal barrier can be understood in one of two ways: (i) In a *nonequilibrium trapping model*, once the recombination energy converts the metastable defect, it could be considered inactive and "stuck" (unable to relax further or anneal because the temperature is too low) with a large number of relatively unstable, non-equilibrium metastable defects, as is the case in the hydrogen-trapping model.[8] (ii) Alternately we can use a *"high-temperature" equilibrium model,* and allow the defect further reconfiguration as if the athermal excitation that converted it in the first place were equivalent to a very high-temperature source with an equivalent kT of 0.5eV. Defects included within the quasi-Fermi level splitting can convert and equilibrate. Sometimes the defects would anneal back with another recombination event, but the saturated result would be a large number of the defect states converted. These metastable states would have higher formation energies and saturate with broader distributions having more, relatively unstable, metastable defects. These configurations would fill higher states in the "well" as shown in Fig. 5.

## CONCLUSIONS

I found a defect formation energy of 0.35 eV and an anneal activation energy of 2.1 eV in a 60 μm undoped a-Si:H sample; these energies are higher than previously reported values and may be indirectly due to the low spin density of the sample. The generation of thermal defects is very slow indicating that the formation barrier could be higher than 2.1 eV.

For defect generation β is much closer to unity than for annealing. Consequently, the source of the metastable spin defect may lie at a well defined energy and require only a single reconfiguration to generate a spin while anneal of a spin may require many dispersive steps. The "source" density is of the order $10^{19}$ $cm^3$.

Extra excess defects introduced at a higher equilibration temperature anneal faster, thereby supporting a model in which metastable states with higher formation energies have smaller annealing activation energies. During annealing, the value of β depends more on the difference between the quench and annealing temperatures than on the annealing temperature alone. $\beta \neq T_A/T_o$

Light-induced defects could then be characterized as a very "high-temperature" distribution of athermal defects similar to that which might be quenched in as a result of $kT \cong 0.5$ eV.

## ACKNOWLEDGEMENTS

I am grateful to Mr. K. Sadlon for EPR interfacing and software, to Dr. J. Xi and Glasstech Solar, Inc. for the a-Si:H sample; and to Drs. R. Crandall, H. Branz, M. Vanecek and B. vonRoedern for discussions and critical reading of the manuscript. This work was supported under DOE Contract No DE-AC02-83CH10093.

## REFERENCES

1. D.G. Ast and M.H. Brodsky, in *The Fourteenth International Conference of Physics of Semiconductors*, B.H.L. Wilson (ed.), Institute of Physics Conference Series No. 43 (Hilger, London, 1979), P. 1159
2. R.A. Street, J. Kakalios and T.M. Hayes, Phys. Rev. B34, 3030 (1986)
3. Z E. Smith, S. Aljishi, D. Slobodin, V. Chu, S. Wagner, P.M. Lenahan, R.R. Arya and M.S. Bennett, Phys. Rev. Lett. 57, 2450 (1986)
4. Z.E. Smith and S. Wagner, Phys. Rev. B32, 5510 (1985)
5. M. Stutzmann, W. B. Jackson and C.C. Tsai, Appl. Phys. Lett. 45, 1075
6. R.A. Street, J. Kakalios, C.C. Tsai and T.M. Hayes, Phys. Rev. B35, 1316 (1987)
7. X. Xu, A. Morimoto, M. Kumeda and T. Shimizu, Appl. Phys. Lett. 52, 622 (1988)
8. R.A. Street and K. Winer, Phys. Rev. B40, 6236 (1989)
9. S. Zafar and E.A. Schiff, J. Non-Cryst. Solids 114, 618 (1989)
10. T.J. McMahon and R. Tsu, Appl. Phys. Lett. 51, 412 (1987)
11. K. Shepard, Z.E. Smith, S.Aljishi and S. Wagner, Appl. Phys. Lett. 53, 1644 (1988)
12. H.R. Park, J.Z. Liu and S. Wagner, Appl. Phys. Lett. 55, 2658 (1989)
13. C. Lee, W.D. Ohlsen and P.C. Taylor, Phys. Rev. B36, 2965 (1987)
14. S. Zafar, PhD. dissertation Syracuse Uuiversty (1990)
15. R.G. Palmer, D.L. Stein, E. Abrahams and P.W. Anderson, Phys. Rev. Lett. 53, 958 (1984)
16. D. Adler, Sol. Cells 9, 133 (1983)

# INVESTIGATION OF DEFECT REACTIONS INVOLVED IN METASTABILITY OF HYDROGENATED AMORPHOUS SILICON

J. David Cohen and Thomas M. Leen

Department of Physics, University of Oregon, Eugene, OR 97403, U.S.A.

## ABSTRACT

We have characterized metastable states in n-type doped a-Si:H due either to fast (quench) cooling or by light-soaking followed by a series of partial dark anneals. These studies have been carried out for samples with a large range of $PH_3$ doping levels using drive-level capacitance profiling to deduce the number of occupied bandtail states ($N_{BT}$) and transient photocapacitance spectroscopy to measure the density of deep defects ($N_D$). By comparing the changes in these quantities we are able to distinguish among many of the metastable defect reactions that have been proposed. In particular we have found conclusive evidence that quench cooling, although increasing $N_{BT}$ typically by a factor of 2, does not modify $N_D$ in any of our samples. Marked changes in $N_{BT}$ with negligible changes in $N_D$ are sometimes also obtained for some of the partial anneal steps following light-soaking. These results are discussed in terms of possible defect reactions which can activate a phosphorous dopant atom without simultaneous formation of silicon dangling bonds.

## INTRODUCTION

There has been increasing evidence that doped a-Si:H exhibits types of metastable behavior not observed in intrinsic samples. The possibility of this was noted several years ago in that the usual mechanism proposed for the creation of light-induced defects by the breaking of weak Si-Si bonds,[1,2]

$$2e^- + 2Si_4{}^0 \rightleftharpoons 2Si_3{}^- \tag{1}$$

could be extended to include bond switching reactions with the phosphorous dopants:[3,4,5]

$$Si_4{}^0 + P_3{}^0 \rightleftharpoons Si_3{}^- + P_4{}^+ \tag{2}$$

$$Si_4{}^0 + P_4{}^+ + 2e^- \rightleftharpoons Si_3{}^- + P_3{}^0 \tag{3}$$

All of the above proposed reactions involve a change in density of dangling bonds which, because we are considering n-type doped material, appear in a negative charge state ($Si_3{}^-$). The presence (or absense) of extra electrons on one side of the reaction implies a shift of the Fermi level ($E_F$) to change the occupation of states. One can immediately recognize, therefore, that a *quantitative* comparison between the changes in occupation of states near $E_F$ and the density of deep defects should be able to identify which of these reactions dominates a given change in metastable state. In moderately doped n-type material this comparison is particularly easy to make since changes in occupation near $E_F$ occur in the conduction bandtail and are easily distinguishable experimentally from changes in the density of deep states, $N_D$.

Our recent studies[6] using this approach have led us to propose a new defect (half) reaction which can activate or de-activate a dopant atom with no change in $N_D$:

$$P_3^0 \rightleftharpoons P_4^+ + e^- \qquad (4)$$

In this paper we shall examine the evidence for the existence of such a dopant activation mechanism for a range of samples with dopant levels between 10 and 300 Vppm. This evidence has been obtained primarily by experimental studies of metastable changes that occur either with by prolonged light exposure followed by partial annealing or by rapid (quench) cooling from a 200°C anneal temperature.

## EXPERIMENTAL METHODS

All of our a-Si:H films were deposited on $p^+$ crystalline silicon substrates by the glow discharge decomposition of silane. Doping was carried out using $PH_3$ admixtures during deposition with concentrations varying from 10-300 Vppm. Semitransparent contacts of chromium or palladium were evaporated onto the samples to obtain Schottky barrier devices for study using junction capacitance techniques.

Samples were prepared in a light soaked state (state B) by exposure to 400 mW/cm$^2$ 1.9 eV light at room temperature for periods not less than 20 hours. Samples were also examined in several partial anneal "states" obtained by elevating the state B sample temperature in the dark for 15 minutes, first at 350K followed by successively higher temperatures in 20-30K increments up to a final temperature of 470K. Our slow-cooled state (state A) were dark annealed at 200°C for 30 minutes and then cooled at a controlled linear rate of 0.1K/second. The quench-cooled metastable state was produced by cooling at a rate of 7K/second from 200°C.

Techniques employed to determine the occupation bandtail states are based on the much faster response of electrons in these states compared to the deeper gap state electrons. Such methods have included the thermally stimulated capacitance (TSCAP) method[7], the carrier sweep-out method[3], and the drive-level capacitance profiling (DLCP) method.[8] We prefer the last technique since one can also confirm the spatial uniformity of any metastable changes that occur within the sample. As previously discussed in some detail,[4,8] this method yields directly an integral over the density of states, g(E); namely

$$N_{DL} = \int_{E_C-E_e}^{E_C-E_F^0} g(E) \; dE \qquad (5)$$

where $E_F^0$ is the position of the Fermi level in the (neutral) bulk, $E_C$ is the position of the conduction band mobility edge, and where the lower energy limit $E_e$ can be selected experimentally by the temperature and frequency of the measurement. For the current studies we employed a measurement frequency of 1 kHz and temperature of 240K which gave a lower energy cutoff near the bottom of the conduction bandtail at $E_C - 0.4$ eV. Under these conditions $N_{DL} \cong N_{BT}$. Some examples of drive-level profiles for several metastable states of our 10 Vppm $PH_3$ sample are given in Fig. 1.

The deep defect density, $N_D$, is determined by the voltage pulse photocapacitance technique. This capacitance transient technique is used to observe the sub-band-gap optical excitation of electrons deep in the gap by detecting

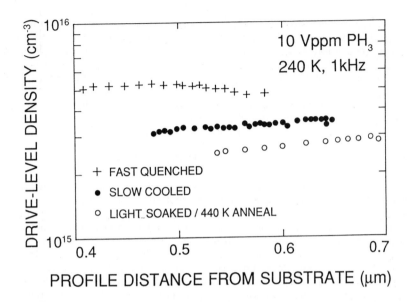

Fig. 1. Spatial profiles of the occupied bandtail states obtained by the drive-level capacitance profiling method for the 10 Vppm PH$_3$ doped sample. Profiles for three metastable states are compared: the slow-cooled state (state A), the quench cooled state, and a 440K partial anneal of the light-soaked sample.

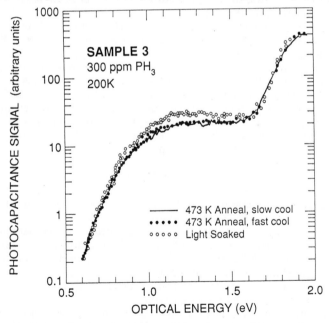

Fig. 2. Voltage pulse transient photocapacitance spectra taken for the 300 Vppm PH$_3$ doped sample for state A, following light-soaking, and following quench cooling.

the charge released to the conduction band. This method has also been described in detail previously.[9] The resultant spectra appear quite similar to sub-band-gap optical absorption data (see Fig. 2 for several examples from our 300 ppm sample) but share the advantages of other junction capacitance techniques that they are preferentially sensitive to the deep bulk region of our samples. The signal level near an optical energy, $E_{opt}$, of 1.5 eV relative to that of the valence bandtail (1.6 eV $< E_{opt} <$ 1.8 eV) is to a good approximation proportional to an integral over the density of states for gap states lying roughly in the energy range $E_C$ - 0.6 eV $< E < E_V$ + 0.4 eV. We identify this integral with $N_D$.

EXPERIMENTAL RESULTS

Our results for the changes in $N_{BT}$ and $N_D$ with metastable treatment for our 3 most heavily doped samples have been plotted in Fig. 3 in a manner which can be compared directly with the predictions of the proposed defect reaction models (Eqs. 1-4). The horizontal position indicates the DLCP determined value of $N_{BT}$ with a vertical line indicating its value in state A. The vertical axis shows the deviation in the value of $N_D$ from its state A value and is plotted in the same density units as $N_{BT}$. A zero value for $N_{BT}$ signifies that $E_F$ is deeper than 0.4 eV below $E_C$ so that there are *no* occupied conduction bandtail states. Open circles denote metastable states obtained by light-soaking and partial annealing; the filled squares denote the quench-cooled metastable states.

We observe that many metastable treatments lead to a change in $N_{BT}$ without a change of comparable magnitude in $N_D$. This is, in fact, the case for *all* of the quench-cooled metastable states. A representative comparison of the actual photocapacitance spectra indicating the changes in $N_D$ for some of these metastable states is given for one sample in Fig. 2.

In Fig. 4 we summarize the total $N_D$ *vs.* $PH_3$ doping level for the fully light-soaked (state B) and state A measurements for all 5 samples studied. We note that the relative increase in deep states after light-soaking approaches a factor of 2 for the most lightly doped samples but that this ratio *decreases* as the doping level *increases*. We also plot $N_{BT}$ *vs.* doping level and compare the values of $N_{BT}$ between the slow-cooled states (state A) and the quench-cooled metastable states. Here we also observe a factor of two increase in $N_{BT}$ in many of the samples studied.

The straight lines included in this figure are for comparison purposes to indicate the approximate power law dependence of $N_D$ and $N_{BT}$ on $PH_3$ doping level. These lines have slopes of 1.0 and 1.5, respectively.

DISCUSSION

The correlations between $N_{BT}$ and $N_D$ vary markedly among the proposed defect reactions. The predictions resulting from reactions (1), (2), and (4) are indicated schematically in Fig. 3(a). Only the most heavily doped sample appears to possibly be consistent with a reaction of type (2). That reaction maintains a balance between the densities of $P_4^+$ and $Si_3^-$ so that one could expect significant changes in $N_D$ with very little shift in $E_F$ (and hence little change in $N_{BT}$). In general, however, the metastable changes generated by light soaking and partial annealing do not seem to obey reactions of strictly a single type.

On the other hand, there is very consistent behavior for the change in properties associated with quench cooling. Here we observe large changes in $N_{BT}$ without any observable change in $N_D$. This result appears to contrast markedly with reported results for the quench cooling of *intrinsic* a-Si:H samples.[10] Such a

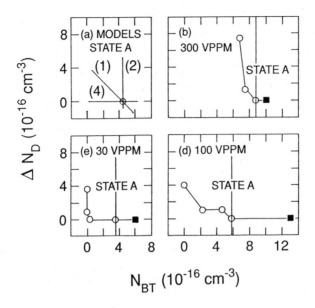

Fig. 3.   Change in the deep defect densities ($N_D$) as a function of the occupied bandtail state density ($N_{BT}$) for (a) the proposed defect reactions (1), (2), and (4) discussed in the text, (b) the 300 Vppm PH$_3$ doped sample, (c) the 100 Vppm PH$_3$ doped sample, and (d) the 30 Vppm PH$_3$ doped sample.    Open circles denote metastable states obtained by light-soaking and partial annealing, and the filled squares indicate quench-cooled metastable states.   The vertical lines denote the $N_{BT}$ values of the sample in the slow-cooled state (state A).   Error bars are comparable to the data point sizes.

Fig. 4.  A summary of the total deep defect densities ($N_D$) vs. PH$_3$ doping level for the fully light-soaked state (state B) and state A measurements.    Also plotted are the values of conduction bandtail occupation ($N_{BT}$) vs. doping level for state A and the quench-cooled metastable states.   The straight lines indicate the approx-imate power law dependence of $N_D$ and $N_{BT}$ for state A on PH$_3$ doping level and have slopes of 1.0 and 1.5, respectively.

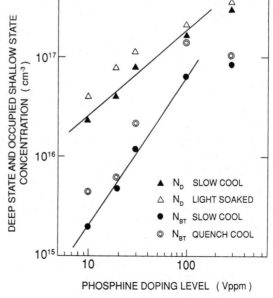

result can only be accounted for by a process like reaction (4) in which phosphorus changes its doping coordination independent of bond switching with silicon atoms. Although one might attempt a composition of reactions of types (1) and (2) to accomplish the same net result as reaction (4), we do not believe this a reasonable alternative because it would require that a second dangling bond be fortuitously located very close to the $Si_3^-/P_4^+$ pair created by reaction (2), a situation that must be considered extremely unlikely.

There has been a great deal of discussion of hydrogen diffusion mechanisms accounting for metastable behavior in a-Si:H, and we recently proposed a couple hydrogen bond switching reactions which could result in the activation of a donor without changes in $N_D$:[6]

$$P_3^0 + Si:H \rightleftharpoons P_4^+ + Si_4^0 + e^- + H^* \qquad (6)$$

$$P:H + Si_4^0 \rightleftharpoons P_4^+ + Si:H + e^- \qquad (7)$$

where :H represents a passivated bond on an otherwise 3-fold coordinated atom and H* is an interstitial hydrogen.   Such proposed reaction are also quite consistent with recent studies of hydrogen in *crystalline* silicon where evidence for a mobile species of negatively charged hydrogen has been found.[11]   These researchers have proposed the reaction

$$(P-H)^0 \rightleftharpoons P_4^+ + H^- \qquad (8)$$

where $(P-H)^0$ is a neutral hydrogen-dopant complex. In a-Si:H, the existence of such a a reaction could quite naturally account for metastable changes observed in n-type material a-Si:H under bias annealing[12] where $N_{BT}$ has been observed to increase dramatically while, as our more recent studies of this effect have shown, there is no observable change in $N_D$.

The observations of the increased doping effects of reaction (8) have been made through the imposition of an applied electric field to remove the H⁻ species from one region of the sample.[11] However, if one speculates that reaction (8) might be followed by the ultimate conversion of H⁻ into a neutral interstitial species this could lead to an increase in $N_{BT}$ in a manner functionally equivalent to that given in reaction (6).

Finally, we should comment on the nearly linear dependence of $N_D$ on phosphorus doping suggested in Fig. 4.   Previously it has been argued that $N_D$ scales as $[P]^{1/2}$.[3]  Because our only knowledge of the phosphorus content in these samples is based on gas phase concentrations, it is not entirely certain that we have a linear dependence on actual phosphorus incorporation.  However, it has been found elsewhere that [P] generally *does* scale linearly with $PH_3$ levels present during growth.[13]  Also, our own data suggests the usual sublinear dependence of $N_D$ with [P] at levels above 100 Vppm $PH_3$.

Theoretically, one understands the 1/2 power law dependence as a consequence of reaction (2) in the equilibration process of dangling bonds together with the requirements of charge neutrality so that $[P_4^+] \cong [N_D^-]$. However, given that alternative defect reactions may actually dominate metastable processes for activating $P_4$ dopants and that an H⁻ species could alter the charge balance condition, it is not unreasonable to speculate that a linear dependence of $N_D^-$ on [P] could occur in the more lightly doped samples.

A simpler explanation might be that silicon-dopant complexes actually form in the plasma-phase during film growth and that these dissociate upon deposition but do not react in solid-phase thermal cycling. This would result in a linear dependence of $N_D$ on the gas phase $PH_3$ concentration and still provide a mechanism for self-compensation. Within such a picture phosphorous dopants and deep defects would each equilibrate *independently* (as our results indicate); however, the two processes would often be correlated due to the effect the Fermi level position has on each of them. Indeed, convincing evidence that deep defect formation responds to $E_F$ position independent of any dopant bond switching reaction has recently been obtained through post-growth interstitial lithium doping.[14]

Finally, it has also been suggested that dopant impurities account directly for both stable and light-induced deep defects through the formation of dopant complexes similar to the DX center in III-V semiconductors.[15] Such a model would quite naturally lead to the observed linear dependence of deep defect with dopants in the low concentration regime.

The correct explanation will hopefully also ultimately account for the observed power law dependence of $N_{BT}$ on $PH_3$ doping. Indeed, the previously proposed explanations for a 1/2 power law of $N_D$ on doping also lead to a 1/2 power for $N_{BT}$, which is totally inconsistent with our own data and the previous studies which obtained $N_{BT}$ using the carrier sweep-out technique.[3] Further experiments are now underway to better distinguish the defect reactions that are controlling the behavior of $N_{BT}$ with doping and metastable treatment.

## ACKNOWLEDGEMENTS

Research at the University of Oregon was carried out under SERI Subcontract XM-8-18061-1.

## REFERENCES

1.  H. Dersch, J. Stuke, and J. Beichler, Appl. Phys. Lett. 38 (1981).
2.  M. Stutzmann, W.B. Jackson, and C.C. Tsai, Phys. Rev. B32, 23 (1985).
3.  R.A. Street, D.K. Biegelsen, W.B. Jackson, N.M. Johnson, and M. Stutzmann, Phil. Mag. B52, 235 (1985).
4.  K.K. Mahavadi, K. Zellama, J.D. Cohen, and J.P. Harbison, Phys. Rev. B35, 7776 (1987).
5.  M. Stutzmann, Phys. Rev. B35, 9735 (1987).
6.  T.M. Leen, J.D. Cohen, and A.V. Gelatos, Mat. Res. Soc. Symp. Proc. 192, 707 (1990).
7.  D.V. Lang, J.D. Cohen, and J.P. Harbison, Phys. Rev. B25, 5285 (1982).
8.  C.E. Michelson, A.V. Gelatos, and J.D. Cohen, Appl. Phys. Lett. 47, 412 (1985).
9.  For a good review see J.D. Cohen and A.V. Gelatos, in Advances in Amorphous Semiconductors, ed. by H. Fritzsche (World Scientific, Singapore, 1988), pp. 475-512.
10. Z E. Smith, S. Aljishi, D. Slobodin, V. Chu, S. Wagner, P.M. Lenahan, R.R. Arya, and M.S. Bennett, Phys. Rev. Lett. 57, 2450 (1986).
11. J. Zhu, N.M. Johnson, and C. Herring, Phys. Rev. B41, 12354 (1990).
12. D.V. Lang, J.D. Cohen, and J.P. Harbison, Phys. Rev. Lett. 48, 421 (1982).
13. See, for example, M. Stutzmann, Phil. Mag. Lett. 53, L15 (1986).
14. K. Pierz, W.Fuhs, H. Mell, Philos. Mag. B63, 123 (1991).
15. D. Redfield and R.H. Bube, Phys. Rev. Lett. 65, 464 (1990).

# LIGHT–INDUCED CHANGES IN COMPENSATED a–Si:H FILMS

J.K. Rath*, B. Hackenbuchner, W. Fuhs and H. Mell
Fachbereich Physik und Wissenschaftliches Zentrum für Materialwissenschaften,
Universität Marburg, Renthof 5, D–3550 Marburg, FRG

## ABSTRACT

We report on concomitant measurements of the metastable light–induced excess conductivity $\sigma_e$ and of the defect density $N_d$ in two series of n–type a–Si:H films doped with 30 or 300 ppm $B_2H_6$, respectively, and various volume parts of $PH_3$. The data for $\sigma_e$ are analysed in terms of the light–induced shift of the Fermi level, $\Delta E_f$. This quantity is determined by the density $N_e$ of light-induced centers providing the negative excess charge (E–centers) and by the density of states at the Fermi level, $g(E_f)$. The changes of $N_e$ and $g(E_f)$ give rise to the complicated dependence of $\Delta E_f$ on the exposure time $t_e$ and on the Fermi–level position, $E_c-E_f$, in the films. We also present data on the formation and annealing kinetics of the E–centers. Their possible nature is discussed in the light of recent ideas on the mechanism of metastable phenomena in a–Si:H films.

## INTRODUCTION

The observation of a light–induced metastable increase of the dark conductivity in compensated a–Si:H films was first reported in 1983 [1]. Since then a number of studies have dealt with this phenomenon [2-6] but its microscopic origin and mechanism is poorly understood. A photo–induced excess conductivity $\sigma_e$, often denoted as persistent photoconductivity (PPC), has also been observed in doping modulated multilayer films (npnp or nipi structures) [7-10]. Also the nature of this effect is still debated and it is not yet clear whether the two excess conductivities have the same origin [2,3,5,6,11]. The aim of the present study has been to improve the knowledge on the $\sigma_e$ effect in compensated a–Si:H films. To this end we have varied the Fermi level position in the films in a wide range by changing the volume ratio of $PH_3$ and $B_2H_6$. The samples were characterized by concomitant measurements of the excess conductivity $\sigma_e$ and of the defect density $N_d$. We analyse the data in terms of the light–induced shift of the Fermi level, $\Delta E_f$, which is directly related to the density of centers providing the negative excess charge (E–centers). The quantity $\Delta E_f$ is also used to study the formation and annealing kinetics of the E–centers. Finally we discuss possible mechanisms for the creation of E–centers.

## EXPERIMENTAL DETAILS

The compensated a–Si:H films used in this study were prepared by r.f. glow–discharge decomposition of undiluted silane in a capacitively coupled

*Permanent address: Energy Research Unit, Indian Association for the Cultivation of Science, Jadavpur, Calcutta 700032, India

plasma reactor. The deposition conditions were as follows: $T_s = 230°C$, $p = 0.3$ mbar, $F = 10$ sccm, $P \simeq 20$ mW/cm$^2$, $r \simeq 2.5$ Å/s. The substrate was corning 7059 glass and the film thickness ranged from 0.8 to 1.1 $\mu$m. Several series of compensated films were prepared by premixing the SiH$_4$ with various volume parts v of PH$_3$ and B$_2$H$_6$. In each series v(B$_2$H$_6$) was fixed while v(PH$_3$) was varied. Here we report on two series of n–type samples with v(B$_2$H$_6$) = 30 and 300 ppm, respectively.

Mg contacts having a length of 6mm and a gap of 0.3 mm were deposited on top of the samples for conductance measurements. The defect density $N_d$ in the volume of the films was infered from the defect related sub–band–gap absorption determined by photoconductance spectroscopy (CPM) [12]. To eliminate effects of adsorbates and previous light exposures the samples were first annealed at 200∘C in vacuum (state A). Except where otherwise noted, the excess conductivity $\sigma_e$ is obtained by illumination with heat–filtered white light from a tungsten–halogen lamp of intensity $F = 50$mW/cm$^2$. The value of $\sigma_e$ is defined as the conductivity measured 15 min after the light has been turned off. A photo–induced excess conductivity was also observed in a–Si:H films doped singly with B$_2$H$_6$ [13]. It could be traced to the presence of a surface–oxide layer. To check for this possibility we measured the value of $\sigma_e$ for several samples before and after a 2 min etch in 5% HF and found only small changes ($< 10\%$) which could be attributed to the slow decay of $\sigma_e$ occuring at room temperature. We conclude from this result that the excess conductivity observed in our films is a bulk effect.

## 3. RESULTS AND DISCUSSION

Fig. 1 shows the dependence of the light–induced excess conductivity $\sigma_e$ at $T = 300$K and of the defect density $N_d$ on the exposure time $t_e$ for a sample doped with 30 ppm B$_2$H$_6$ and 300 ppm PH$_3$. Also indicated is the behavior of the photoconductivity $\sigma_{ph}$ at $T = 300$K. While this quantity varies only by about 10% up to $t_e = 80$ h, $\sigma_e$ exceeds the dark conductivity $\sigma_d$ by a factor of 10 already for $t_e = 1$s. Up to 10 s, $\sigma_e$ increases almost proportional to t. For longer exposure times the enhancement becomes increasingly weaker. Near t = $10^4$s, $\sigma_e$ reaches a maximum of 1100 $\sigma_d$ and decreases for t $> 5$h. The defect density $N_d$, within experimental accuracy, does not change up to 100s but thereafter increases continously up to a factor of 3.7 for $t_e = 80$ h. This increase of $N_d$ is similar in magnitude to that in undoped a–Si:H films with a similar Fermi level position, $E_c$–$E_f$. We attribute it to the light–induced formation of dangling–bond defects [14,15].

The behavior of $\sigma_e$ and $N_d$ in Fig. 1 suggests that there is no well–defined correlation between these two quantities. This is more clearly revealed by Fig. 2 in which $\sigma_e$ is plotted as a function of $N_d$. Curve 1 shows the correlation for the transition from the annealed state A to the light–soaked state B ($t_e = 80$ h). The reverse transition is characterized by curve 2. It was obtained by annealing the sample at increasingly higher temperatures from 320K to 470K for 5 minutes. The correlation of $\sigma_e$ and $N_d$ for the transition B to A differs completely from that for the transition A to B. In both cases $\sigma_e$ changes most strongly when $N_d$ is almost constant.  This behavior is at variance with the parallel changes of $\sigma_e$ and $N_d$ reported by Yoo et al [6].

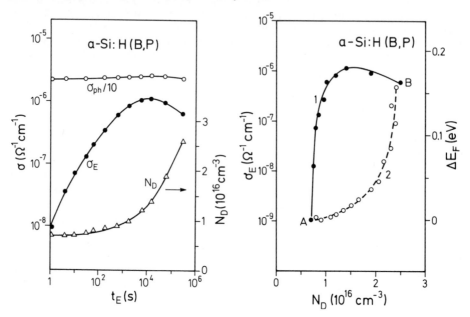

Fig. 1
Dependence of the photoconductivity $\sigma_{ph}$, excess conductivity $\sigma_e$ and defect density $N_d$ on the exposure time $t_e$ for a sample doped with 30 ppm $B_2H_6$ and 150 ppm of $PH_3$.

Fig. 2
Correlation of $\sigma_e$ and $N_d$ for the transition A to B (curve 1, light soaking) and for the reverse transition B to A (curve 2, annealing), respectively, measured for the sample in Fig. 1.

A straightforward interpretation of the $\sigma_e$ effect is that illumination increases the electron concentration in our n–type compensated $a$-Si:H films. This increase can be attributed to special centers (E–centers) [9] which either release an electron by optical excitation or which capture a hole, thereby preventing its recombination with an electron.  The excess negative charge, $-eN_e$, will reside predominantly in gap states above the dark Fermi level $E_f$ in state A causing a shift of the quasi–Fermi level $E_{fn}$ in the dark, $\Delta E_f$. Using for the dark conductivity $\sigma_d$ the expression

$$\sigma_d = \sigma_0 \exp \left[ -(E_c-E_f)/kT \right] \tag{1}$$

and
we obtain

$$\sigma_e = \sigma_0 \exp \left[ -(E_c-E_f-\Delta E_f)/kT \right] \tag{2}$$

$$\Delta E_f = kT \ln (\sigma_e/\sigma_d) \tag{3}$$

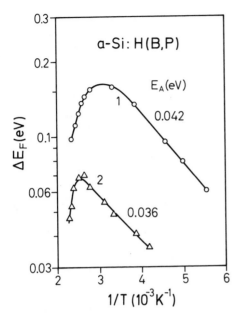

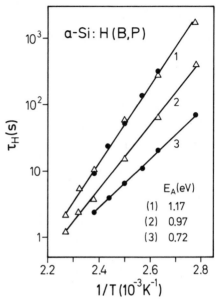

Fig. 3
Temperature dependence of the Fermi-level shift $\Delta E_f$ induced by light exposure. Curve 1: 50 mW/cm², 15 min. Curve 2: 0.5 mW/cm², 10 s.

Fig. 4
Temperature dependence of the relaxation time $\tau_h$ of $\sigma_e$ (curves 2 and 3) and of $\Delta E_f$ (curve 1) for the two exposures in Fig. 3. Circles: 50 mW/cm². Triangles: 0.5 mW/cm².

Assuming an energy independent density of states near $E_f$ of value $g(E_f)$ we find the relation

$$N_e = g(E_f)\,\Delta E_f \tag{4}$$

Similar expressions were proposed by Kakalios [9] for the $\sigma_e$ effect observed in npnp doping modulated a–Si:H films but were not used for the analysis of his experimental data. The right–hand scale in Fig. 2 indicates the values of $\Delta E_f$ observed for the sample in Figs. 1 and 2. In terms of relation (4) the results in these two figures can be understood as follows: For short exposure times, when the defect density $N_d$, and thereby $g(E_f)$, does not deviate from its value in state A, $N_d^a$, the creation of E–centers causes a steep increase of $\Delta E_f$. For $t_e >$ 100s, when $N_d$ exceeds $N_d^a$, the increase of $\Delta E_f$ is more and more reduced and finally, for $t > 10^4$s the value of $\Delta E_f$ decreases, since the relative increase of $N_d$ and $g(E_f)$ is stronger than that of $N_e$. The steep decrease of $\Delta E_f$ during the transition from state B to state A (curve 2 in Fig. 2) is due to the fact that for low annealing temperatures the metastable light–induced defects do not anneal, i.e. $g(E_f)$ remains constant, whereas the density of E–centers decreases appreciably.

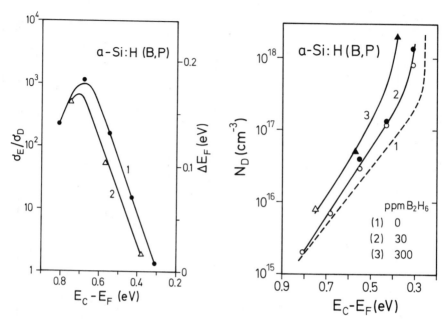

Fig. 5
Dependence of $\Delta E_f$ on the
Fermi–level position $E_c$–$E_f$ for
samples doped with 30 (curve 1) and
300 ppm $B_2H_6$ (curve 2), respectively.

Fig. 6
Dependence of the defect density $N_d$
in the annealed state A on $E_c$–$E_f$.
Curve 1: behavior of films doped
singly with $PH_3$ (from Ref. 12).

Using the quantity $\Delta E_f$ we have studied the kinetics of defect creation and
annealing in more detail. Curve 1 in Fig. 3 shows the temperature dependence
of $\Delta E_f$ induced by a 15 min illumination. Below room temperature $\Delta E_f$ is
activated with $E_a \simeq 0.04$ eV. Above T $\simeq 350$ K $\Delta E_f$ decreases with increasing
temperature, since the E–centers anneal during illumination. According to
Fig. 1, for $t_e = 15$ min $= 900$ s, the defect density increases by about 30%.
Since this increase may vary with T and so $g(E_f)$, the data in curve 1 of Fig. 3
may not truely reflect the behavior of $N_e$. Therefore we carried out a second
measurement using a much weaker light exposure, namely 10s with monochromatic light ($h\nu = 2$eV) of intensity F $= 0.5$ mW/cm$^2$. The results are
shown by curve 2 in Fig. 3. The maximum of $\Delta E_f$ here is shifted to higher
temperatures but the activation energy $E_a \simeq 0.04$ eV is about the same as for
curve 1. The same small activation energy has been found for the creation rate
of light–induced dangling–bond defects [14,15]. The physical reason for it is not
yet clear.

The annealing kinetics of the E–centers is represented by the data in Fig.
4. To obtain these data we have monitored the decay of $\sigma_e$ after terminating
the light exposure. The relaxation time is represented by $\tau_h$, the time necessary
to decrease $\sigma_e$ or $\Delta E_f$, respectively, to half its value at t $= 0$. In the literature
the relaxation of the $\sigma_e$ effect has been characterized by $\tau_h$ of $\sigma_e$ [1,3,7,8]. We
have determined this time for the two types of light exposure used in Fig. 3.

Curve 3 in Fig. 4 shows the data for white–light illumination and curve 2 that for the monochromatic illumination. The two curves differ appreciably. On the other hand, when the same data are analysed in terms of $\Delta E_f$, the relaxation times for both illuminations nicely fall on the same straight line (curve 1 in Fig. 4). This indicates that during the relaxation of the E–centers $g(E_f)$ does not change, i.e. the values of $\tau_h$ in curve 1 represent the relaxation time of $N_e$. This quantity is activated with $E_a \simeq 1.2$ eV, the extrapolated value for $1/T = 0$ being $\tau_0 \simeq 10^{-13}$s. Extrapolation to room temperature yields $\tau_h(300K) = 5\cdot10^6$s $\simeq 50$d for the density of E–centers. The decay time of $\sigma_e$, on the other hand, is much shorter. Extrapolation of curve 3 in Fig. 4 gives $\tau_h$ $(300K) = 7\cdot10^3$s $\simeq 2$h in accordance with the slow decrease of $\sigma_e$ observed at room temperature.

Fig. 5 displays the dependence of the light–induced excess conductivity on the Fermi level position in the n–type compensated films investigated. $E_c-E_f$ has been calculated from the dark conductivity $\sigma_d$ according to relation (1) using $\sigma_0 = 150\Omega^{-1}cm^{-1}$. Curve 1 shows the behavior of films doped with 30 ppm of $B_2H_6$ and 90 to 900 ppm of $PH_3$. The samples in curve 2 were doped with 300ppm of $B_2H_6$ and 1500 to 9000 ppm of $PH_3$. The excess conductivity, in each case, was determined after a 15h illumination with white light of 50 mW/cm$^2$. Qualitatively the decrease of $\Delta E_f$ with decreasing $E_c-E_f$ can be explained by the assumption that $N_e$ is constant within one series of films and $g(E_f)$ increases. The strong increase of the defect density $N_d$ with decreasing $E_c-E_f$ (curve 2 or 3, respectively, in Fig. 6) even suggests that $N_e$ may increase somewhat with decreasing $E_c-E_f$.

The decrease of $\Delta E_f$ for $E_c-E_f > 0.7$ eV presumably is related to the fact that the $\sigma_e$ effect is small or even does not exist in p–type compensated a–Si:H films [1]. As indicated by curve 2 in Fig. 5 the quantity $\Delta E_f$ in the 300 ppm–$B_2H_6$ films is slightly smaller compared to the 30 ppm films. Since $N_d$ is somewhat larger in the more highly doped films (curve 3 in Fig. 6) the density of E–centers seems to be roughly the same in the two series of films. One may be tempted to conclude from this result that the E–centers are not related to the boron atoms in the films. However, the creation rate of E–centers may be proportional to the photoconductivity in the films as in the case of light–induced Staebler-Wronski defects [15,16]. For the 300 ppm films $\sigma_{ph}$ is about an order of magnitude lower compared to the 30 ppm samples. This decrease may balance the increase of the boron concentration.

We now discuss the possible microscopic mechanism of E–center formation. The first suggestion [1] was that in n–type films the $\sigma_e$ effect could be due to the slow trapping of holes in centers which are in poor communication with the rest of the material. Since the excess conductivity is not observed in singly doped material, except for boron–doped films with a surface–oxide layer [13], the presence of both B and P seems to be a necessary prerequisite. This has led to the suggestion that the E–centers might be boron–phosphorus (B–P) complexes [2,4] or B–P–H complexes [3,5]. In our view it is rather unlikely that such centers can form at the lowest $B_2H_6$ and $PH_3$ doping levels used in this study. Otherwise we would expect a strong increase of $N_e$ with increasing B and P content which is not indicated by our results in Fig. 5. We also consider it very unlikely that the E–centers release electrons by optical excitation since we have found a sub–linear increase of $\Delta E_f$ with the light intensity.

Kakalios and Street [11] first proposed that holes could be captured in a process which involves a change of the defect or dopant bonding configuration, e.g. the transition of a fourfold coordinated active acceptor ($B_4^-$) into a threefold coordinated, electronically inactive boron atom ($B_3^o$). Such reaction changes the doping efficiency of boron and is assumed to occur during equilibration after a boron doped sample is annealed in the dark above 450K and rapidly cooled to room temperature. In this model [11] internal potential fluctuations play an essential role. They separate spatially the photo–excited electrons and holes which then can equilibrate with the dopants and defects. Since holes equilibrate faster than electrons, a negative excess charge remains. The faster increase of $\sigma_e$ with $t_e$ in doping–modulated samples has been attributed to the presence of p–type layers in which the equilibration is particularly fast because the diffusion coefficient of hydrogen is high.

A similar model which is also based on the faster speed of structural changes in p–type a–Si:H films was proposed by Hamed and Fritzsche [10] for doping modulated samples. Brief light exposures create predominantly light-induced defects in the p–type layers making them more intrinsic. This in turn pushes $E_f$ in the whole sample closer to the conduction band. We believe that this model is not applicable to our compensated films. There may be p–type regions in nearly intrinsic samples but it appears unlikely in films with $E_c$–$E_f$ < 0.5 eV. We also believe that potential fluctuations do not play an important role in the formation of E–centers in our compensated films. In the presence of significant potential fluctuations we would expect differences in the formation and annealing kinetics for high– and low– intensity illumination which however is not observed in Figs. 3 and 4. Therefore, we favor an interpretation of the $\sigma_e$ effect in compensated films which does not require potential fluctuations.

Branz [17] discussed in detail a class of bistable charge trapping defects which may cause the metastable phenomena observed in doped a–Si:H films. There are several reactions involving the capture of holes. As an example we consider two of them

$$h^+ + P_3^o + Si_3^- \longleftrightarrow P_4^+ + Si_4^o + e^- \tag{5}$$

$$h^+ + B_4^- + Si_3^{o(-)} \longleftrightarrow B_3^o + Si_4^o (+ e^-) \tag{6}$$

Eq. (5) is a bond–formation reaction in which not only a hole is captured but also an electron emitted. Eq. (6), on the other hand, is a bond–switching reaction. In this case it appears unlikely that the dangling bond is negatively charged ($Si_3^-$), since the Coulomb interaction with the $B_4^-$ neighbor may shift the energy level beyond $E_f$. Yoo et al. [6] have used other hole trapping reactions which lead to the formation of $Si_3^o$ without a $B_4^-$ neighbor. In n–type films these centers would capture an electron and there would be no net change of charge.

Hole trapping reactions similar to (5) and (6) can also be obtained by taking into account the motion of hydrogen [17,18]. All reactions can also proceed to the left when an electron is captured, and there are other possible reactions by which electrons can be captured. Thus the question arises: why is the capture of holes favored? The reason presumably is that, in n–type films, the hole concentration during illumination deviates much more from its (equilibrium) dark value than the electron concentration [17]. By that, reactions (5) and (6) may be driven further to the right than corresponding reactions in which electrons are captured. All these reactions will also occur in singly doped material and indeed light–inducated changes of the doping efficiency have been reported [17]. However, the effect on the conductivity is small and therefore masked by other effects, since the high density of defects in singly doped films allows only a small shift of $E_f$. In compensated films, on the other hand, the defect density and thereby $g(E_f)$ is much lower, so that according to (4) a large shift of $E_f$ is induced.

The present data do not allow to decide whether the $\sigma_e$ effect is due to the activation of donors $(P_4^+)$ or to the deactivation of acceptors $(B_4^-)$. Further studies are necessary to clarify the detailed microscopic mechanism of the metastable capture of holes in compensated a–Si:H films.

## ACKNOWLEDGEMENTS

We thank W. Beyer for helpful discussions. One of us (J.K.R.) gratefully acknowledges the award of a fellowship by the United Nationsl Industrial Development Organization (UNIDO). This work was supported by the Bundesminister für Forschung und Technologie (BMFT).

## REFERENCES

1.  H. Mell and W. Beyer, J. Non–Cryst.Solids 59/60, 405 (1983)
2.  S.C. Agarwal and S. Guha, Phys. Rev. B 32, 8469 (1985)
3.  S.–H. Choi, B.–S. Yoo and C. Lee, Phys. Rev. B36, 6479 (1987)
4.  Y.–F. Chen and Y.–S. Huang, J. Appl. Phys. 62, 2578 (1987)
5.  D.H. Zhang and D. Haneman, J. Appl. Phys. 63, 1591 (1988)
6.  B.–S. Yoo, Y.–H. Song and C. Lee, Phys. Rev. B 41, 10787 (1989)
7.  J. Kakalios and H. Fritzsche, Phys. Rev. Lett. 53, 1602 (1984)
8.  M. Hundhausen and L. Ley, Phys. Rev. B32, 6655 (1985)
9.  J. Kakalios, Phil. Mag. B 54, 199 (1986)
10. A. Hamed and H. Fritzsche, Phil. Mag. Lett. 60, 171 (1989)
11. J. Kakalios and R.A. Street in Disordered Semiconductors, Eds. M.A. Kastner, G.A. Thomnas and S.R. Ovshinsky, Plenum, New York (1987) p. 529
12. K. Pierz, W. Fuhs and H. Mell, Phil. Mag. B63, 123 (1991)
13. B. Aker, S.–Q. Peng, S.–Y. Cai and H. Fritzsche, J. Non–Cryst. Solids 59/60, 509 (1983)
14. H. Dersch, J. Stuke and J. Beichler, Appl. Phys. Lett. 38, 456 (1981)
15. M.Stutzmann, W.B. Jackson and C.C. Tsai, Phys. Rev. B32, 23 (1985)
16. D. Staebler and C. Wronski, J. Appl. Phys. 51, 3262 (1980)
17. H.M. Branz, Phys. Rev. B38, 7474 (1988)
18. K. Winer and W.B. Jackson, Phys. Rev. B40, 12558 (1989)

# Accelerated stability test of a-Si:H by defect saturation

M. Isomura and S. Wagner
Department of Electrical Engineering, Princeton University
Princeton, NJ, 08544

## Abstract

The temperature dependence of the saturated light-induced defect density ($N_{sat}$) of a-Si:H was measured while precisely controlling sample temperature during illumination. $N_{sat}$ decreases if the sample temperature is held above 75°C to 100°C, and increases below 50°C. At intermediate temperature $N_{sat}$ levels off. The two thermally activated regions of $N_{sat}$ suggest that the activation energy of the annealing process has a distribution with two peaks. In experiments geared to raising the stability of a-Si:H, we found that $N_{sat}$ is reduced to the lowest value measured so far by thermal annealing at 190°C for 3 to 19 hours prior to light-soaking. This result suggests that the microscopic structure of a-Si:H plays on important role in stability.

## Introduction

Light-induced defects in a-Si:H pose a serious problem to device applications, and their understanding is of fundamental importance to a-Si:H. Much attention has been drawn to the Staebler-Wronski effect and many researchers are working on the phenomenon, but we do have neither a clear understanding of its mechanism nor an unequivocal method for its evaluation. Therefore, it is quite important to acquire precise quantitative information about the effect. The rate of buildup of the defect density is, however, affected very much by the conditions of light-soaking and obtaining information about stability usually takes a long time.

We recently introduced a new method for reaching saturation of the light-induced defect density ($N_{sat}$) within a few hours by soaking with a Kr-ion laser whose light ($\lambda$=647nm) is absorbed nearly uniformly and at 3 Wcm$^{-2}$ produces a carrier generation rate of about $3 \times 10^{22}$cm$^{-3}$s$^{-1}$ [1]. Because of saturation, the measured defect density $N_{sat}$ is independent of spatial or temporal variations of the light intensity. We also found that $N_{sat}$ has a clear correlation with the defect growth rate at typical sunlight intensity, and that we can estimate the defect growth rate dNs/dt once we have determined $N_{sat}$ [2]. Therefore, the light-soaking history of a material may be evaluated in a few hours, and feedback on stability can be provided quickly to the deposition laboratory.

Some reports about the saturated light-induced defect showed that the saturation value is affected by thermal annealing.[3,4] Thermal annealing underscores the question for the mechanism for defect saturation in our method, which could be exhaustion of sites which are convertible to defects, or attainment of a steady state between annealing and generation of defects. To obtain an answer, we measured the temperature dependence of $N_{sat}$ while precisely controlling the temperature of the a-Si:H films during light-soaking.

In samples obtained from many different laboratories and deposited by different techniques, $N_{sat}$ rises with the Tauc gap ($E_{opt}$) or E04 and with the total hydrogen concentration ($c_H$), but is not correlated with the initial defect density or the Urbach energy.[5] The $N_{sat}$ values measured so far lie between $4 \times 10^{16}$ and $2 \times 10^{17}$cm$^{-3}$ with apparent lower limits that depend on $E_{opt}$ and $c_H$ [2]. Therefore, $E_{opt}$ and $c_H$ likely are important for stability. However, the $N_{sat}$ values scatter considerably and may be influenced by other factors, too. Therefore, we attempted to change the "structure" of a-Si:H by long anneals at temperature near the substrate temperature during deposition.

106

Such anneals are not expected to change $E_{opt}$ and $c_H$ substantially, and could help identify other variables that affect $N_{sat}$.

## Samples and experimental techniques

The samples were prepared from silane gas on Corning 7059 glass or quartz substrates by the radio-frequency (13.56MHz) glow discharge method. Sample S28 for the temperature-dependence measurement of $N_{sat}$ was deposited on a thin (0.3mm) quartz substrate for better control of the temperature because of its high thermal conductance and low heat capacity. We list the preparation temperature and samples properties in table 1. The defect density ($N_S$) was measured with the constant photocurrent method (CPM) [6] by converting the integrated subgap absorption intensity to $N_S$ using a calibration factor.[7] $N_S$ of the samples before light-soaking lies between $9 \times 10^{14}$ and $5 \times 10^{15} cm^{-3}$. In a series of experiments done on the same sample *relative* values of $N_S$ are determined by overlaying the CPM spectra. We estimate the accuracy of such relative $N_S$-data to be $\log(N_S)$ ±0.05, or $N_S$ ±15%. The scatter in absolute values of $N_S$, as used for comparing results from different samples, is $N_S$ ±30% if the same skilled operator evaluates the CPM spectra. [8]

For light-soaking the samples were mounted on a thermoelectric element whose reference-temperature side is attached to a fan-cooled aluminum heat sink, as shown in fig. 1. We measured the sample temperature by the calibrated reflectance of a low intensity He-Ne laser beam off the surface of the a-Si:H films during Kr-ion laser soaking. Fig. 2 shows the reflectance of the He-Ne laser with and without illumination from the Kr-ion laser as a function of the temperature of the thermoelectric element. The U-shaped reflectance plot is shifted to the left by Kr-ion laser illumination, and the temperature of the a-Si:H films can be seen to rise by 10°C to 15°C above that set by the thermoelectric element.

## The temperature dependence of $N_{sat}$

How does thermal annealing during high-intensity light-soaking affect the $N_{sat}$ value? If it does, the $N_{sat}$ should be sensitive to temperature. To find the answer, we conducted saturation experiments with samples held at temperatures between 25°C and 130°C. The sample temperature was controlled as shown in fig.1. We show the result in fig. 3. Samples S28-1,2,3,4 and 7, which which are the pieces from the same substrate and have identical sample histories, were used for the different temperatures. For each temperature, we measured the defect density after illumination with the Kr-ion laser for 5, 8 and 13 hours. The $N_{sat}$ values are not affected by temperature between 50°C and

Table 1

| Sample No. | $T_S$ (°C) | d (μm) | $c_H$ (%) | $E_0$ (eV) | $E_u$ (meV) | Annealing treatment $T_A$ (°C) | time (hr) |
|---|---|---|---|---|---|---|---|
| S28 | 180 | 1.65 | 16.5 | 1.75 | 48-50 | 190-265 | 3-19 |
| (S28-1-4,7) | 180 | 1.65 | 16.5 | 1.75 | 49-50 | 190 | 3 |
| (S28-8,9) | 180 | 1.65 | 16.5 | 1.75 | 52 | 190 | 6 |
| H18 | 170 | 1.10 | 19.5 | 1.77 | 51-53 | 185 | 6 |
| S14 | 180 | 0.55 | 16.0 | 1.72 | 47-50 | 220 | 6 |

$T_S$=deposition temperature, d=thickness, $c_H$ in at.%, $E_0$=Tauc gap, $E_u$=Urbach energy

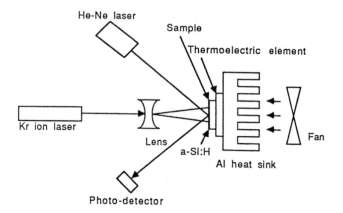

Fig.1  Schematic diagram of the sample configuration for Kr-ion laser soaking at controlled temperature.

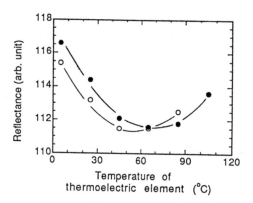

Fig.2  Reflectance of a He-Ne laser beam off an *a*-Si:H sample as a function of the temperature of the thermoelectric element (fig.1) when the Kr-ion laser is on (O) and off (●).

100°C. However, the $N_{sat}$ value reached at 130°C is low compared to the values reached at the lower temperatures, probably because of thermal annealing. The $N_{sat}$ value of sample S28-1, which had been low at 130°C, after additional illumination at 100°C for 3 hours rose nearly to the $N_{sat}$ value measured in other samples tested at 50°C to 100°C. Therefore, the drop of $N_{sat}$ above 100°C must be caused by temperature. On the other hand, at 25°C the defect density did not reach saturation within 13 hours and increased up to a slightly higher value than the $N_{sat}$ reached above 25°C. Thus, at 25°C, illumination for longer than at 50 to 130°C does not saturate the defect density, plus $N_{sat}$ is likely to lie higher at 25°C than at 50 to 130°C. S28-9 was also used to test the effect of sequential light-soaking at decreasing temperature, as shown in fig.4. S28-9 saturated at 100°C within one hour. After saturation at 100°C the samples was saturated to a higher $N_{sat}$

value at 75°C by additional 3 hours illumination. Subsequent exposure for 3 hours at 50°C did not raise $N_{sat}$ further. Fig.4 also shows a separate sample S28-9 from the same substrate as S28-8, which has the exactly same history. S28-9 was soaked only at 25°C, but did not saturate within 28 hours. Fig.4 shows the plateau between ~50°C and ~75°C that also appears in fig.3. We estimate the scatter in $N_{sat}$ of samples from the same substrate with the same history is at most $N_s \pm 30\%$.

The observation of a plateau of $N_{sat}$ in figs.3 and 4, flanked by two thermally activated regimes, suggests that the distribution of thermal activation energies has two peaks. The most simple phenomenological explanation is a three-level system as depicted in fig.5. A group of metastable states peaked at A has such a low activation energy that, under our high-intensity illumination, all A-states are annealed out at 50°C. Another group of metastable states peaked at B can anneal out only at high temperature, so that the $N_{sat}$ values on the plateau are determined by the number of B-states. We believe that all of our earlier $N_{sat}$ data [1,2,5] were obtained on the plateau. The plateau is the valid region for using $N_{sat}$ as a stability criterion. It retains the advantage of $N_{sat}$ that it

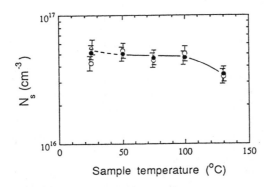

Fig.3  Defect density $N_S$ after Kr-ion laser soaking for 5hrs(O), 8hrs(●) and 13hrs(Δ), as a function of sample temperature.

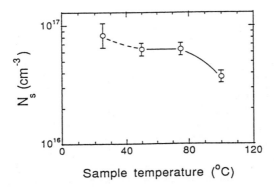

Fig.4  Defect density $N_S$ after Kr-ion laser soaking as a function of sample temperature for S28-8(●) and S28-9(O).

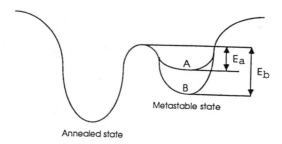

Fig.5  A simple energy configuration model with two groups of metastable defects.

is insensitive to important experimental conditions like light intensity and sample temperature. Furthermore, the plateau luckily coincides with the operating temperature of solar cells and thus lends more credence to the use of $N_{sat}$ data. The question about the difference between the A and B states remains to be answered. One possibility is tied to the participation of hydrogen motion in the annealing process [9]: state A may require only a single jump of an hydrogen atom for annealing; state B may require hydrogen motion over distances of several interatomic spacings.

The sample temperature during light-soaking affects the *rate* of defect growth, too. Fig.6 shows the density of defects $N_S$ induced in two samples taken from the same deposition run. S28-8 was light-soaked in steps at 25°C, and S28-9 at 100°C. Note two effects, one on growth rate, the other on saturation. A given $N_S$ is reached at 100°C about five times faster than at 25°C. Bandgap narrowing [10] contributes only 20% to the observed increase in defect growth rate so that raising the temperature must accelerate the intrinsic rate of defect growth. However, more measurements should be made to confirm this enhancement of the defect growth rate by increased temperature. The slopes of $N_S$ versus illumination time at both temperatures are the same and lie close to

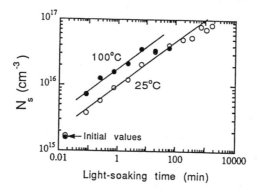

Fig.6 Defect density $N_S$ as a function of Kr-ion laser soaking time for sample temperatures of 25°C (O) and 100°C (●).

1/3, in accordance with the recombination model. These identical slopes appear to disagree with the models in which the slope changes with temperature, e.g., because of dispersive hydrogen diffusion.[11,12] The data of fig.6 also show that saturation is reached at 100°C but not at 25°C.

## Reduction of $N_{sat}$ value by thermal annealing

We recently found that $N_{sat}$ can be reduced by thermal annealing *before* light-soaking. We annealed for several hours at temperatures near the deposition temperature. Fig. 7 shows the $N_{sat}$ values of three different samples which were annealed before light-soaking for 6 hours, as functions of the annealing temperature. $N_{sat}$ was measured after Kr-ion laser-soaking at 50°C. The initial $N_S$ and the $N_{sat}$ of each sample were reduced by the thermal annealing. For sample S28-8, annealing around 190°C reduced $N_{sat}$ the most. The data are shown in fig.8, where $N_{sat}$ versus the annealing time at 190°C. $N_{sat}$ is reduced to half of the as-deposited value within 3 to 6 hours, with longer annealing not having much further effect. Thus the stability of a-Si:H is improved by annealing close to the deposition temperature. This annealing treatment promises to be useful for the improvement of devices, because the annealing temperature is not too high for spoiling the device performance.

In fig.9, we plot $N_{sat}$ of the 190°C-annealed samples as a function of their Tauc gap together with earlier data of standard annealed (170°C, 90min) or as-deposited samples which we have reported before.[2,5] In commenting the earlier data we drew a line which denotes the lowest observed values of $N_{sat}$ versus $E_{opt}$.[2] The 190°C-anneal data clearly fall below this line. Note that relative values of ns were determined in these experiments. This means that a-Si:H can be made more stable than we thought earlier.

Another important conclusion is that the microscopic "structure" does play an important role in stability. Earlier [2,5] we had identified correlations of the band gap (fig.9), and of hydrogen content, with $N_{sat}$. The 190°C-anneals do not change $E_{opt}$ nor are they expected to change the hydrogen concentration. This evidence leaves the microscopic structure as a possible explanation for the reduction of $N_{sat}$. The long-term anneals may be the equivalent of low-temperature deposition which is expected to produce a more relaxed network.

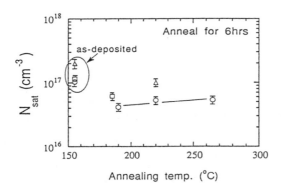

Fig.7 Saturation value $N_{sat}$ as a function of pre-light-soak annealing temperature for samples S28(O), H18(□) and S14(Δ).

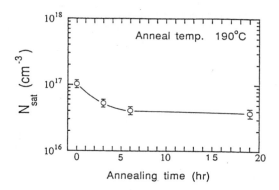

Fig.8 Saturation value $N_{sat}$ as a function of pre-light-soak annealing time at 190°C.

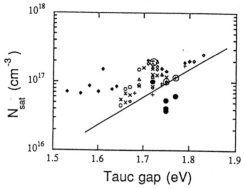

Fig.9 Saturation value $N_{sat}$ as a function of the Tauc gap. (●)pre-soak annealed samples, (⊕)initial samples (+)UHV RF(13.56MHz)-GD, (×)RF(13.56MHz)-GD, (△)RF(70MHz)-GD, (◆)DC magnetron sputter, (O)triode DC-GD and (◇)triode DC-GD fluorine.

## Summary

We found that at high-intensity light-soaking the saturated defect density exhibits a plateau in the range of ~50°C to ~75°C. $N_{sat}$ drops to lower values during light-soaking above 75°C-100°C. During light-soaking at 25°C $N_S$ rises to higher values, and we have not observed saturation up to 28 hours. These observations suggest that the distribution of activation energies for defect annealing peaks at two values. We found preliminary evidence for an acceleration in the growth rate of the defect density when the sample temperature during light-soaking is raised.

In experiments on improving the stability of a-Si:H we found that long anneals near deposition temperature, carried out before light-soaking, reduced $N_{sat}$ to the lowest values we have measured so far. We surmise that the stability improvement has a microscopic origin. On account of the low annealing temperature we expect the treatment to be applicable to raising the stability of completed devices.

## Acknowledgments

We thank S.A. Lyon for making his Kr-ion laser available to us, C. Peterson for instructing us in its operation and maintenance, and N. Hata, N.W. Wang and J.H. Yoon for helpful discussion. This work is supported by Thin-Film Solar Cell Program of the Electric Power Research Institute.

## References

1.  H.R. Park, J.Z. Liu and S. Wagner, Appl. Phys. Lett. 55, 2658 (1989)

2.  M. Isomura, H.R. Park, N. Hata, A. Maruyama, P. Roca i Cabarrocas, S. Wagner, J.R. Abelson and F. Finger, Proc. 5th International PVSEC, Kyoto, Japan, 71 (1990)

3.  M. Stutzmann, W.B. Jackson and C.C. Tsai, Phys. Rev. B 32, 23 (1985)

4.  H. Ohagi, J. Nakata, A. Miyanishi, S. Imao, M. Jeong, J. Shirafuji, K. Fujibayashi and Y. Inuishi, Jpn. J. Appl. Phys. 27, 12, 2245 (1988)

5.  H.R. Park, J.Z. Liu, P.Roca i Cabarrocas, A.Maruyama, M.Isomura, S.Wagner, J.R.Abelson and F.Finger, Appl. Phys. Lett. 57, 1440 (1990)

6.  M. Vanecek, J. Kocka, J. Stuchlik, Z. Kosicek, O. Stika and A. Triska, Solar Energy Materials 8, 411 (1983)

7.  Z E. Smith, V. Chu, K. Shepard, S. Aljishi, D. Slobodin, J. Kolodzey, S.Wagner and T.L. Chu, Appl. Phys. Lett. 50, 1521 (1987)

8.  N.W. Wang, X. Xu and S. Wagner, this volume

9.  W.B. Jackson and J. Kakalios, Phys. Rev. B 37, 1020 (1988)

10.  G. Weiser and H. Mell, J. Non-crystalline Solids 144, 29 (1989)

11.  D. Redfield and R.H. Bube, Appl. Phys. Lett. 54, 1037 (1989)

12.  W.B. Jackson, Philosophical Magazine Lett. 59, 2, 103 (1989)

# AN EXPERIMENT TO DISTINGUISH BETWEEN BIMOLECULAR AND SINGLE-CARRIER DRIVEN MODELS OF METASTABLE DEFECT GENERATION

S.J. Fonash, J.-L. Nicque, J.K. Arch, S.S. Nag, and C.R. Wronski
Center for Electronic Materials and Processing
The Pennsylvania State University, University Park, PA 16802

## ABSTRACT

The computer program AMPS has been used to explore experimental situations for testing the validity of bimolecular and single-carrier driven models of metastable defect generation in a-Si:H. In this study, AMPS was used to evaluate the effect of blue laser soaking on the experimentally-measured quantum efficiency of a-Si:H Schottky barrier structures. Results of these simulations indicate that neither a surface degradation nor a bulk degradation model alone can satisfactorily explain the experimentally observed degradation in quantum efficiency due to blue laser soaking. These results suggest that neither a bimolecular nor single-carrier-driven defect generation mechanism can clearly account for the degradation experimentally observed.

## INTRODUCTION

A number of attempts have been made to try to distinguish among the different possible mechanisms that have been proposed to be the cause of metastable defect formation in a-Si:H. These efforts have included analyzing light induced creation rates[1] as well as investigating the effects of current injection.[2-3] However, no clear picture has yet emerged to allow the assessment of the role of recombination or of single carrier defect generation mechanisms.

In this report we present an up-date on our ongoing efforts to use the AMPS (Analysis of Microelectronic and Photonic Structures) transport simulation programs to design and analyze experiments which can help to differentiate among metastable defect creation mechanisms which are driven by the presence of an excess of both carriers, by the presence of an excess of a single carrier, or by some other mechanism.[4-6]

## THE EXPERIMENT

The experiment reported on here involves blue laser soakings of Schottky barrier solar cell structures and device evaluations using quantum efficiency (QE) measurements. Three different device lengths (L) were used: 0.8 $\mu$m, 1.7 $\mu$m, and 3.5 $\mu$m. In each case the metallization used was Pt and in each case the structure had an $n^+$ back contact. The QE evaluations were done with and without bias light in the annealed state for each length. The bias light QE used a red light bias ($\lambda \geq 600$ nm; flux = 3 x $10^{15}$ photons/cm$^2$-sec). After the blue laser soakings, QE evaluations were again done with and without this same bias light. These blue laser soakings ($\lambda = 450$ nm) utilized a flux of 6 x $10^{18}$ photons/cm$^2$-sec and were done for 10 minutes.

The experimental QE's for these three structures (L = 0.8, 1.7, and 3.5 μm) are shown in Figs. 1, 2, and 3 prior to any laser exposure. Each figure shows the experimental QE for a given device in the annealed state with and without bias illumination. Also shown in these three figures are the corresponding AMPS computer fits to the experimental data. The parameters used for these fits to the annealed QE with and without bias light are seen in Table I. As may be noted, these fits for QE with and without bias light are achieved using essentially the same material parameters for each of the three lengths. The density of states (DOS) picture used by the AMPS modeling to describe these three devices in the annealed state is shown in Fig. 4. As can be seen from this figure, this DOS picture includes exponential tail states and Gaussian deep-level (dangling bond) states. In addition, we point out that we needed to include $8 \times 10^{14}$ cm$^{-3}$ fully-ionized donor states to obtain a good match to the experimental data for all three thicknesses. These donors could arise from n-layer phosphorous contamination or from some other impurity or defect level.

---

### Table I.  Input parameters for annealed case.

Device thicknesses:  0.8, 1.7, and 3.5μm
Schottky barrier height:  1.1eV
All devices have a 200Å n$^+$ back layer ($N_D = 10^{18}$cm$^{-3}$).

$\mu_n = 10$ cm$^2$ V$^{-1}$ s$^{-1}$; $\mu_p = 1$ cm$^2$ V$^{-1}$ s$^{-1}$; $N_c = N_v = 2 \times 10^{20}$ cm$^{-3}$

Tail states (exponential distribution)
Density of states at the optical edges = $10^{21}$ cm$^{-3}$eV$^{-1}$

Valence tail:     $E_d = 50$ meV; $\sigma_{charged} = 10^{-15}$ cm$^2$; $\sigma_{neutral} = 10^{-19}$ cm$^2$

Conduction tail: $E_a = 27$ meV; $\sigma_{charged} = 10^{-15}$ cm$^2$; $\sigma_{neutral} = 10^{-19}$ cm$^2$

Dangling bonds states (Gaussian distribution)
(a) Donor:       $E_{peak}-E_v = 0.9$ eV; $N_{total} = 6 \times 10^{14}$ cm$^{-3}$
(b) Acceptor:    $E_{peak}-E_v = 1.4$ eV; $N_{total} = 6 \times 10^{14}$ cm$^{-3}$

For both Gaussians:  Standard deviation = 0.15eV
$\sigma_{charged} = 5 \times 10^{-14}$ cm$^2$ ; $\sigma_{neutral} = 5 \times 10^{-16}$ cm$^2$

In addition, $8 \times 10^{14}$cm$^{-3}$ fully-ionized (i.e., located in energy near the conduction band edge) donors are included throughout the bulk.

---

After the QE evaluations of the annealed state seen in Figs. 1, 2, and 3, the structures were subjected to the blue laser soakings at open circuit. These blue laser exposures were selected because our AMPS modeling showed that recombination varies significantly across the device in this case as seen in Fig. 5. Hence the recombination, as calculated by AMPS, present during the blue laser exposures of the annealed samples (defined by Table I) is shown for the three different lengths. Also shown for comparison is the recombination that would be present had equivalent (same flux density) red laser exposure been used. Fig. 5 shows that, if recombination causes degradation of a-Si:H, then we should expect the blue laser soaking to produce a degraded surface layer. This layer should be the same according to Fig. 5 for L = 0.8, 1.7, and 3.5 μm devices.

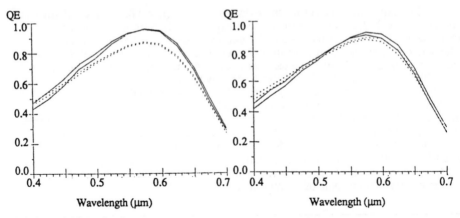

Fig. 1. Computer match ( - - - ) to experimental QE (——) for L = 0.8 μm. Upper curve in blue wavelengths in each pair is QE under light bias; lower is QE in the dark. Matching parameters are identical with those used in Fig. 2 and very similar to those in Fig. 3.

Fig. 2. Computer match ( - - - ) to experimental QE (——) for L = 1.7 μm. Upper curve in blue wavelengths in each pair is QE under light bias; lower is QE in the dark. Matching parameters are identical with those used in Fig. 1 and very similar to those in Fig. 3.

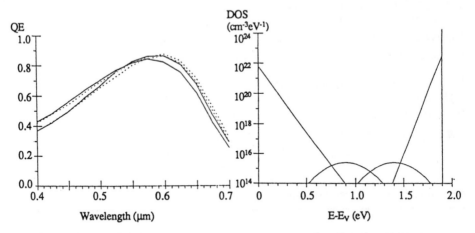

Fig. 3. Computer match ( - - - ) to experimental QE (——) for L = 3.5 μm. Upper curve in blue wavelengths in each pair is QE under light bias; lower is QE in the dark. Matching parameters are identical with those used in Fig. 1 and 2 except Gaussians have a factor of 3 more states.

Fig. 4. Density of states used by AMPS to model the annealed state for all three device lengths.

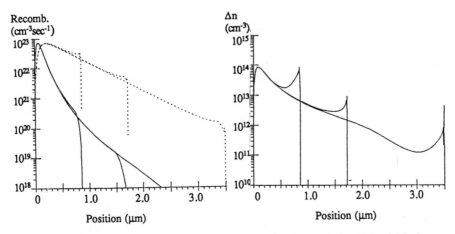

Fig. 5. Recombination in the three annealed devices as a function of position in the device during the blue laser soaking ( —— ) and during an equivalent red laser soaking ( - - - ). Annealed material parameters are used.

Fig. 6. Electron population n during blue laser soaking minus electron population $n_0$ at thermodynamic equilibrium for all three device thicknesses. Annealed material parameters are used.

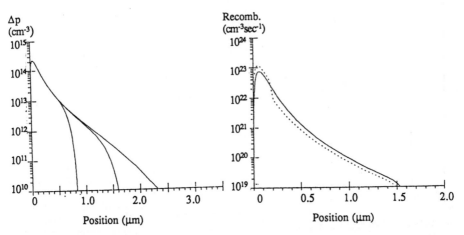

Fig. 7. Hole population p during blue laser soaking minus hole population $p_0$ at thermodynamic equilibrium for all three device thicknesses. Annealed material parameters are used.

Fig. 8. Recombination in the 1.7 µm device as a function of position during blue laser soaking using annealed material parameters ( —— ) and surface degraded material parameters ( - - - ).

Continuing to examine the AMPS-predicted degradation based on the models proposed for defect creation, we plot in Fig. 6, for all three devices, the change in electron population between that present during the blue laser soaking and that present in thermodynamic equilibrium. The corresponding change in hole population under the blue laser exposures is plotted in Fig. 7. In the case of electrons, Fig. 6 suggests that a degradation mechanism driven by electrons would produce similar degraded layers in the 1.7 and 3.5 μm devices. Fig. 6 suggests that an electron-driven mechanism would yield a more uniform degradation across the 0.8 μm device. In the case of holes, Fig. 7 suggests that a hole-population driven degradation mechanism would yield identical degraded surface layers for all three device lengths.

Figures 5-7 show recombination, change in electron population, and change in hole population under open circuit blue laser exposure. They are calculated using the material parameters of the annealed state given in Table I. Since the material parameters are actually evolving during the blue laser exposure, we recalculated these quantities assuming a 2000Å degraded layer at the front surface. This degraded layer is simply the same as the bulk except the total number of states in each Gaussian has been increased by a factor of 100. Figure 8 compares the recombination present during blue laser soaking using both the annealed parameters and the surface-degraded parameters. As can be seen from this figure, the evolution of material parameters during laser soaking does not substantially change the picture of Figure 5. Similar comparisons show that the free electron and free hole profiles do not change significantly either. These profiles of recombination, change in free electron population, and change in free hole population also do not change during degradation if a bulk degradation model is assumed.

To summarize, the picture is the following. If degradation is driven by recombination, then this modeling indicates an identical degraded surface layer should be produced in all three device lengths after the blue laser exposure. If degradation is driven by the presence of excess holes, then this modeling also indicates an identical degraded surface layer should be produced in all three device lengths after the blue laser exposure. If degradation is driven by the presence of excess electrons, then this modeling suggests that the degradation should be relatively uniform across the 0.8 μm device; however, it will result in degraded surface layers for the 1.7 and 3.5 μm devices.

## QE RESULTS AND DISCUSSION

After the blue laser exposure, each of the three devices was recharacterized using QE measurements in the dark and under bias light. The experimental results are shown in Figs. 9 (dark) and 10 (bias light). The degradation in the QE that resulted from the open-circuit blue laser soakings may be ascertained for each device by comparing the data of Figs. 9-10 with the experimental (solid) curves of Figs. 1-3.

Just as we had matched the QE seen in Figs. 1-3 for the annealed devices, we undertook the matching of the QE data of Figs. 9-10. Both the recombination-driven and hole-driven models, as noted, suggest that this blue laser soaking degradation should result in a surface degraded layer. The electron driven model, except for the 0.8 μm device, also suggests a surface degraded layer should result. Hence we have expended a great deal of effort in trying to match the experimental QE data of Figs. 9-10 with a degraded surface layer. The best fit obtained to date, that can be used for all three lengths, is shown in Fig. 11 for the case of dark QE (corresponding to Fig. 9) and in Figure 12 for the case of bias light QE (corresponding to Fig. 10). The surface degradation model used the same material parameters as the annealed case with the exception of a 2000Å front degraded layer. The only difference in this layer is that the Gaussians of Fig. 4 have their state densities increased by a factor of ~70.

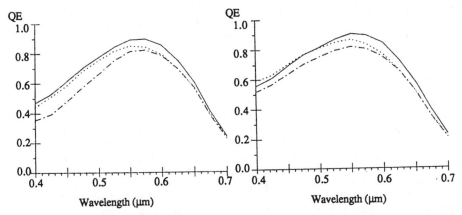

Fig. 9. Experimental dark QE after 10 minutes blue laser soaking.
[—— 0.8μm; - - - - 1.7μm; —·— 3.5μm.]

Fig. 10. Experimental bias light QE after 10 minutes blue laser soaking.
[—— 0.8μm; - - - - 1.7μm; —·— 3.5μm.]

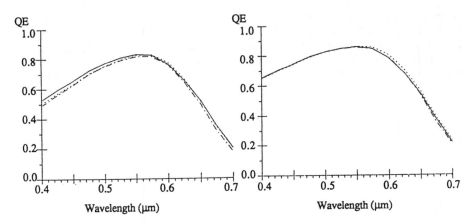

Fig. 11. Computer match to experimental dark QE after 10 minutes (Figure 9) as a function of thickness using surface-degraded parameters (identical for all thicknesses).
[—— 0.8μm; - - - - 1.7μm; —·— 3.5μm.]

Fig. 12. Computer match to experimental bias light QE after 10 minutes (Figure 10) as a function of thickness using surface-degraded parameters (identical for all thicknesses).
[—— 0.8μm; - - - - 1.7μm; —·— 3.5μm.]

A comparison of the trends with device thickness seen experimentally in Figs. 9-10 and those seen in the computer match of Figs. 11-12 reveals that the surface-degraded model is unable to predict the complete experimental thickness dependence of QE after 10 minutes blue laser soaking. Fig. 11 shows that the surface-degraded model is successful in predicting the drop in long-wavelength QE. Figures 11 and 12 show that matching the long-wavelength QE, in both the light and dark, with a surface-degraded model necessitates a strongly-degraded surface. In fact, the surface is so degraded that the short-wavelength QE loses the sensitivity to device length that is seen experimentally. Detailed computer analysis using AMPS shows that, in the degraded surface model, the short-wavelength QE is controlled by a large front-surface field. This electric field, which is established by positive space charge residing in the large number of surface-layer states used in the surface-degraded model, is found to be nearly the same for all three device thicknesses.

Because of these problems in attempting to use a simple surface-degraded model to explain degradation by blue laser soaking, we have also explored what we term here a bulk degradation model. In reality it is a hybrid surface/bulk degradation model which assumes a surface layer degradation and uniform additional degradation throughout the device. Compared to the annealed state picture of our material, the "surface" portion of our hybrid "bulk" degradation model has a 500Å surface layer with new donor-like states in it of density $10^{16}$ cm$^{-3}$, while the "bulk" degradation consists of enhanced Gaussians (those seen in Fig. 4) with state densities increased to $4 \times 10^{15}$ cm$^{-3}$. The bulk and surface parameters used in the fits of this bulk degradation model are the same for each device length.

Figs. 13-14 plot the computer "match" to the experimental light-soaked data of Figs. 9-10 using this "bulk" degradation model. These figures show that this bulk degradation model does a better job predicting the experimentally-observed thickness dependence of short-wavelength QE, both with and without bias light. In addition, reasonable fits to the experimental data are obtained using lower defect state densities ($4 \times 10^{15}$ cm$^{-3}$ versus $4 \times 10^{16}$ cm$^{-3}$) in the bulk degradation model than were needed to obtain reasonable QE's using the surface degraded model. These results suggest that there is some bulk component to degradation even in the thickest device (L = 3.5 μm)

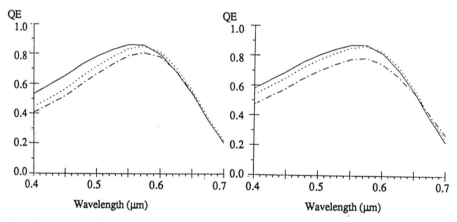

Fig. 13. Computer match to experimental dark QE after 10 minutes (Figure 9) as a function of thickness using bulk-degraded parameters (identical for all thicknesses).

[—— 0.8μm; - - - - 1.7μm; —·— 3.5μm.]

Fig. 14. Computer match to experimental bias light QE after 10 minutes (Figure 10) as a function of thickness using bulk-degraded parameters (identical for all thicknesses).

[—— 0.8μm; - - - - 1.7μm; —·— 3.5μm.]

subjected to the blue laser soaking. However, this bulk-degraded model also does not predict completely the trends in QE with device thickness. In particular, the bulk model is not entirely successful in predicting the drop in long-wavelength QE with increasing device thickness that is seen experimentally in Figs. 9-10. These results suggest that neither a simple surface model nor a simple bulk model alone can completely explain the degradation in QE that is seen experimentally. In general, these results imply that neither the recombination-driven, electron-driven, nor hole-driven models of degradation are completely satisfactory.

## SUMMARY

We have explored an experimental, light-soaking situation where recombination-driven, hole-driven, and electron-driven (at least for L = 1.7 and 3.5 μm) defect creation models all suggest front surface layer degradation should occur. In this straightforward situation of blue laser exposures we surprisingly find that it has been difficult to match the degradation (for all three device lengths used) across the QE-mapped wavelength range using a simple, surface degraded model. Such a degraded surface layer successfully lowers the red wavelengths QE by field redistribution. However, the surface-degradation model fails to predict the experimental drop in QE with device thickness after 10 min. blue laser soaking. The large front-surface field created by these many surface states results in a short-wavelength QE that is relatively insensitive to device thickness. The bulk model is more successful in predicting the experimental trends in short-wavelength QE after blue laser soaking; however, the model is less successful in predicting the long-wavelength QE trends.

In our first report of this need for some bulk component to degradation even for blue laser soaking we believed this correlated with the presence of excess electrons.[7] However, we only used the thinnest device (L = 0.8 μm) at that time. Fig. 6 shows this correlation is not present in the other two lengths. Hence, we now find that our results to-date suggest that neither recombination-driven, electron-driven, nor hole-driven defect generation mechanisms clearly account for observed degradation.

## ACKNOWLEDGEMENTS

This work is supported by the amorphous silicon thin film photovoltaics program of the Electric Power Research Institute.

## REFERENCES

1. W.B. Jackson, Philo. Mag. Letts., 59 103 (1989).
2. W. Kruhler, H. Pfleiderer, R. Plattner, and W. Setter, "Optical Effects in Amorphous Semiconductors," (AIP Conf. Proc. No. 120), edited by P.C. Taylor and S.G. Bishop (NY American Institute of Physics, 1984) p. 311.
3. D.L. Staebler, R. Crandall, and R. Williams, Appl. Phys. Lett., 39 733 (1981).
4. D. Redfield, Appl. Phys. Lett., 52 492 (1988).
5. M. Stutzmann, Appl. Phys. Lett., 56 2313 (1990).
6. R.A. Street, Phys. Rev. Lett., 49 1187 (1982).
7. S.J. Fonash, S.S. Nag, P.J. McElheny, J.K. Arch, J.-L. Nicque, and C.R. Wronski, Record of the 5th International Photovoltaic Science and Engineering Conference (Kyoto, Japan, 1990).

# THE STAEBLER-WRONSKI EFFECT - A FRESH ASSESSMENT

Bolko von Roedern, Solar Energy Research Institute
Golden, CO 80401-3393, USA

## ABSTRACT

Arguments are presented that the performance of amorphous silicon solar cells is inadequately described by the density of deep defects in the material. In addition to the density of deep defects, carrier extraction (or fill factors) is limited by the carrier mobilities. Carrier mobilities are determined by the presence of charged defects, which will affect the bandmobility via potential-fluctuations of the band edges. The primary degradation reaction is postulated to be the generation of additional charged defects which will affect the solar cell performance. Because temperature will establish an equilibrium between charged and neutral defects, the commonly observed increase of the spin densities or midgap defect densities occurs. However, this increase is not the major cause for the inferior performance after light-soaking. The model presented here is not a radical departure from the models used to describe the electronic properties, but rather an attempt to provide one consistent description using well established concepts.

## INTRODUCTION

The modeling of the electronic properties of amorphous silicon (a-Si:H) based materials and the Staebler-Wronski effect has been based on the generally accepted theory that the density of neutral dangling bond defects ($T_3^0$ defect with a g=2.0055 paramagnetic resonance value) is the major parameter in determining the electronic behavior of a-Si:H based materials and devices. On the other hand, there is evidence that the density of midgap states or recombination is not limiting the electronic properties once the density of midgap defects is below $5 \times 10^{16}$ cm$^{-3}$eV$^{-1}$ [1]. The latter observation is often confused by the empirical observation that in many instances the materials development was sufficiently guided by the quest to minimize this value. When the method (of minimizing the initial $T_3^0$ concentration) is applied to glow discharge or photo-CVD produced i-layers, it has resulted in material yielding the best initial and stabilized performance in solar cells. This has led many groups to believe that the increase in $T_3^0$ defects in poorer quality samples was the direct cause for the lower electronic performance. Recently, von Roedern and Madan [2] presented evidence that the electron bandmobility also varies in materials of different quality and has to be invoked to explain differences observed in different a-Si:H based materials.

I will provide a microscopic model showing how the density of charged $T_3^{+/-}$ dangling bond defects can be used to account for electronic differences in different samples and, most important, develop a self-consistent description for the Staebler-Wronski degradation. Surprisingly, the model borrows concepts well established in the literature (i.e., the concept of Branz and Silver, who proposed that the charged state of a dangling bond defect would be determined by local potential fluctuations) [3]. In addition to the defect site (or local potential) considerations, a thermal equilibrium between neutral and charged dangling bonds is established. There is strong evidence, however, that the charged state of a dangling bond alone cannot account for all phenomena observed. Additional defects are being generated during light-soaking. When a light- or otherwise induced recombination process generates an additional Staebler-Wronski defect, it is plausible that in the vicinity of a strained bond, impurity atom, or structural inhomogeneity the generated dangling bond would be charged rather than neutral. While there is no evidence that the generation of the light-induced defects can be avoided, it becomes understandable that in certain instances materials with low stabilized dangling bond densities can be obtained, that exhibit nevertheless very poor performance in solar cells.

### BASIS FOR MODEL

The possibility of the presence of charged dangling bonds has long been postulated. However, most electronic models do not specifically involve them, as there is no clear experimental feature that could provide an easy quantitative measure of their presence. In chalcogenide glasses, charged defects have long been suspected to cause poor mobilities, which prevent these materials from being used in devices such as field effect transistors or solar cells. In recent years, growing evidence has been presented that charged and neutral dangling bond defects coexist in amorphous silicon [4], and that under certain circumstances these defects rather easily alter their charged state [5]. This latter fact has lead to the idea that a "defect pool" is responsible for the electronic properties [6]. The concept of a defect pool can be used to account for the fact that solar cell parameters or the photoconductivity behavior can be modeled in terms of midgap density of states as the major input parameter determining these properties. I will show that this relationship is not a phenomenological one but an accidental consequence of the defect pool.

The most overwhelming evidence that something besides the midgap density of states controls the electronic and solar cell performance can be found in examples of a-Si:H based alloys or in a-Si:H samples with established low midgap densities of states, yet poor solar cell performance. As has been pointed out by von Roedern and Madan [2], for a-SiC:H alloys it has been reported that the mobility recombination-lifetime ($\mu\tau$) product of electrons decreases by orders of magnitude upon alloying with carbon, and that the field effect mobility fully accounts for the observed drop of $\mu\tau$. This is direct evidence that the understanding of the photo-transport properties has to include and account for variations of the bandmobility in these materials. In Table 1, data reported by Catalano et al. [7] substantiate this conjecture. Many groups have now reported that the inverse proportionality of the photoconductivity and the midgap density of states ($\mu\tau \sim \sigma_l \sim 1/N_D$) is not observed in several instances, and should therefore not be regarded as the dominant or sole mechanism which determines the $\mu\tau$ products. Unfortunately, for many other materials and solar cell devices only drift-mobility results are available. These results are commonly interpreted to arise from multiple trapping of carriers in shallow states (normally exponential tail states). Within the framework of this paper, it will become apparent that multiple trapping is not the major mechanism determining the time of flight transients. Of greater importance is the effective mobility at the extended-states transport path.

**Table 1:** *Field Effect Mobilities and $\mu\tau$ products of electrons for a series of a-SiC:H Alloys Note: Data in parenthesis are extrapolated from Figures of ref. [7].*

| Methane fraction | Bandgap $E_g$ (eV) | Field effect $\mu_e$ (cm$^2$V$^{-1}$s$^{-1}$) | $(\mu\tau)_e$ (cm$^2$V$^{-1}$) |
|---|---|---|---|
| 0 | 1.70 | 0.4 | $5 \times 10^{-7}$ |
| 0.4 | 1.80 | 0.1 | $2 \times 10^{-9}$ |
| 0.5 | 1.89 | $7 \times 10^{-3}$ | $\{3 \times 10^{-10}\}$ |
| 0.76 | 2.19 | $5 \times 10^{-6}$ | $\{ < 10^{-11}\}$ |

Changes of the band mobility have long been invoked as being present in a-Si:H based materials [8]. These earlier conclusions were derived from an analysis of thermopower and conductivity data. It was found that there is a difference between the activation energy and the thermopower. This difference is normally parameterized as a quantity $E_Q$, which characterizes the difference between a random mobility and a mobility in which the carriers can follow an external electric field. In fact, already in 1982 has it been reported that light-soaking increases $E_Q$ [9]. Since $E_Q$ is derived from the difference of the temperature dependence of thermopower and conductivity, the influence of gap states is eliminated and a change in the mobility at the transport level has to be invoked. It is found that in good quality a-Si:H, $E_Q$ may be as small as 50 meV, while in the light-soaked state $E_Q$ increases 150 meV or more. While these observations in 1982 were correctly deduced for free carriers present in the dark, these findings were never consequently applied to photo-carriers, which are affected similarly. In Figure 1, a simple model of the band edge in an amorphous semiconductor is presented in spatial and energy coordinates. Potential fluctuations lead to spatial variations of the band edges. The minimum energy at which continuous transport is possible is the conduction band edge $E_C$. However, when an external electric field is applied, the ensemble of dark or photo-carriers with a density $n_0$ follows the applied field

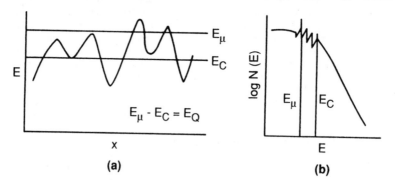

(a)                                                    (b)

**Figure 1.**    *Spatial (a) and energetic (b) representation of the band edge and transport path in the presence of long-range potential fluctuations.*

with an activated mobility behavior with $\mu(T) = \mu_o exp(-E_Q/kT)$. The energy level at which the carriers can move in an external electrical field can be termed "average transport level" or, for historical reasons "quasi-mobility edge," $E_\mu$. It lies above the minimum continuous conduction band edge, which is consistent with recent internal photoemission results that postulate that the "mobility gap" is larger than the optical bandgap [10]. Only at low temperatures (T << 300 K) will we observe a regime where low experimental mobilities are deduced because of transport paths which are much longer than the geometrical distance [11].

While the above is consistent with the experimental findings, a concrete source for potential fluctuations has to be identified. Small local potential fluctuations will have a tendency to increase the abundance of charged dangling bonds. The approach taken here is to assume that the long-range potential fluctuations originate from charged dangling bonds, and then to point out that there are no inconsistencies with this assumption. In this context it is important to remember that following the arguments of Branz and Silver [3] the charged and neutral dangling bonds appear to have a negative correlation energy. Since the charged state of a dangling bond will depend on its interactions with the surrounding atoms of the lattice, it follows that the negatively charged dangling bond ($T_3^-$) will be located below midgap and the positively charged dangling bond defect ($T_3^+$) above midgap. An increase of the stable or metastable $T_3^{+/-}$ densities is the controlling mechanism for the poor performance of inferior samples, such as those that are heavily doped, those containing fluorine, oxygen or nitrogen, and Ge and C alloys, and those with structural inhomogeneities or hydrogen clustering ($SiH_2$ bonds), or after light-soaking any good sample. From the results of the photoluminescence, it can be argued that recombination, which is commonly assumed to limit the performance, only poses a limit when the midgap density of states exceeds $10^{17}$ $cm^{-3}eV^{-1}$ [1].

## MODEL FOR THE STAEBLER-WRONSKI EFFECT

If the foregoing is accepted as a plausible basis, the existing models for the Staebler-Wronski effect should be revised. As the material with the best initial properties appears to have low densities of both neutral and charged defects, the mechanism proposed by Branz and Silver [3] for the degradation is eliminated, i.e. that the total number of dangling bonds is conserved as light-induced recombination converts charged $T_3^{+/-}$ defects into neutral $T_3^o$ defects. In my view the temperature dependence of the Staebler Wronski degradation suggests the following reaction path:

$$(1) \quad T_4^* \xrightarrow{\text{(recomb.)}} T_3^{+/-} \xrightarrow{(kT)} T_3^o$$

Here, $T_4^*$ is what is commonly referred to as a "weak bond". However, the term "weak bond" should not be limited to a strained Si-Si bond. It could be any susceptible site due to the presence of an irregularity such as an impurity atom, internal surface, $SiH_2$ group etc. In any case, following the argument of Branz and Silver [3], in the vicinity of such a disturbance, the local potential fluctuation associated with the disturbance will be causing a tendency for the resulting dangling bond to be charged rather than neutral. However, at a given temperature, a thermodynamic equilibrium will control the charged status of the dangling bond defects, with a characteristic energy of 0.35 eV.

The kinetics of how the experimentally observed neutral dangling bonds are being created will thus depend on 2 factors:
(1) The number of susceptible sites $(T_4^*)$ which are available to be converted.
(2) the local potential fluctuations dictated by microstructure, impurities, and H-clusters.
This, along with temperature, will determine the ratio established for a given material between charged and neutral dangling bonds.

This suggests that the observed decay of the photoconductivity in poorer quality material is primarily determined via a decreased mobility due to an increased density of $T_3^{+/-}$ defects, rather than by the $T_3^0$ defects. I believe that this model can explain all degradation phenomena observed to date, as well as "mysteries" such as the original findings of Han and Fritzsche [12], who reported that when samples are light-soaked at low temperatures the decrease of the photoconductivity is not accompanied by an increase of the sub-bandgap absorption.

It is well known that the generation of recombination-induced defects is self-limiting at approximately the $10^{17}$ cm$^{-3}$eV$^{-1}$ level. I would like to suggest that the electron transport itself provides the mechanism for this limitation. As can be seen in Figure 1, most free carriers will be confined to regions with a low conduction band energy. Consequently, trapping and recombination will be limited to regions within the bulk of the semiconductor. From extrapolation of defect densities or from experiments using very high generation rates the saturation level for the metastable defect density may be on the order of $>10^{18}$ cm$^{-3}$eV$^{-1}$ [13,14]. A saturation at $10^{17}$ cm$^{-3}$eV$^{-1}$ at typical 1-sun intensities is a consequence of the spatial band edge fluctuations.

The saturation levels, as well as the generation rate in the "reserve" regime (where in this picture $T_4^*$ sites remain available and accessible to be converted), appear to be dictated by the volume-dose. The resulting increase of defect densities in this regime is likely to be proportional to $t^\alpha$ ($\alpha \approx \frac{1}{3}$), which is often experimentally observed.

The "incubation" time (i.e. the region where the measured $T_3^0$ density does not appear to grow upon <u>initial</u> illumination) is due to the fact that in most materials the initial frozen-in $T^3_0$ defect density has a greater than the equilibrium $T_3^0$ density, so that while $T_3^{+/-}$ defects are being generated, an increase of the $T_3^0$ density is not observed initially. With respect to the midgap density of states, higher temperatures will decrease the incubation time, lower temperatures will increase it. The opposite behavior is observed in solar cells, where a higher temperature will cause the more detrimental $T_3^{+/-}$ defects to be converted into $T_3^0$ defects according to equation (1). Different materials have different propensities for the $T_3^{+/-}$ or $T_3^0$ defects to form. For example, material deposited at higher deposition rates degrades faster than does standard glow-discharge produced material [15]. Fluorinated material has been reported to have inferior properties when fluorine is incorporated into the films [16]. This can be explained by the electro-negative fluorine atoms significantly enhancing the $[T_3^{+/-}]/[T_3^0]$ ratio. While materials with this feature may show reasonably high photoconductivities and large photo-to-dark conductivity ratios [17], their incorporation into solar cells has not yet been reported to result in higher stabilized efficiencies. Similar results were obtained on samples prepared by other deposition methods, such as by sputtering. The stability of material compensated with phosphorous and boron is easily explained by my model, as compensation will lead again to an increase of the $[T_3^{+/-}]/[T_3^0]$ ratio. In the context of this model the presence of a higher initial number of $T_3^{+/-}$ defects will result in increased stability; in reality, however, this is accomplished by a mechanism equivalent to pre-degradation.

## A CRUCIAL TEST FOR THE MODEL'S VALIDITY

A crucial test for the validity of this model lies in the fact that it correctly predicts the different temperature dependencies of the degradation of the midgap density of states and solar cell performance. It has been reported that up to about 70° C, the saturated defect densities as measured by the constant photocurrent method (CPM) are independent of the temperature [18]. On the other hand, the stabilized performance of solar cells shows significant dependencies on the operating temperature, in the laboratory as well as in amorphous silicon solar array installations [19,20]. The differences in temperature behavior of the saturation values and degradation rates cannot be reconciled with a model that only invokes one parameter, the midgap density of states. In Figure 2, I have plotted the degradation of a solar cells and midgap density of states as a function of temperature at which the samples or solar cells were light-soaked. At present, I have sketched the performance gathered from the literature [19,21]. The important point to note is the opposite temperature dependence of the degradation rate of the solar cell and materials properties. While consistent with the model presented here, this feature is inconsistent with most models presented to date, which rely only on recombination losses and ignore changes of the carrier mobilities upon light-soaking. Work at SERI is in progress to better quantify the opposite trend in the degradation rates in solar cells and i-layer properties [22].

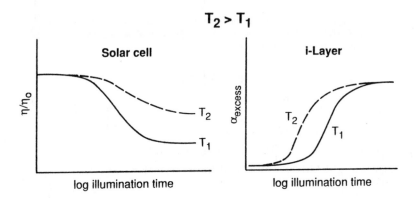

**Figure 2.**   *Differences in the temperature dependence of the degradation of solar cell efficiencies and the excess absorption measured by CPM.*

It is also possible that the effect of fluctuating band edges has been neglected in the analysis of polycrystalline semiconductors. Only gallium arsenide is optimized by maximizing the mobility; most other semiconductors are optimized by minimizing recombination by reducing gap state densities. There is growing evidence, however, that even these materials effects of fluctuating band edges are observable and should be considered when data are analyzed. However, the magnitude is smaller than in the amorphous materials.

## CONSEQUENCES TO BE CONSIDERED

1.  The neutral dangling bond is not the most detrimental defect in a-Si:H based solar cells for concentrations of $10^{17}$ cm$^{-3}$eV$^{-1}$ or less.

2.  Mobility changes, presumably caused by charged dangling bonds, are important and have been largely overlooked to date.

3.  When "more stable" materials are synthesized, their stability is gained at the expense of tolerating lower initial mobilities of the material. Lower initial and stabilized solar cell performances will result when these materials are incorporated into solar cells.

4.  Research which focusses solely on producing materials with low stabilized dangling bond densities may be unproductive, as obtaining a low dangling bond density is not a sufficient condition for good solar cell performance.

5.  The validity of results obtained from photocurrent and capacitance measurements should be re-examined. Presently, it is assumed that these quantities are affected by midgap states only. If the bandmobility and/or transport level are altered (e.g. by light-soaking), analyses of the experimental results have to consider these changes.

## ACKNOWLEDGEMENT

This work is supported by the U.S. Department of Energy under Contract No. DE-AC02-83CH10093. The author would like to thank Tom McMahon, Dick Crandall and Howard Branz for stimulating discussions.

## REFERENCES

1.  R.W. Collins, P. Viktorovitch, R.L. Weisfield, and W. Paul, Phys. Rev. B 26, (1982), 6643.

2.  B. von Roedern and Arun Madan, Phil. Mag. B 63, 1991, 293.

3.  H.M. Branz and M. Silver, Phys. Rev. B 42, 1990, 7420.

4.  S. Yamasaki, H. Okushi, A. Matsuda, K. Tanaka, and J. Isoya, Phys. Rev. Lett. 65, (1990), 756.

5.  J.M. Essick and J.D. Cohen, Phys. Rev. Lett. 64, (1990), 3062.

6.  Z.E. Smith, in Advances in Amorphous Semiconductors, edited by H. Fritzsche (World

Scientific, Singapore, 1988), 409.

7.  A. Catalano, J. Newton, and A. Rothwarf, IEEE Transactions on Electron Devices, 37, (1990), 391.

8.  H. Overhof and W. Beyer, Phil. Mag. B 47, (1983), 377.

9.  D. Hauschildt, W. Fuhs, and H. Mell, Phys. Stat. Solidi (b) 111, (1982), 171.

10. S. Lee, D. Heller, and C.R. Wronski, Materials Research Society Symposium Proceedings, (ed. Taylor, Thompson, LeComber, Hamakawa, Madan), 192, (1990), 89.

11. C. Cloude, W.E. Spear, P.G. LeComber, and A.C. Hourd, Phil. Mag. B 54, (1986), L113.

12. D. Han and H. Fritzsche, J. Non-Cryst. Solids, 59&60, (1983), 397.

13. T.J. McMahon, these proceedings.

14. U. Schneider, J. Sopka, B. Schröder, and H. Oechsner, Proc. 20th IEEE PV Specialists Conference, (1988), 346.

15. H. Curtins, M. Favre, Y. Ziegler, N. Wyrsch, and A.V. Shah, Materials Research Society Symposium Proceedings, (ed. Madan, Thompson, Taylor, LeComber and Hamakawa), 118, (1988), 159.

16. A.A. Langford, A.H. Mahan, M.L. Fleet, and J. Bender, Phys. Rev. B 41, (1990), 8359. It should be noted that fluorine in the plasma may be beneficial (e.g., by etching the growing surface it may help to control the hydrogen content or microstructure).

17. H. Matsumura, H. Ihara, H. Tachibana, and H. Tanaka, J. Non-Cryst. Solids 77&78, (1985), 793.

18. H.R. Park, J.Z. Liu, and S. Wagner, Appl. Phys. Lett. 55, (1989), 2658.

19. H.S. Ullal, D.L. Morel, D.R. Willett, D. Kanani, P.C. Taylor, and C. Lee, Proc. 17th IEEE PV Specialists Conference, (1984), 359. Note: replotting the data of Figure 6 on a semi-logarithmic scale substantiates a lower degradation rate for solar cells at higher temperatures.

20. S.L. Hester, T.U. Townsend, W.T. Clements, and W.J. Stolte, Proc. 21st IEEE PV Specialists Conference, (1990), 937.

21. P. Santos and W.B. Jackson, private communication.

22. B. von Roedern and T.J. McMahon, to be published Proc. 10th European Photovoltaic Solar Energy Conference, Lisbon, April 1991.

# SCENARIOS OF DEFECT GENERATION IN a-Si:H MATERIAL FOR VERY LONG-TERM OR VERY INTENSE IRRADIATION

M.Gorn, B. Scheppat, and P. Lechner
Phototronics Solartechnik GmbH, Hermann-Oberth-Str. 9,
D-8011 Putzbrunn, GERMANY

## ABSTRACT

A complete investigation of the kinetics for defect generation by Stutzmann, Jackson, and Tsai (SJT) necessitates numerical
solution of the rate equations. The time where saturation of defects occurs, depends critically on temperature and activation
energy for annealing. This saturation time can vary by orders of
magnitude according to different experimental conditions. It is
shown how high-intensity light or 20 keV-electron irradiation data
can be explained in terms of the SJT model.

## INTRODUCTION

Material characterization in terms of stability comprises
measurements of the time development of the total defect density
under prolonged light or electron beam irradiation for different
external parameters such as intensity and temperature and internal
material parameters. In order to draw significant conclusions from
these experiments, a good understanding of the kinetics is imperative. Using the complete formulae of the work by Stutzmann, Jackson, and Tsai (SJT) [1], the rate equation for defect generation,
including annealing, is numerically solved in order to allow for
checks with experiments in regimes where explicit analytical formulae do not exist: Non-negligible initial defect density; room
temperature, where saturation occurs at very long times (depending
on activation energy for annealing); high intensities, where the
approximate formulae leading to the well-known $t^{1/3}$-dependence
cannot be used. In fact, as shown in this paper, there is a whole
variety of super $t^{1/3}$-laws up to the linear behaviour observed
in 20 keV-electron irradiation, for instance.

## BASIC PARAMETERS OF THE SJT MODEL

The SJT model [1] is based on the assumption that new defects
(N) are created by non-radiative tail-to-tail recombination
between optically excited electrons (n) and holes (p), described
by the equation

$$\frac{dN}{dt}_{ill} = c_{sw} A_t\, n\, p \tag{1}$$

where $A_t$ is the tail-tail transition probability and $c_{sw}$ the
efficiency of such a transition to create a new dangling bond N
during the illumination time $t_{ill}$. The problem is the calculation of n and p, as in photoconductivity for instance, and, in

130

principle, many material parameters enter, leading to a large variety of effects in the time dependence, not fully investigated up to now. Therefore, we shall restrict ourselves to the somewhat simplified solutions given in ref. 1., where after consideration of all possible transitions involving positively, negatively charged and neutral dangling bonds, effective transition rates of excess carriers into dangling bonds (without further specification of their charge) are defined, leading to effecitve transition probabilities $A_n$ and $A_p$ into dangling bonds. These transitions compete with the direct tail-tail recombination (probability $A_t$). Taking the exact (but charge averaged) solutions of ref.1, G being the generation rate (intensity), equation (1) yields the defect generation rate,

$$\frac{dN}{dt}_{ill} = c_{sw} \frac{A_t G^2}{A_n A_p N^2} \frac{4}{\left\{1 + \sqrt{1 + \frac{4A_t G}{A_n A_p N^2}}\right\}^2} \tag{2}$$

having two important limits, according to whether the critical parameter

$$\xi = 4 A_t G/(A_p A_n N^2) \tag{3}$$

is small or large. In case $\xi$ is small, the well-known low-intensity limit results where the defect generation rate is proportional to $1/N^2$,

$$\frac{dN}{dt}_{ill} = c_{sw} \frac{A_t G^2}{A_n A_p} \frac{1}{N^2} \tag{4a}$$

The second limit is obtained for very high intensities, or more correctly, if $\xi$ is very large, leading to a defect generation rate independent of N and proportional to G,

$$\frac{dN}{dt}_{ill} = c_{sw} G \tag{4b}$$

The low-intensity limit as found from ref.1, is the case for usual light soaking below 700 mW/cm$^2$ corresponding to a generation rate G = $2 \times 10^{22}$ cm$^{-3}$sec$^{-1}$ for uniformly absorbed light of 1.9 eV in a 1 um thick film, for instance, and leads to the well-known $t^{1/3}$-law (if the initial defect density $N_0$ can be neglected).

The high-intensity limit leads to a linear time dependence, eq. 4b. Consequently, there is a priori no need to invoke a new defect creation mechanism for 20 keV irradiation as proposed by Gangopadhyay, Schröder, and Geiger [2]. In contrast to their conclusions, the SJT model is therefore capable of explaining also electron-induced effects. However, the generation rate G must be extremely large in order to obtain a linear behaviour if the probability values of ref. 1, $A_t = 3 \times 10^{-11}$ cm$^3$sec$^{-1}$, $A_p = 0.8 \times 10^{-9}$ cm$^3$sec$^{-1}$, and $A_n = 2.7 \times 10^{-9}$ cm$^3$sec$^{-1}$, or the ratio

$$A_t/(A_p A_n) = 1.4 \times 10^7 \text{ cm}^{-3}\text{sec} \tag{5}$$

are taken, then $G > 10^{26}$ cm$^{-3}$sec$^{-1}$ in order to have $> 1$ with defect densities between $10^{16}$ and $10^{17}$ cm$^{-3}$. However, the probabilities can depend on the generation rate itself and it is very likely for $A_t$ to be enhanced with respect to $A_p$ and $A_n$ at higher intensities. We will elaborate on this further below.

The complete rate equation also holding for long-term irradiation is obtained by adding the thermal annealing term,

$$\frac{dN}{dt}_{ann} = -v_o (N-N_o) \exp(-E_a/kT) \tag{6}$$

with $E_a$ the activation energy for annealing of defects, $N_o$ the initial defect density, T the temperature, $v_o$ the attempt to escape frequency ($10^{10}$ sec$^{-1}$ assumed after ref. 1), and k Boltzmann's constant. In fact, there is a whole spectrum of activation energies but for simplicity we take some representative value, lying around 1.0 to 1.1 eV for not too intense light soaking, after ref. 1, and for 20 keV electron irradiation between 0.8 and 0.9 eV after Schneider [3]. 

All solutions for the time dependence of defect generation are numerically calculated using the total rate equation. In some cases analytical statements can be made.

## SATURATION OF DEFECT DENSITY

The annealing term always leads to a steady-state equilibrium of the defect density, here called saturation, which can be expressed analytically for the two limits, eqs. 4,
a) $t^{1/3}$-law:

$$N_{sat} = (c_{sw} A_t/(A_p A_n))^{1/3} \, G^{2/3} \, v_o^{-1/3} \, \exp(3E_a/kT) \tag{7a}$$

b) $t^1$-law:

$$N_{sat} = c_{sw} \, G \, v_o^{-1} \, \exp(E_{act}/kT) \tag{7b}$$

(initial $N_o$ dropped, for simplicity). The saturation time, defined at the point where the extrapolated and unsaturated defect density increasing without annealing equals the steady-state saturation value, turns out to be independent of intensity. Numerical examples are shown in fig. 1. In case of the two limits, eqs. 4,

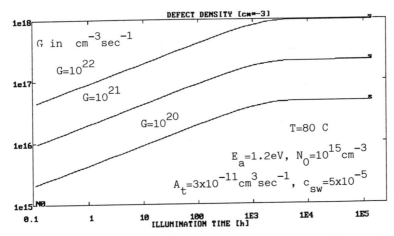

Fig. 1: Numerical solutions of complete rate equations of the STJ model. Saturation time turns out to be independent of irradiation intensity (see text).

the saturation times can be expressed analytically (again neglecting the initial defect density $N_0$, for simplicity),

a) $t^{1/3}$ - law:     $t_{sat}' = 1/(3v_0) \, \exp(E_{act}/kT)$     (8a)

b) $t^1$ - law:     $t_{sat} = 1/v_0 \, \exp(E_{act}/kT)$     (8b)

    It is important to state that the fact of the saturation times being completely independent of intensity and of the constant $c_{sw}$ raises doubts on the significance of fast aging experiments with high intensities, unless the STJ model is incorrect.

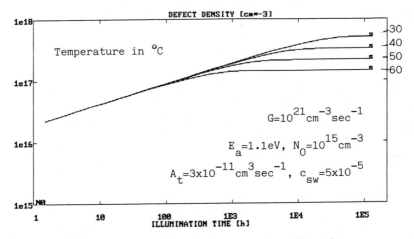

Fig. 2a: Influence of temperature on saturation time

An earlier saturation, in the SJT kinetics, means either higher temperature or lower activation energy. The drastic changes in the time development due to small changes in the two parameters are shown in figs. 2a and 2b. Indeed,

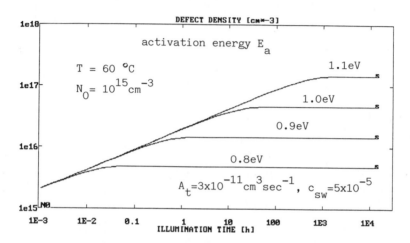

Fig. 2b: Influence of activation energy on saturation time

in 20 keV-electron irradiation, activation energies as low as 0.8 eV are observed, corresponding at T=30°C to a saturation time of some ten minutes, eq. 8b. Though not yet measured, it is likely that also high intensities of light produce less stable defects with lower activation energies in between the values for low intensity light and 20 keV-electron irradiation (e.g. 0.85 to 0.9 eV, saturating at around 1 hour at 30°C).

## DISCUSSION OF HIGH INTENSITIES

At higher intensities also the ratio between $A_t$ and $A_p$, $A_n$ might change with respect to the standard value, eq. 5, because of tail-tail recombination dominance. This leads to a deviation from the usual $t^{1/3}$-law as shown in fig. 3, due to the more appropriate formula in eq. 2 for defect creation.

The transition to a super $t^{1/3}$-law for higher intensities (G) occurs only if the critical parameter in eq. 3 is large enough. This depends therefore on the value of $A_t$ or, more correctly, on the ratio $A_t/(A_n A_p)$. There is no such deviation from $t^{1/3}$ with $G = 10^{22}$ cm$^{-3}$sec$^{-1}$ using the set of probability values from ref. 1, given above, as can be seen in fig. 1.

Recent high intensity light soaking data by Park, Liu, and Wagner [4] indeed exhibit a steeper rise, as seen in fig. 4, before the saturation. A good fit is obtained by the combination $A_t = 5 \times 10^{-8}$ cm$^3$sec$^{-1}$ and $c_{sw} = 3 \times 10^{-8}$. This differs from the values of ref. 1 but it does not change the essence of this work since the product $c_{sw}A_t$ = $1.5\times10^{-15}$ $cm^3sec^{-1}$ has remained the same. As stated already in ref. 1, only this pro- duct can be determined from the $t^{1/3}$-regime, also to be seen from eq. 4a (if $A_p$ and $A_n$ are known). We are therefore in a position to draw conclusions on the value of $A_t$ itself, which could not be done from the low intensity data in ref. 1.

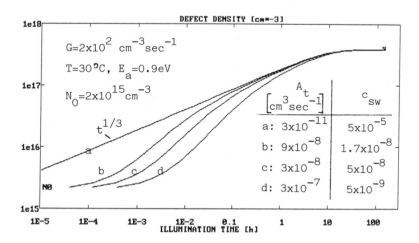

Fig. 3: Super $t^{1/3}$-behaviour by enhancement of tail-tail re-combination

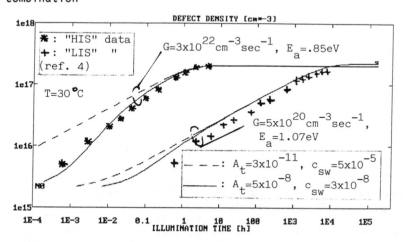

Fig. 4: Comparison of data (ref. 4) with the SJT kinetics. High intensity ("HIS") is well reproduced by an enhanced tail-tail recombination. For the low intensity ("LIS") the difference when using the standard values for $A_t$ and $c_{sw}$ is not so large.

The new values of $A_t$ and $c_{sw}$ reproduce all high and low intensity data of ref. 4, as seen in fig. 4. Of course, the fits to the low intensity data by Stutzmann et al., ref. 1, remain unaffected.

We have therefore found a universal set of parameters reproducing high- and low-intensity light soaking data. (For bandgaps of around 1.9 eV, the issue of this dependence has to be addressed in further investigations.)

Now turning to 20 keV-electron irradiation, we encounter the problem of guessing the generation rate. In fact, there are no numbers in the literature. A simple estimate for the number of e-h pairs can be obtained from the picture developed in ref. 1, where it is established that the SWE is independent of photon energy down to 1.3 eV. This may be explained by assuming that e-h pairs with higher energy first thermalize down to the tail separation of 1.3 eV before undergoing tail-tail recombination. For an intensity of $I = 170$ mWcm$^{-2}$ as given in ref. 2, an e-h density of $8 \times 10^{17}$ cm$^{-2}$sec$^{-1}$ would result or a generation rate of $8 \times 10^{21}$ cm$^{-3}$sec$^{-1}$, respectively, in a 1 um thick film, if absorption is homogeneous. The linear rise of the defect density with time, with such a moderate G, only arises for a strongly enhanced $A_t = 3 \times 10^{-4}$cm$^3$sec$^{-1}$, as shown in fig. 4. The data by

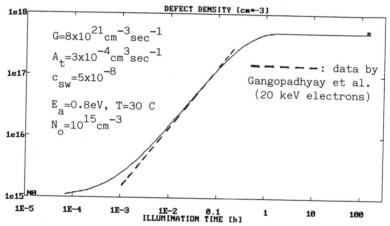

Fig. 4: Fit of 20 keV-electron irradiation data with SJT kinetics and strongly enhanced tail-tail recombination

Gangopadhyay et al. (ref. 2) are well reproduced using the new value $c_{sw} = 3 \times 10^{-8}$ obtained in connection with fig. 4, the high intensity light soaking data. Another scenario for obtaining the linear rise is to assume higher generation rates. This implies, however, a new assumption, the Staebler-Wronski susceptibility $c_{sw}$ to be reduced at highter intensities. A curve almost identical to the one in fig. 4 can be obtained, for example by the combination $G = 8 \times 10^{25}$ cm$^{-3}$sec$^{-1}$ and $c_{sw} = 6 \times 10^{-12}$, and so on, but with the same tail-tail recombination as in connection with high-intensity light soaking, fig. 4, namely $A_t = 5 \times 10^{-8}$ cm$^3$sec$^{-1}$. Intermediate scenarios are also possible.

## SUMMARY AND CONCLUSIONS

We have found that existing high- and low intensity light soaking data on a-Si:H (for constant bandgap) can be parameterized in terms of the Stutzmann-Jackson-Tsai kinetics. From the super $t^{1/3}$-behaviour at high light intensity, a new estimate of the tail-tail recombination probability $A_t$ could be obtained whereas in the $t^{1/3}$-regime only the product $A_t c_{SW}$ ($c_{SW}$ being the Staebler-Wronski susceptibility) has been determined previously. Our new combination of $c_{SW}$ and $A_t$ reproduces very well all high- and low-intensity light soaking data, if the following properties, not yet measured in detail, turn out to be true:

- Higher intensities create more unstable defects with lower activation energies leading to earlier saturation times.

- The model parameters depend on the bandgap (issue not addressed in this paper).

For 20 keV-electron irradiation, the tail-tail recombination must be largely enhanced with respect to recombination at defects or the electron-hole pair generation rate must be much larger than estimated assuming homogeneous absorption and a reduced value for $c_{SW}$, in this case. Both scenarios lead to a defect generation linear in time before the saturation as experimentally observed.
A convenient feature of the SJT-kinetics is that its parameters can be checked by more detailed experiments and measurements and by more detailed modelling of recombination.

## REFERENCES

1. M. Stutzmann, W.B. Jackson, and C.C. Tsai, Phys. Rev. B32, 23 (1985)
2. S. Gangopadhyay, B. Schröder, and J. Geiger, Phil. Mag. B 56, 321 (1987)
3. U. Schneider, PhD Thesis, Kaiserslautern, Germany (1989).
4. H.R. Park, J.Z. Liu, and S. Wagner, Appl. Phys. Lett. 55(25), 2658 (1989).

# DEGRADATION RATE AND SATURATION DEFECT DENSITY IN a-Si:H AS A FUNCTION OF TEMPERATURE AND LIGHT INTENSITY

M. Grimbergen, L. E. Benatar, A. Fahrenbruch, A. Lopez-Otero,
D. Redfield, and R. H. Bube
Department of Materials Science & Engineering
Stanford University, Stanford, CA 94305
TELEPHONE: (415) 723-1946; FAX: (415) 725-4034

## ABSTRACT

Defect density as a function of light exposure time and the value of the saturation defect density in undoped hydrogenated amorphous silicon (a-Si:H) have been measured. We varied conditions of temperature and light intensity to test the predictions of both a stretched exponential description of the kinetics and the steady-state solution to the underlying rate equation. A stretched exponential fits the time dependence, with a stretching parameter that increases and a time constant that decreases with temperature over the range 270 to 360K. The saturation density of light-induced defects is both temperature and intensity dependent in the range 420-480K for generation rates $6 \times 10^{20}$ to $2 \times 10^{22}$ cm$^{-3}$ sec$^{-1}$. The observed decrease in saturation density with increasing temperature and increase in saturation density with increasing intensity are consistent with the model predictions.

## INTRODUCTION

Behavior of the degradation with respect to light intensity, temperature and time are all important attributes of instability in a-Si:H. Of the many descriptions that have been proposed for light-induced degradation in amorphous silicon, only one predicts a temperature dependence for degradation rates [1,2]. This description yields a stretched exponential function for the time dependence of defects, and has been markedly sucessful in describing published degradation data.[2-4]. In this paper we examine the temperature dependence in greater detail.

The existence and importance of saturation in light-induced generation of defects has become well established, both theoretically[2,4,5] and experimentally[6-8]. It has been shown that the temperature dependence of saturation density of light induced defects should exhibit distinctly different behaviors in three regions of temperature. Confirmation of this behavior will provide a major test of the theoretical description of degradation, as well as an aid in evaluating some physical parameters of the defects.

Experimentally, however, there has been one characterization of the temperature dependence of saturation defect density to date[6], and this study did not include intensity dependence. We have undertaken a study of both temperature and light intensity dependences, using measurements relatively insensitive to surface states.

In this paper, we review briefly the stretched exponential formulation before describing the time-dependent and steady-state experiments. Experimental results are then discussed and compared with predictions from the stretched exponential formulation.

## STRETCHED EXPONENTIAL DESCRIPTION

A rate equation has been given for the formation and anneal of defects in hydrogenated amorphous silicon which includes both optical and thermal terms for

138

both processes.[4] This formulation may be taken further if similar dispersive processes control the rates in both directions, generation and anneal, in which case the final form of the solution to the rate equation is a stretched exponential (SE)[2]. A stretched exponential has been used to fit the annealing kinetics of defects[9] as well as the kinetics of defect generation[3]. However, explicit forms of the parameters in the equation and their dependences on temperature, light intensity and perhaps sample condition have not been explored in depth.

Specifically, the stretched exponential description of the defect density is[2]

$$N(t) = N_s - (N_s - N_o) \exp[-(t/\tau)^\beta] \tag{1}$$

where $N_s$ is the saturated defect density, $N_o$ is the initial defect density, $\tau$ is the time constant, and $\beta$ is the stretch parameter. The steady-state defect density, $N_s$, takes the form[4]

$$N_s = N_T (C_1 G + v_1) / [(C_1 + C_2)G + v_1 + v_2] \tag{2}$$

where G is the excitation rate, C's represent the effectiveness of recombination processes, v's are thermally activated frequencies (subscripts 1 and 2 denote generation and anneal, respectively), and $N_T$ is the total density of potential defects in the material. Numerical values of these parameters used later in the modeling of steady-state effects are given in Table II.

By measuring the kinetics of defect generation at a variety of temperatures for known excitation rates, we observe the behavior of the stretch parameter and time constant with temperature. In separate experiments, the value of the saturation defect density was measured as a function of temperature and, for the first time, excitation rate.

## EXPERIMENT

Two series of experiments were designed to measure the transient and steady-state defect response to degradation in undoped a-Si:H samples grown by glow-discharge at 230°C. Samples were approximately one micrometer thick and deposited onto metal-patterned fused silica in a coplanar geometry. Kinetics experiments were performed using several samples from the same slide, while steady-state measurements were made on samples from a different deposition run.

In the first series, the samples were held in an evacuated, temperature-controlled (±2K) Joule-Thompson refrigerator for in-situ degradation at temperatures between 90 and 360K with a nearby filtered light source. The source consisted of a focused beam from a 250 W quartz-halogen lamp with a 660 nm bandpass filter. Light intensities obtained from this system were 1-2 W/cm$^2$ with the filter and 3-4 W/cm$^2$ without the filter. Care was taken to assure degradation beam uniformity across the sample to better than 10%. Heating of the sample due to absorbed light was also closely monitored.

After each degradation, the Constant Photocurrent Method (CPM) was used to measure $\alpha(E)$ at room temperature[10]. The CPM set-up is on the same optical table as the degradation apparatus so that the sample can be left inside its chamber throughout a degradation experiment, minimizing uncertainty due to repositioning or making of electrical contacts. Defect densities were calculated from the integration of subgap absorption[11].

In the second series of measurements, samples were degraded in a nitrogen atmosphere at elevated temperatures under illumination until saturation was reached for each temperature and intensity. The saturated density was then retained by quenching the sample in less than half a second in water at 300K. The degrading source was similar to that already described. Increases in film temperature due to absorption of high light intensity were monitored as calibrated changes in band gap transmission at 650 nm. Saturated defect densities were then measured at room temperature by CPM, using the procedure described above.

## KINETICS: RESULTS

Kinetics behavior above room temperature was studied by degrading one sample at 360K until N(t) stabilized to within 5%, which is within our experimental error, indicating saturation (see Figure 1). After annealing, the same sample was degraded at a lower temperature (320K). The incident light intensity was adjusted so that more points could be measured in the steepest portion of the curve--the region which provides the most accurate data for obtaining fitting parameters. This shift also positioned the initial rise on the scale of several minutes to ensure that the sample temperature (influenced by both the incident light as well as heating or cooling from the sample stage) had equilibrated. In this manner we were able to attain high confidence in the sample temperature during degradation.

The stretched exponential fits (equation 1) to both curves are shown in Figure 2 and the values of parameters obtained from the fits are given in Table I. The principal finding was a noticeable drop in the stretch parameter, $\beta$, with decreasing temperature. It is also important to note from the figure that N(t) in a single sample shows two different apparent slopes; the smaller range ($N_s$-$N_o$) and lower temerature are both thought to contribute to a decrease in slope.

An additional finding relates to the annealing of defects: after saturation at 360K and before the 320K run, the sample was annealed for 5 hours at 180°C in a nitrogen atmosphere, but the defect density measured after the anneal was found to be almost twice that of the sample in its original (as received) condition. Other tests of annealing showed that for samples which had not been degraded to saturation the same annealing procedure returned the defect density to its initial value. From these preliminary findings we consider that the last defects to form may be the most difficult to anneal. Further tests of annealing are planned.

A second sample was used for measuring kinetics at temperatures below room temperature. Of interest was a noticeable shift to longer times for the data taken at low temperatures (90K and 270K). For example, data for defects generated at a higher G ($1.6 \times 10^{22}$cm$^{-3}$s$^{-1}$) but lower temperature (270K) overlapped with data for defects generated at a lower G ($.68 \times 10^{22}$cm$^{-3}$s$^{-1}$) but higher temperature (320K). The only shift in time expected was that due to G, which would have placed the higher-G data at shorter times. We interpret these preliminary findings to be an increase in the time constant, $\tau$, with decreasing temperature. Figure 3 shows the kinetics data for the 270K degradation--the shift to longer times led us to use white light (~4 W/cm$^2$) to achieve saturation. The stretch parameter, $\beta$, again drops with temperature (see Table I).

## STEADY-STATE: RESULTS

Predicted saturation behavior using equation (2) and values from Table II shows three general regions, depending on the relative strengths of the thermal and light contributions. Figure 4 demonstrates that at the highest temperatures, thermal annealing dominates, while at the lowest temperatures, light generation and annealing dominates. It is in the intermediate range of temperature where most of the experimental data gathered here lie.

Measured saturation defect densities resulting from quenching after degradation at 420-480K (150°C to 200°C) for a single sample under several light intensities are shown in Figure 5. The range of degrading light intensities used corresponds to generation rates between mid-$10^{20}$ and low $10^{22}$ cm$^{-3}$sec$^{-1}$. The lowest curve shows the density in the dark at two temperatures, with an activation energy of 150 meV.

The curves show several important trends, which are expected when both light generation and annealing processes operate concurrently. Increasing temperature results in reduced saturation density as annealing during degradation becomes more important. Similarly, reduced intensity at a given temperature results in lower saturation density. This is in contrast, however, to behavior near room temperature and below, where thermal annealing is insignificant and no intensity dependence is expected.

The measured steady-state behavior matches qualitatively that predicted by the simple time-independent solution presented earlier in Figure 4. Extrapolation of the equilibrium data to very high temperatures intercepts the zero ordinate at $1/2N_T$, as predicted by the model. The calculated curves in Figure 4 are for each measured generation rate in Figure 5 using the values listed in Table II. The calculated curve for $G = 1.8 \times 10^{22}$ cm$^{-3}$ sec$^{-1}$ approximately matches saturation data taken at this generation rate. For lower generation rates, the temperature dependence appears more gradual than that predicted by calculated curves in Figure 4. The actual low-temperature saturation value may be slightly higher than $6 \times 10^{16}$ cm$^{-3}$ which was assumed in the calculation.

Further measurements will determine whether some of the model constants should have different values or show some temperature dependence. For example, both thermal pre-exponential coefficients have been assigned constant values based on annealing lifetimes observed in doped material.[12] The coefficients for light generation and annealing are assumed equal and temperature independent. The goal of ongoing work is to evaluate these assumptions.

## CONCLUSIONS

Measurements of kinetics of defect generation on two undoped samples for temperatures between 270K and 360K show a decrease in the stretch parameter, $\beta$, and an increase in the time constant, $\tau$, with decreasing temperature. These results are predicted by the SE description of defect kinetics.[2] The fits to the data provide further evidence that a SE description of defect generation is valid; continued experiments will be carried out to increase our confidence in this description and to obtain more complete data for determining $\beta(T)$ and $\tau(T,G)$. In addition, we will be testing the application of the SE to annealing of light-induced defects as a further test of the similarity between the rate-controlling processes for generation and anneal.

The steady-state data show clearly that the saturation defect density is dependent upon degradation temperature above 400K in undoped material. The data also show that, at elevated temperatures, the saturation density is intensity dependent. Although both effects have been predicted,[4] this is the first time these effects have been quantified using uniformly absorbed light and CPM.

By considering both the optical and thermal effects on the generation of metastable defects in a-Si:H and their anneal, the rate equation for the defect population adopted here ties together these two processes which define the stability of the material. Studies of the parameters in both the time-dependent solution to the rate equation--the stretched exponential--and its solution in the steady state serve as guidelines for development of a microscopic model, an important step towards improving the material's stability.

## ACKNOWLEDGEMENTS

This work was supported by the Electric Power Research Institute. The authors gratefully acknowledge provision of samples by Murray Bennett of Solarex, Inc.

## REFERENCES

1. R. H. Bube, and D. Redfield, J. Appl. Phys. **66,** 820 (1989)
2. D. Redfield and R. H. Bube, Appl. Phys. Lett. **54**, 1037 (1989)
3. R. H. Bube, L. Echeverria, and D. Redfield, Appl. Phys. Lett. **57**, 79 (1990)
4. D. Redfield, Appl. Phys. Lett. **52**, 492 (1988)
5. D. Redfield and R. H. Bube, in *Amorphous Silicon Technology-1990*, Matl. Res. Symp. Proc. **192**, 273 (1990)
6. C. Lee, W. Ohlson, P. Taylor, H. Ullal, and G. Ceasar, Phys. Rev. B **31**, 100 (1985)
7. H. Curtins, et al., in *Amorphous Silicon Technology*, Matl. Res. Soc. Proc. **118**, 159 (1989)
8. H. Park, J. Liu, and S. Wagner, Appl Phys. Lett. **55**, 2658 (1989)
9. W. B. Jackson, Phys. Rev. B **41**, 1059 (1990)
10. M. Vanecek, J. Kocka, J. Stuchlik, Z. Kosisek, O. Stika, and A. Triska, Sol. Energy Mater. **8**, 411 (1983)
11. Z E. Smith, V. Chu, K. Shepard, S. Aljishi, D. Slobodin, J. Kolodzey, S. Wagner, and T. L. Chu, Appl. Phys. Sett. **50**, 1521 (1987)
12. J. Kakalios, R. Street, and W. Jackson, Phys. Rev. Lett. **59**, 1037 (1987)

Table I   Kinetics parameters deduced from data of Figures 1 and 2

| Sample | Degradation temperature T (K) | Stretch parameter β |
|---|---|---|
| 2Aa2 | 360 | 0.66 |
| 2Aa2 | 320 | 0.50 |
| 2Ab1 | 270 | 0.39 |

Table II   Parameter values used in Eqn. 2 for saturation defect density plotted in Figure 4, where k is the Boltzmann constant.

| Recombination effectiveness coefficients (cm$^{-3}$) $C_1 + C_2$ | Thermally activated generation frequency $v_1$ (sec$^{-1}$) | Thermally activated anneal frequency $v_2$ (sec$^{-1}$) | Total potentail defect density $N_T$ (cm$^{-3}$) |
|---|---|---|---|
| $3.5 \times 10^{-25}$ | $5 \times 10^9 \exp(-1.25/kT)$ | $5 \times 10^9 \exp(-1.10/kT)$ | $1.2 \times 10^{17}$ |

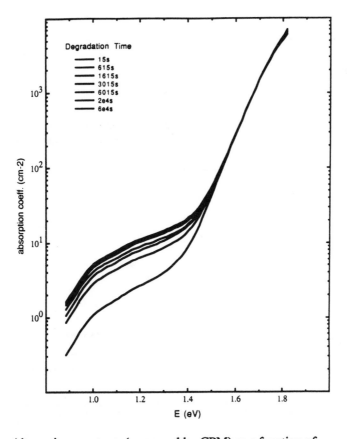

Figure 1.   Absorption spectrum (measured by CPM) as a function of degradation time at 360K under 1 W/cm$^2$ illumination (G=2x10$^{22}$cm$^{-3}$s$^{-1}$).

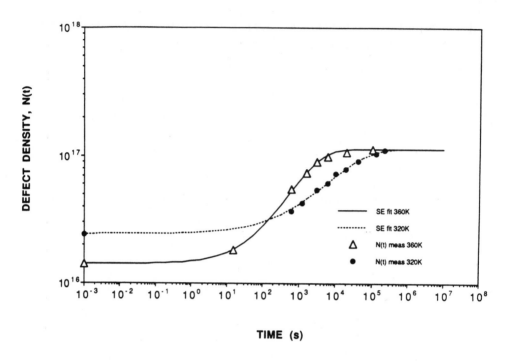

Figure 2.   Defect density vs. light exposure time for
sample 2Aa2 degraded at 320 and 360K.

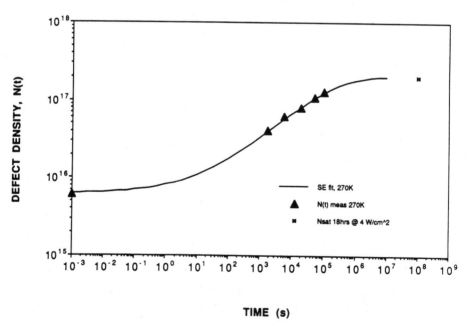

Figure 3.   Defect density vs. light exposure time
for sample 2Ab1 degraded at 270K

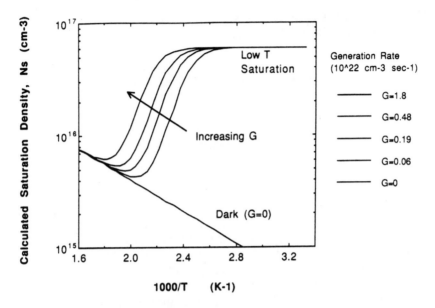

Figure 4.    Saturation defect density calculated from
equation (2) using values from Table II.

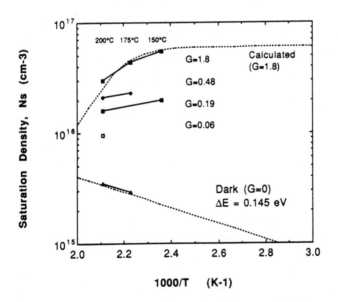

Figure 5.    Measured saturation density as a function
of temperature and generation rate, G (x $10^{22}$ cm$^{-3}$s$^{-1}$).

# LIGHT-INDUCED CHANGES IN SUBBAND ABSORPTION IN A-Si:H USING PHOTOLUMINESCENCE ABSORPTION SPECTROSCOPY

S.Q. Gu and P.C. Taylor
Department of Physics, University of Utah
Salt Lake City, UT 84112

S. Nitta
Department of Electrical Engineering
Gifu University, Gifu, Japan

## ABSTRACT

We have used the photoluminescence (PL) generated in a thin-film sample of a-Si:H to probe low absorption levels by measuring the absorption of the PL as it travels down the length of the film in a waveguide mode. This technique, which we have called PL absorption spectroscopy or PLAS, allows the measurement of values of the absorption coefficient $\alpha$ down to about 0.1 cm$^{-1}$. Because this technique probes the top and bottom surfaces of the a-Si:H sample, it is important to separate surface from bulk absorption mechanisms. An improved sample geometry has been employed to facilitate this separation. One sample consisted of an a-Si$_{1-x}$N$_x$:H/a-Si:H/ a-Si$_{1-x}$N$_x$:H/NiCr layered structure where the silicon nitride layers served as the cladding layers for the waveguide. In a second sample the a-Si:H layer was interrupted near the middle for two separate, thin (100 Å) layers of a-Si$_{1-x}$N$_x$:H in order to check for the importance of the absorption at the silicon/silicon nitride interfaces in these PLAS measurements. Changes in the below-gap absorption on light soaking were examined using irradiation from an Ar$^+$ laser (5145 Å, ~200 mW/cm$^2$ for 5.5 hours at 300 K). The silicon/silicon nitride interface is responsible for an absorption which has a shoulder near 1.2 eV while the bulk a-Si:H absorption exhibits no such shoulder. The metastable, optically-induced increase in the below gap absorption appears to come entirely from the bulk of the a-Si:H. These low temperature PLAS measurements are compared with those obtained at 300 K by photothermal deflection spectroscopy.

## INTRODUCTION

The below gap absorption in a-Si:H is related to the defect states, which can significantly affect the optical and electronic properties. Therefore, the precise measurement of the subband absorption is very important. However, this absorption is very weak, typically 1-10 cm$^{-1}$, and the thicknesses of the samples of a-Si:H are on the order of 1 $\mu$m. These parameters yield an optical path $\alpha d$ on the order of 10$^{-4}$ to 10$^{-3}$, which is too small to be measured by conventional transmission techniques.[1]

This limitation on $\alpha d$ has resulted in the development of many techniques other than the conventional one to probe this weak absorption. For example, Loveland et al.[2] have measured the photoconductivity to probe the absorption while others have used photothermal spectroscopic techniques such as photothermal deflection spectroscopy (PDS)[1] and photoacoustic spectroscopy (PAS).[3] All these techniques use the assumption that the nonradiative quantum efficiency $\eta(\lambda)$ or the lifetime of the excited carriers $\tau(\lambda)$ is not sensitive to the exciting-photon energy, an assumption which is still controversial.

Therefore, it is desirable to develop some techniques which can probe the low absorption without using any photothermal effect which is related to $\eta(\lambda)$. Photopyroelectric spectroscopy (PPES) was developed by Coufal[4] to measure the absorption coefficient $\alpha$ and $\eta$ separately. Fan et al.[5] reported that a minimum at 1.7 eV for $\eta$ has been found for a-Si:H by PPES. Ranganathan and Taylor used a technique called photoluminescence absorption spectroscopy (PLAS)[6] to measure the subband absorption. For PLAS, instead of using the thickness of the sample as the optical path d, we use the PL generated in the sample as the light source and the length of the sample as d, so we can increase $\alpha d$ from $10^{-4}$ to the order of 1. In addition, as this technique probes both the front and the back surfaces of the sample, one must separate the surface and the bulk contributions to the absorption.

The PLAS technique can also be used to study the metastable states in a-Si:H. Many results have been published[7] using PDS to measure the effect of light soaking at room temperature, but no low temperature results are available. Since PLAS can measure the absorption at 77 K or even lower temperatures, such measurements help in understanding the metastable states at low temperature.

In this paper we present an improved PLAS technique and report some light-induced changes in the subband absorption. In addition, we show that surface effects can be probed by this technique.

## EXPERIMENTAL ASPECTS

In order to suppress the scattered PL which does not travel in a waveguide mode within the film, we developed an improved geometry for the samples, as shown in Fig. 1(b). Sample 1 consists of NiCr/a-Si$_{1-x}$N$_x$:H(0.4$\mu$m)/ a-Si:H(1.0$\mu$m)/a-Si$_{1-x}$N$_x$:H (0.4$\mu$m), where the nitride layers serve as the cladding layers for the waveguide. Sample 2 has a similar structure except that two 100 Å a-Si$_{1-x}$N$_x$:H layers were inserted within the a-Si:H layers and each separate a-Si:H layer is about 0.5 $\mu$m. The two thin nitride layers will increase the contribution to the surface absorption. The samples were prepared as follows: the thick layer of NiCr was deposited on top of the 7059 Corning glass by sputtering; the a-Si$_{1-x}$N$_x$:H and a-Si:H were deposited by glow discharge deposition. For a-Si:H, the deposition system operated with a flow rate of 33 sccm, the pressure was approximately 210 m torr, the rf power density was 17 mw/cm$^3$, the substrate temperature was about $T_S = 250°C$, and the growth rate for a-Si:H was 1.6 Å/sec. For a-Si$_{1-x}$N$_x$:H, the deposition system was operated with a flow of SiH$_4$:N$_2$:NH$_3$ at rates of 200:50:10 sccm, respectively, a substrate temperature of $T_S = 250$ C,

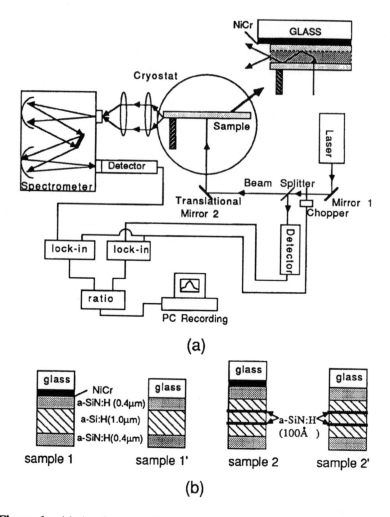

**Figure 1.**  (a) A schematic diagram for PLAS; (B) the geometries of the samples.

a pressure 210 m torr, an rf power density of 6 mw/cm$^3$, and a growth rate of 1.1 Å/sec.  In order to form a sharp edge to the waveguide end which provided the PL emission, a photoresist was deposited on the surface, and the unwanted sample was etched away by plasma etching.

A schematic diagram of the PLAS set-up is shown in Fig. 1(a).  By moving mirror 2 translationally, one can move the laser spot on the sample and, hence, adjust the length of the sample along which the PL travels.  A baffle was mounted on the top of the sample near the emitting edge to block the scattered PL which is not traveling within the sample in a waveguide mode.  The PL was detected by a sensitive Ge p-i-n detector (North Coast, model 817) and a lock-in amplifier. The instability of the laser was corrected using a reference circuit. The final PL spectrum was recorded on a computer. The absorption coefficients were obtained by calculating the ratios of the PL spectra for the laser hitting the sample at different positions.

## RESULTS AND DISCUSSION

After the plasma etch, the samples were annealed at 206°C for about 30 minutes, then were measured at 77 K. Without moving the sample, it was light-soaked using an $Ar^+$ laser (5145 Å, ~200 mW/cm$^2$ for 5.5 hours at 300 K), after which the absorption was again measured using PLAS. Figure 2(a) shows the absorption of sample 1 after annealing and light soaking. We found that the absorption after annealing increased very slowly from 1.0 eV to about 1.5 eV. After light soaking the absorption increased. The difference in the absorption $\Delta\alpha$ between the light soaked film and the annealed film is shown in Fig. 4.

In order to compare with

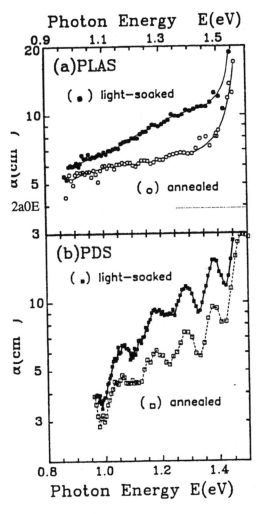

**Figure 2.** (a) The absorption spectra using PLAS at 77 K of sample 1 after annealing (○) and light soaking (•); (b) the absorption spectra using PDS at 300 K of sampel 1' after annealing (□) and light soaking (■).

PDS measurements, we performed PDS measurements at 300 K on sample 1' which was prepared in the same run as sample 1 except that there was no NiCr layer on the glass. Figure 2(b) shows the absorption using PDS for sample 1' after light soaking and before light soaking. The slope of the below-gap absorption is greater for PDS than for of PLAS in Fig. 2(a), and the absolute value is lower than that of PLAS in the low energy region (~ 0.9 eV). In a-Si:H the energy gap increases by about 1.0 eV when the temperature cools down from 300 K to 77 K. Comparing the absorption[3] using PAS at 50 K with our results of PLAS, we found that the $\alpha$ determined by PLAS drops much more slowly than that of PAS for an annealed sample in the energy region below 1.3 eV. This result may imply that the PLAS technique probes different absorption processes than PAS.

Figure 3(a) shows the absorption measured by PLAS in sample 2 after annealing and light soaking. Obviously there is a shoulder around 1.2 eV in the absorption spectra, which is not found in sample 1. As there are two a-Si:H and a-SiN:H interfaces within the a-Si:H layer, the shoulder may originate from the silicon/silicon nitride interfaces. There was a peak at 1.15 eV in the absorption spectra in a-Si:H found by Ranganathan et al.[6] earlier, but we did not find this peak in our sample geometry. We suggest that the peak may originate from the surface absorption of the a-Si:H/SiO$_2$ interface. Figure 3(b) shows the absorption using PDS at 300 K of a sample which has

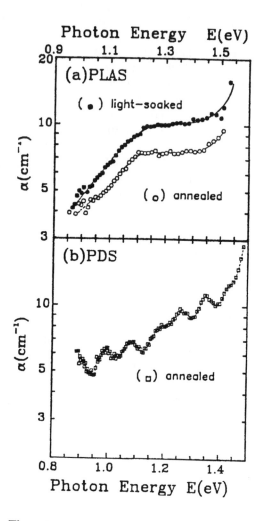

**Figures 3.**  (a) The absorption spectra using PLAS at 77 K of sample 2 after annealing (○) and light soaking (●); (b) the absorption spectra using PDS at 300 K of sample 1' after annealing (□) and light soaking (■).

the same structure as sample 2 except no NiCr on top of the glass. By scaling the PDS absorption spectrum with the known temperature dependence of the optical energy gap ($\sim$ 0.1 eV) in a-Si:H between 300 K and 77 K, one can compare the PDS and PLAS results. In this way we find that the magnitudes of the absorption measured by PDS and PLAS are essentially the same within $\sim$ 10% error.

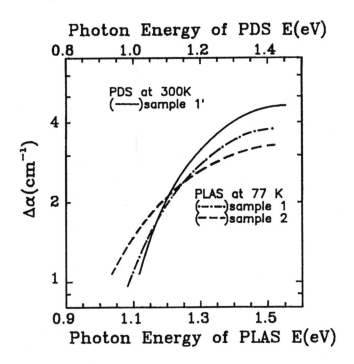

**Figure 4.** The light induced changes of the absorption
$\Delta\alpha$ for smaple 1' using PDS at 300 K; sample 1 using
PLAS at 77 K; sample 2 using PLAS at 77 K.

However, a shoulder is found using PLAS in sample 2, which suggests that the
PLAS may probe more surface state contributions.

Combining the light induced changes of the absorption $\Delta\alpha$ of sample 1 using
PLAS, PDS and of sample 2 using PLAS, we find that they are all quantitatively
the same within experimental error. All drop very fast when the absorption energy
decreases below about 1.3 eV at 77 K or 1.2 eV at 300 K, as shown in Fig. 4.
In this figure the absorption spectra are from two different samples, measured
using two different techniques, and show different absorption shapes, but the light
induced absorption spectra $\Delta\alpha$ are essentially the same. This fact suggests that the
light induced changes are independent of the differences between the PDS and
PLAS techniques. Since these differences probably involve the relative importance
of surface absorption, we can conclude that the light induced changes are a bulk
effect. As we have corrected for changes of the energy gap with temperature, the
energy of the sharp rise in $\Delta\alpha$ appears to scale with the energy gap. This means
that the light induced changes in the absorption may arise from transitions between
the optically-induced localized states and one band edge.

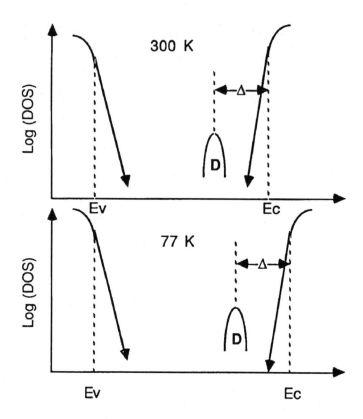

**Figure 5.** Schematic density of states (DOS) as a function of energy at 77 K and 300 K.

Measuring the light induced changes of the absorption in the a-Si$_{1-x}$C$_x$:H alloys and varying the energy gap by changing the composition x, Bennett et al.[8] found that the greatest change in absorption on light soaking occurred at an energy which was 0.55 eV less than the energy of the band gap. In our case, the energy gap at 300 K is about 1.8 eV($\alpha$(Eg) $= 10^4$ cm$^{-1}$), and the difference between the energy of the band gap and the energy of the sharp rise in $\Delta\alpha$ is about 0.6 eV, which is very close to the value given by Bennett et al.[8] This agreement exists not only for the PDS measurements at 300 K but also for the PLAS experiments at 77 K. In our earlier results on the PL excitation spectra,[9] we found that the PLE spectrum for a good sample of a-Si:H shows a sharp drop around 1.3 eV at 77 K, which we explained in terms of a two step excitation process. This explanation requires that one mid-gap state has an energy difference of about 1.3 eV from the band edge at 77 K. Combining these results with our light induced $\Delta\alpha$ measurements, we propose that the light soaking increases the density of metastable mid gap states, where the difference of the defect energy (D$^0$ or D$^-$) to the band edge is independent of the temperature and is about 0.6 eV as shown schematically in Fig. 5. Because of the energy at which the sharp rise in $\Delta \alpha$ occurs ($\sim$ 1.3 eV at 77 K), the most probable process for the light induced absorption is from the valence band to the D⁻ states.

## CONCLUSION

We have developed an improved sample geometry for the PLAS technique to measure low absorption. Absorption spectra for two a-Si:H samples with different numbers of silicon and silicon nitride interfaces have been measured using PLAS at 77 K and PDS at 300 K. Comparison of the absorption obtained by PLAS and PDS shows that they are all essentially of the same order. The PLAS absorption spectrum of the a-Si:H sample which has more a-Si:H/a-SiN:H interfaces shows a shoulder at about 1.2 eV. The light induced changes $\Delta \alpha$ for both the PLAS and PDS measurements for the two samples are essentially the same. The sharp rise in $\Delta \alpha$ occurs at the same energy with respect to the energy gap in both techniques. These results suggest that only the bulk absorption contributes to the light induced absorption and that the energy difference from the conduction band edge to the metastable states induced by light soaking is independent of the energy gap.

## ACKNOWLEDGEMENTS

The authors are grateful to K. Gaughan and S. Hershgold for making the samples and to B. Moosman for assistance in depositing the NiCr films. We thank Z.H. Lin for valuable discussions. This research is supported by SERI under subcontract No. XM-91-81413.

## REFERENCES

1.  N.M. Amer, W.B. Jackson, in *Semiconductors and Semimetals*, J.I. Pankove, ed. (Academic, New York, 1984), Vol. 21B, p. 83.
2.  R.J. Loveland, W.E. Spear and A. Sharbaty, J. Non-Cryst. Solids **40**, 474 (1973).
3.  S. Yamasaki, Philos. Mag. **B56**, 79 (1987).
4.  H. Coufal, Appli. Phys. Lett. **45**, 516 (1984).
5.  J. Fan, J. Kakalios, in *Amorphous Silicon Technology - 1991* (Materials Research Society, Pittsburgh, 1991), to be published.
6.  R. Ranganathan, P. C. Taylor, J. Non-Cryst. Solids **97+98**, 707 (1987).
7.  A. Skumanich, N.M. Amer and W.B. Jackson, Bull. A.P.S. **27** 146 (1982).
8.  M.S. Bennett, S. Wiedeman and K. Rajan, in *Amorphous Silicon Technology* (Materials Research Society, Pittsburgh, 1989), Vol. 149, p. 577.
9.  S.Q. Gu and P.C. Taylor, in *Amorphous Silicon - 1990* (Materials Research Society, Pittsburgh, 1990) Vol. 192, p. 107.

# CAPACITANCE STUDIES OF BIAS-INDUCED METASTABILITY IN p-TYPE HYDROGENATED AMORPHOUS SILICON

Richard S. Crandall, Kyle Sadlon, Stanley J. Salamon, and Howard M. Branz
Solar Energy Research Institute, Golden, CO, 80401

## ABSTRACT

We report experimental studies of metastability in the p-layer of hydrogenated amorphous silicon p-n junction devices. After reverse bias annealing, we observe a completely reversible metastable increase in the capacitance caused by hole emission from defects in the p-layer. Hole recapture over an energy barrier during subsequent zero-bias annealing restores the initial capacitance. The hole emission has a thermal activation energy of $1.3 \pm 0.2$ eV and the hole capture has an activation energy of $0.89 \pm 0.1$ eV. We discuss microscopic models of the metastable defect and relate our findings to other carrier-induced metastability in hydrogenated amorphous silicon.

## INTRODUCTION

The stability of hydrogenated amorphous silicon (a-Si:H) against changes in structure caused by non equilibrium carrier densities is one of the most important research topics in amorphous solids. Deleterious carrier-induced metastabilities are observed in solar cells, thin film transistors, and other a-Si:H devices. Consequently, the problem has technological as well as intellectual interest. Despite research into these metastabilities for fifteen years, there is still no universally accepted theory. There is considerable evidence[1] that metastable effects in undoped a-Si:H are caused by defects that trap charge and metastably reconfigure. Junction capacitance experiments in both p-i-n and undoped Schottky barrier solar cells [2,3] have been an important technique in these studies. Such charge-trapping metastabilities are also observed by capacitance and other studies of n-type a-Si:H. Increased room temperature capacitance of n-Schottky barriers results from annealing above 420K and then cooling the device under reverse-bias.[4] These changes can be removed by annealing the device without the bias. Few careful studies of carrier-induced metastability in p-type a-Si:H and its kinetics have been published. Crandall et al.[5] metastably trapped charge in the p-layer of a p-n junction by illumination and by forward-bias injection. They used capacitance techniques to determine the annealing activation energy for electron emission. Street et al.[6] produced increases in sweep-out current by reverse bias annealing the p-layer of i-p-i structures and studied the kinetics by a sweep-out measurement. There are also kinetic studies of light-induced metastability by Jang et al.[7] These effects may be related to metastable changes produced in p-type a-Si:H films by quenching from high temperature.[8,9]

In this paper, we describe the first detailed capacitance study of metastable changes in a-Si:H p-layers. We produce the metastable change by reverse-bias annealing p-n junction devices having thin, heavily-doped n-layers. After cooling under applied reverse bias, we observe a significant increase in the capacitance. This metastable change anneals out at zero bias above about 370K. We measure the characteristic times and activation energies for both production and annealing of these metastable changes. Our measurements show that charge trapping is an integral part of the metastability in p-type a-Si:H. It is clear that reverse-bias annealing removes holes from a metastable defect that can only anneal by hole capture over a barrier.

## EXPERIMENTAL DETAILS

We study radio-frequency glow-discharge deposited p-n junction structures:

$$glass/SnO_2/7800 \text{ Å p-type a-Si:H}/300 \text{ Å n-type a-Si:H}/Al.$$

Details of the deposition conditions are published elsewhere.[10] The n layer
($PH_3/SiH_4=2\%$) is thin and heavily doped to ensure the p-layer depletion dominates the
measured capacitance. The p layer is moderately doped at $BH_2/SiH_4=0.013\%$.
Devices are mounted in an evacuated liquid nitrogen dewar capable of maintaining a
stable temperature (T) between 80K and 550K. We measure the junction capacitance
between 100 and 20,000 Hz by a standard lock-in technique. We measure capacitance
either as a function of frequency ($\omega$) and temperature at constant voltage (C$\omega$T) or
isothermally after an applied voltage pulse. All capacitance values reported in this work
are measured at an applied dc bias ($V_b$) of -3V.

We make constant-voltage measurements of the capacitance (C) at four
frequencies as we increase the temperature from 200 to 500K at 10 K/min. From these
C$\omega$T measurements, we can determine[11] the hole-occupied density of states (DOS) in
the p-layer between the Fermi level and the demarcation level, $E_d = k_BT \ln(\nu/\omega)$ -$E_v$.
Here $k_B$ is Boltzmann's constant, $\nu$ is the "attempt-to-escape" frequency, and $E_v$ is the
energy of the valence-band mobility edge.

We use the C$\omega$T technique to probe metastable changes that we produce by
subjecting the device to bias treatments above 500K for times long enough to ensure
defect equilibration with the bias-induced carrier density.[12] The sample is cooled
without changing the applied bias before the C$\omega$T scans are taken at -3V. The results
are independent of the rate of cooling from high T. However, as described below, they
depend dramatically on the bias treatment at high T.

We also perform isothermal capacitance transient measurements of the
production and relaxation of the metastable changes. We fix the device temperature and
record a capacitance transient, C(t), at -3V after a bias pulse is applied for a time, $t_p$.
The transients reported in this paper are taken after -5V or -1V bias pulses ranging from
5 ms to 500 sec. The measurement temperature is chosen so that near-complete
relaxation occurs in 0.01 to $10^4$ sec. The data are fit by the method of least squares to
the stretched exponential time dependence $C(t) = C(\infty) \pm \Delta C \exp(-[t/\tau_R]^\beta)$, where $\tau_R$
is the relaxation time and $\beta$ is a parameter. Thus, $\tau_R$ is the time for the capacitance to
relax a fraction, $1-e^{-1}$, of the pulse-induced capacitance change.

## EXPERIMENTAL RESULTS

Figure 1 shows typical C(T) data taken at 6310 Hz and -3V following different
bias treatments. The lowest curve (■) is taken after a 20 minute 530K anneal and
subsequent cool to 200K at $V_b = 0V$. The uppermost curve (▲) is taken after an
identical temperature cycle at $V_b = -5V$. The middle curve (●) is taken after a 530K
anneal and subsequent cool to 305K at $V_b = -5V$, followed by a 120 sec bias treatment
at $V_b = 0V$ and subsequent cool to 200 K at -5V. We obtain similar results at
frequencies down to 100 Hz.

The CωT experiments show clearly that reverse bias annealing increases the capacitance above the $V_b = 0V$ anneal value. The temperature above which the capacitance increase is observed decreases monotonically from about 280K at 6310 Hz (see Fig. 1) to about 200K at 100 Hz. The metastable capacitance increase anneals out between about 360 and 400K. The precise annealing T depends on the scan rate but is independent of measuring frequency. High hole densities ($V_b = 0V$) cause a partial annealing of the effect in short times even at about room temperature, as illustrated in Fig. 1.

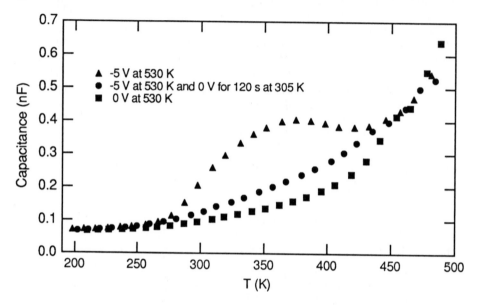

Figure 1. Capacitance versus T measured at 6310 Hz and -3V. The bias-temperature history is summarized in the legend. Bias time at 530K was 20 min.

Figure 2 shows the result of a typical isothermal capacitance experiment. We let the sample reach steady state at 449K and -3V and then apply a -1V bias pulse for a time $t_p = 5$ ms. The pulse admits holes into a previously hole-depleted region of the p-layer. We record the C(t) transient at -3V as the resulting capacitance decrease anneals out by an activated hole emission process. The zero of time is the end of the voltage pulse. The value of $\tau_R$ determined from the stretched exponential fit is independent of the pulse voltage, but there is a weak dependence on $t_p$ (see below). We repeat this procedure at different temperatures to determine $\tau_R(T)$ for the hole emission process. These hole emission data are shown (●) in the Arrhenius plot of Fig. 3.

After a -5V pulse, we measure the isothermal capacitance transient for the hole capture process. The pulse removes holes from the bulk at the edge of the depletion region. The subsequent capacitance decrease at $V_b = -3V$ follows a stretched exponential form as the holes are slowly recaptured. The value of $\tau_R$ is determined at each temperature from the fit to the data. We plot $\tau_R(T)$ for the hole capture process (○) in Fig. 3.

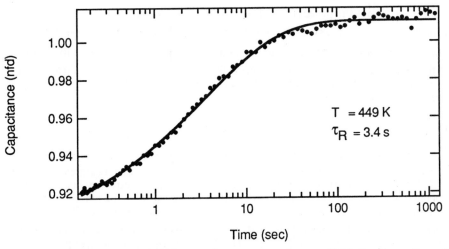

Figure 2. Isothermal (449K) capacitance versus time at -3V following a 5 ms -1V bias pulse. Solid line is the stretched exponential fit to these hole emission data.

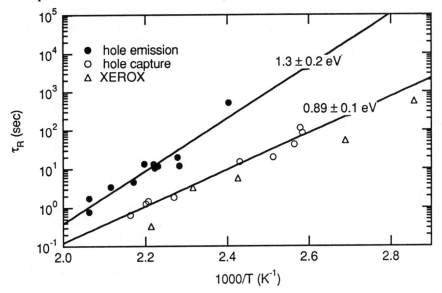

Figure 3. Relaxation times versus 1/T for hole capture (O) and hole emission (●) from isothermal capacitance decays. The lines are Arrhenius fits to our data. The other data (△) are that of Street and coworkers[13, 14] for relaxation of quenched-in defects in p-type a-Si:H.

The solid lines in Fig. 3 are the least-square Arrhenius fits to the isothermal capacitance transient data. We find $E_A = 1.3 \pm 0.2$ eV and $\nu = 1.1 \times 10^{14}$ sec$^{-1}$

for the hole emission process and $E_A = 0.89 \pm 0.1$ eV and $v = 1.3 \times 10^{10}$ sec$^{-1}$ for the hole capture process. Most of the error in $E_A$ comes from the dependence of $\tau_R$ on $t_p$. As previously observed, there is a weak sublinear increase of $\tau_R$ with pulse length.[3] We will describe this phenomenon in more detail in a future publication. In Fig. 3, we plot the data with the least error in the fitting; generally these are the longest pulses at each temperature.

## DISCUSSION

The $C\omega T$ measurement shows the general character of the bias-induced metastability. The dramatic capacitance increase after a -5V bias anneal treatment (see Fig. 1) begins at a temperature that increases with measuring frequency because it is associated with a feature in the hole-occupied DOS. In contrast, the temperature at which the metastable capacitance increase begins to disappear is about 370K, independent of $\omega$. This means that the decrease is associated with the annealing of a metastable defect, rather than a feature in the DOS. The annealing temperature is comparable to that observed in B-doped a-Si:H and about 40K too low to be associated with the P-doped layer.[13] Further, as shown in Fig. 3, there is a remarkable consistency between time constants for annealing of quenched-in defects in B-doped a-Si:H measured by Street et al.[13, 14] and the hole capture times we measure with isothermal capacitance transients. We conclude that our capacitance measurements probe metastability in the p-layers of these p-n devices. Our use of thin, heavily doped n-layers in the devices ensures the p-layer depletion width dominates the measured capacitance changes.

The 0V bias anneal treatment admits holes to regions of the p-layer that are normally depleted at the -3V measurement bias. During the anneal, metastable hole capture decreases the negative charge density in the p-depletion region, widens the depletion region, and reduces the measured capacitance. Conversely, the -5V anneal results in hole emission which increases the negative charge density, shrinks the depletion layer, and increases the capacitance. The hole emission likely also lowers the Fermi level ($E_F$) toward the valence band, thereby increasing the density of holes that can respond to the applied ac voltage and increasing the measured capacitance. The $C\omega T$ data show (see Fig. 1) that at high temperature, the capacitance is independent of the bias-anneal history Above about 450K, relaxation times are so short that the p-layer structure remains in quasi-equilibrium with the measurement bias.

The stretched exponential form $C(t) = C(\infty) \pm \Delta C \exp(-[t/\tau_R]^\beta)$ fits the isothermal capacitance decays quite well, as illustrated in Fig. 2. Stretched exponential relaxation is commonly observed in a-Si:H and other disordered solids. Dispersive hydrogen motion[15] and a distribution of defect conversion barriers[16] have been proposed to explain stretched exponential relaxation in a-Si:H. There is also a model based upon the hierarchy of constraints to relaxation found in glassy materials.[17] Our data are fully consistent with any of these models.

Fig. 3 shows that the activated hole emission is slower than the activated hole capture by one to two orders of magnitude at all temperaures. The hole emission activation energy of $1.3 \pm 0.2$ eV is too large to be simple emission to the valence-band from above $E_F$ It must be associated with emission over a defect reconfiguration barrier, instead. The hole capture also proceeds over a large energy barrier. This barrier may be an actual free energy barrier in configuration space or it may result from the large number of atoms whose position must simultaneously change to lock in the reconfiguration.

Our data strongly support the bistable charge-trapping defect model of metastability in p-type a-Si:H, which was proposed by Branz.[18] Fig. 4 is a configuration-coordinate diagram of the bistable defect. The left-hand well is the ground state of the defect and the right-hand well is the metastable state formed by emission of a free hole. The charge on the defect itself is one electronic charge greater in the ground state than in the metastable state. These experiments cannot determine the absolute charge on the defect, only the change in its charge state.

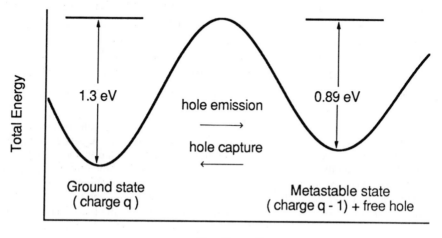

Configuration

Figure. 4 Proposed configuration-coordinate diagram of the metastable defect. Hole emission and capture barriers determined from isothermal capacitance transients are indicated.

The -5V pulses and anneal treatments cause hole emission from the defect and its reconfiguration into the metastable state, as illustrated in Fig. 4. We determine the 1.3 eV energy barrier to the hole emission process from the isothermal capacitance data taken after a -1V filling pulse (Figs. 2 and 3).

The -1V pulses and the 0V anneal treatments cause hole capture and reconfiguration into the ground state. We determine the 0.89 eV energy barrier to the hole emission process from the activated isothermal capacitance data taken after a -5V depletion pulse (Fig. 3).

According to the configuration-coordinate diagram of Fig. 4, rapid quenching leaves the defect in the metastable state and increases the density of free holes. This conductivity increase is observed after p-type a-Si:H is quenched.[8, 9] Our model also predicts that quenched-in metastability must anneal over the same barrier to hole recapture measured by our capacitance experiments. In Fig. 3, we show the coincidence of our hole capture data with the data of Street *et al.*[13, 14] for the annealing of quenched-in conductivity increases in B-doped a-Si:H. This supports the proposal that quenched-in metastabilities in p-type a-Si:H are the same as hole-emission metastabilities.[19]

Branz[19] proposes that the B-dopant is the microscopic defect that captures charge and reconfigures in p-type a-Si:H. Hole capture is accompanied by dopant deactivation, according to $B_4^-$ ($sp^3$) + $h^+$ → $B_3^0$ ($sp^2$). (The superscript is the charge state of the atom, the subscript is its coordination, and the electronic hybridization is in parentheses.) The reverse reaction is dopant deactivation to $B_3^0$. The reaction requires thermal energy to break and reform the bond which accounts for the large measured activation energies.

Another possible microscopic mechanism for the hole capture/emission reaction is $T_3^-$ ($s^2p^3$) + $h^+$ ↔ $T_3^0$ ($sp^3$). This is closely related to the Adler model[1, 20] of light-induced metastability in a-Si:H in which charged dangling-bond defects trap charge, rehybridize and reconfigure. After hole emission, the reconfiguration to $s^2p^3$ hybridization inhibits recapture of the hole at low temperature by changing the bond angles. In homogeneous models of a-Si:H, the (-/0) transition of $T_3^-$ is in the upper half of the gap and acts as a deep hole trap that could not be observed in our capacitance measurement. However, in a model of inhomogenous a-Si:H, Branz and Silver point out that potential fluctuations place the $T_3^-$ level close to the valence band.[21] This hole emission reaction is possible in a lightly doped p-layer, because the number of $T_3^0$ is nearly as high as in undoped a-Si:H.[22]

We note, however, that there are marked parallels between the metastability phenomena in p-type, n-type, and undoped a-Si:H which could suggest a common mechanism. In the present experiment, the hole emission activation energy is about 0.4 eV greater than the hole capture activation energy. Since hole emission and electron capture should stimulate the same metastable charge-trapping reaction,[18] the activation energies can be compared to the annealing energies of electron- and hole-capture-induced metastabilities in undoped and n-type a-Si:H. Crandall[3] finds that the activation energy is consistently about 0.3 eV higher for hole emission than for electron emission, as in the present study.

Another parallel with light-induced metastability[23] in undoped a-Si:H is the ability of free carriers to cause defect reconfiguration even at room temperature and below. If we extrapolate the data for hole capture in Fig. 3 to room temperature, we find $\tau_R \approx 3 \times 10^5$ sec. This time is inconsistent with our CωT observation (Fig. 1) that significant hole-capture defect conversion takes place in just 2 min at 305K. One reason that defect conversion is faster than expected at room temperature might be that some of the energy necessary for the conversion is supplied by the hole capture.

## CONCLUSIONS

Capacitance measurements on a-Si:H p-n junctions clearly show that activated hole trapping and emission cause metastability in p-type a-Si:H. Activation energies for hole capture and emission are $0.89 \pm 0.1$ eV and $1.3 \pm 0.2$ eV, respectively. The relaxation times for hole capture closely match those observed for annealing of quenched-in defects in p-type films of a-Si:H. This supports a bistable charge-trapping defect model of the metastability.[19]

The capacitance changes observed by reverse-bias annealing of p-type a-Si:H are qualitatively similar to the metastabilities observed in n-type[4] and undoped[3] a-Si:H. We suggest that bistable charge-trapping defect models explain most light- and bias-induced metastability in a-Si:H.

## ACKNOWLEDGEMENTS

The authors are indebted to Yueqin Xu for fabrication of device structures. This work was supported by the U.S. Department of Energy under Contract No. DE-AC02-83CH10093.

## REFERENCES

1.  See, e.g, H. M. Branz, R. S. Crandall, and M. Silver, this volume.
2.  R. S. Crandall, Phys. Rev. B **24**, 7457 (1981).
3.  R. S. Crandall, Phys. Rev. B **36**, 2645 (1987).
4.  D. V. Lang, J. D. Cohen, and J. P. Harbison, Phys. Rev. Lett. **48**, 421 (1982).
5.  R. S. Crandall, D. E. Carlson, A. Catalano, and H. A. Weakliem, Appl. Phys. Lett. **44**, 200 (1984).
6.  R. A. Street and J. Kakalios, Phil. Mag. Letters **54**, L21 (1986).
7.  See, e.g., J. Jang, S. C. Kim, S. C. Park, J. B. Kim, H. Y. Chu, and C. Lee, in *Amorphous Silicon Technology*, edited by A. Madan, M. J. Thompson, P. C. Taylor, P. G. LeComber, and Y. Hamakawa (Materials Research Society, Pittsburgh, 1988), p. 189.
8.  J. Kakalios and R. A. Street, Phys. Rev. B **34**, 6014 (1986).
9.  H. M. Branz, Phys. Rev. B **36**, 7934 (1987).
10. Y. S. Tsuo, Y. Xu, R. S. Crandall, H. S. Ullal, and K. Emery, in *Amorphous Silicon Technology-1989*, edited by A. Madan, M. J. Thompson, P. C. Taylor, Y. Hamakawa, and P. G. LeComber (Materials Research Society, Pittsburgh, 1989), p. 471.
11. J. D. Cohen and D. V. Lang, Phys. Rev. B **25**, 5321 (1982).
12. Z. E. Smith and S. Wagner, Phys. Rev B **32**, 5510 (1985).
13. R. A. Street, J. Kakalios, C. C. Tsai, and T. M. Hayes, Phys. Rev. B **35**, 1316 (1987).
14. R. A. Street, M. Hack, and W. B. Jackson, Phys. Rev. B **37**, 4209 (1988).
15. J. Kakalios, R. A. Street, and W. B. Jackson, Phys. Rev. Lett. **59**, 1037 (1987).
16. R. S. Crandall, Phys. Rev. B, Feb. 15, in press (1991).
17. R. G. Palmer, D. L. Stein, E. Abrahams, and P. W. Anderson, Phys. Rev. Lett. **53**, 958 (1984).
18. H. M. Branz, in *Amorphous Silicon Technology*, edited by A. Madan, M. J. Thompson, P. C. Taylor, P. G. LeComber, and Y. Hamakawa (Materials Research Society, Pittsburgh, 1988), p. 199.
19. H. M. Branz, Phys. Rev. B **38**, 7474 (1988).
20. D. Adler, Solar Cells **9**, 133 (1983).
21. H. M. Branz and M. Silver, Phys. Rev. B **42**, 7420 (1990).
22. H. Dersch, J. Stuke, and J. Beichler, Phys. Stat. Solidi (b) **105**, 265 (1981).
23. D. L. Staebler and C. R. Wronski, Appl. Phys. Lett. **31**, 292 (1977).

# WHAT ELECTROLUMINESCENCE AND TRANSIENT SPACE CHARGE LIMITED CURRENTS TELL US ABOUT STAEBLER-WRONSKI DEFECTS

K. Wang, D. Han and M. Silver
Department of Physics and Astronomy
University of North Carolina
Chapel Hill, NC 27599-3255

## ABSTRACT

Steady state electroluminescence depends upon radiative and non-radiative processes often governed by recombination centers. Transient space charge limited currents also depend upon capture by these same deep states although recombination is not involved. We have studied both of these phenomena as a function of photodegradation in order to gain insight into the nature of Staebler-Wronski defects. We have found that photodegradation primarily shortens the non-radiative lifetime but has little effect upon the radiative lifetime. On the other hand, photodegradation seems to have a profound effect upon the transient space charge limited current. We have not yet determined whether this is a junction or a bulk effect.

## INTRODUCTION

It is well-known that photodegradation involves the generation of deep states near the Fermi level in a-Si:H. These states have been associated with neutral dangling bonds[1]. The mechanism for the formation of these Staebler-Wronski defects[2] is still not understood although it is clear that the Fermi level drops toward midgap, lowering the conductivity and reducing the photoconductivity.

Two of the principal experiments used to study the photodegradation mechanism have been photoconductivity[2] and photoluminescence[3] changes brought about by prolonged illumination. However, both of these effects might result in part from changes in geminate recombination. On the other hand, double injection under forward bias in p/i/n structures and the resulting electroluminescence <u>cannot</u> involve any geminate pair recombination[4] because the carriers essentially approach each other from infinity. Consequently only bulk recombination, possibly through deep centers can be involved. Further, transient space charge limited currents in n/i/n devices[5] can yield information about how these localized centers, above the Fermi level, are populated. Even though the quality of the injecting contacts is a factor, we decided to study double injection currents, electroluminescence (EL) and transient space charge limited currents as a function of photodegradation.

In this paper we report on some of our preliminary results. These early results suggest that it is the non-radiative lifetime that is primarily degraded in the Staebler-Wronski effect. Further, we can see from the transient space charge limited current that there is a dramatic increase in the density of deep states and also possibly a decrease in the quality of the n/i injecting contact.

## CONCEPT OF THE EXPERIMENTS

In order to obtain a qualitative picture of what is taking place under double injection, EL and single injection transient space charge limited currents, we will assume that we have an exponential density of states and perfectly injecting contacts. Mark and Lampert[6] have analyzed double injection for perfect contacts and we will use

162

their simple model and simplify the notation by using a "mutual" hole-electron lifetime, $\bar{\tau}$, and a "mutual" hole-electron transit time $t_o$. Mark and Lampert[6] show simply that the excess hole and electron charge (for simplicity assumed to be equal) is

$$Q^{\pm} \cong CV\,\bar{\tau}/t_o \tag{1}$$

With a plethora of localized states in the gap, only a fraction, $\theta$, of $Q^{\pm}$ are in the conduction band. Consequently the forward bias current, $j_F$, is

$$j_F \approx \frac{\theta Q^{\pm}}{\bar{t}_o} \approx \frac{\theta CV\bar{\tau}}{\bar{t}_o^2} \tag{2}$$

Now to evaluate $\theta$, we must assume a density of localized states (DOS) and transitions in and out of these states. Figure 1 shows a schematic of the DOS. and the transitions. The DOS is assumed to have the form $\exp\left[E/kT_o\right]$. The important parameters are: 1) $E_d$ the demarcation level defined as the level where thermal emptying just equals the recombination rate specified by $\tau^{-1}$; 2) $E_L$ the energy from which the radiation primarily emanates and 3) $\tau_L$ the radiative recombination lifetime from these states at $E_L$.

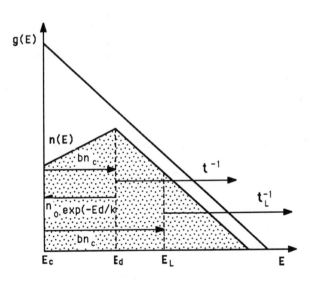

Fig. 1    Schematic diagram of the density of electronic states and their population under doulble injection conditions. Shaded states are filled. The demarcation levels Ed and El are indicated. The transitions into and out of states at Ed and El are indicated by the arrows.

Using these transition rates and equations (1) and (2) we obtain three simple results[7]:

$$j_F = \frac{CV}{Lt_0^2 v} (v\tau)^{T/T_o} \tag{3}$$

$$EL \propto \frac{CV}{t_o} (\frac{\tau}{\tau_L})^{T/T_o} \tag{4}$$

and

$$\frac{EL}{j_F} \propto (v\,t_o)\,(v\tau_L)^{-T/T_o} \tag{5}$$

It should be clear from Eqs. 3, 4, and 5 that measurements of: $j_F$ yield information on $\tau$, EL on the rates of $\tau/\tau_L$ and therefore $EL/j_F$ on $\tau_L$ . Temperature measurements vs degradation will tell us how $\tau$ and $\tau_L$ vary.

Space charge limited current analysis is a little different because single carrier injection is involved and therefore,

$$Q^- \approx CV \tag{6}$$

and

$$j_F \approx \theta \frac{CV}{t_o} . \tag{7}$$

The transient nature is appreciated when $\theta$ and therefore j are functions of time. At early time[8],

$$j_F \propto \frac{CV}{t^{1-\alpha}} \tag{8}$$

where $\alpha = T/T_o$ . The mean time to trap into any deep level is given by

$$\tau = \frac{1}{v} (v\tau_o)^{1/\alpha} \tag{9}$$

where $\tau_o = (bN_R)^{-1}$ . If the density of deep states protrude out significantly from the exponential tail, then for times longer than $\tau$ ,

$$j_F \propto \frac{N_C}{N_R} e^{E_R/kT} \tag{10}$$

where $N_C$ is the density of state at electron band edge $E_C$, $E_R$ is the energy position and $N_R$ is the density of deep states. In Eq. 10 $j_F$ is constant. In principle from Eqs. (9)

and (10) one can determine the energy position and density of deep states. Photodegrada-tion should change $N_R$ and might alter $E_R$ .

## EXPERIMENTAL RESULTS AND DISCUSSION

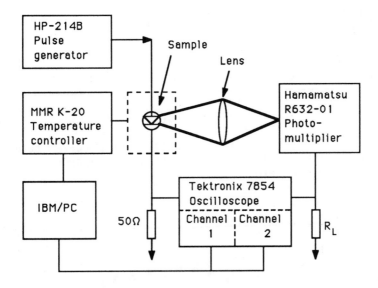

Fig. 2  Schematic diagram of the apparatus used to measure electro-luminescence and forward bias curent in pin devices.

The experimental apparatus for performing EL, $j_F$ and transient space charge limited currents is shown in figure 2. As can be seen, the measuring technique is simple and requires little explanation. A voltage pulse with a rise time of order $10^{-8}$ sec. is applied in forward bias to a p/i/n or an n/i/n sample which is kept at constant temperature $80°$ K $<$ T $< 350°$ . The frequency and width of the applied voltage were 0.1 to 10 Hz and $10^{-2}$ sec. In most cases steady state was achieved within the $10^{-2}$ sec. pulse width. The current and light emission are sampled by a Tektronix dual channel wave form analyzer. Averaging is performed by an IBM/PC. Both p/i/n and n/i/n samples were deposited on ITO-coated glass substrates with a 2μm i layer . The other contact was ZnO for p/i/n and Ag for n/i/n.

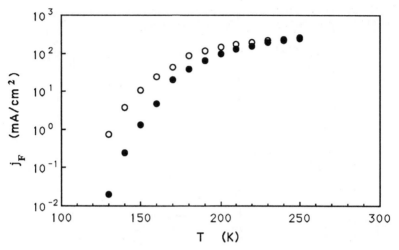

Fig.3  Forward bias current vs. temperature for pin device before
and after light-soaking. Open circles before, dark circles
after degradation.

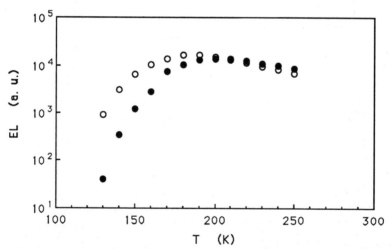

Fig. 4  Electroluminescence intensity vs. temperature before and after light-
soaking for the same pin device under the same conditions as in Fig. 3.
Open circles before, dark circles after degradation.

Photodegradation is achieved by irradiation with a 50 mW/cm$^2$ light for three
hours from a tungsten halogen lamp with an IR cut-off filter. The EL data were taken
under constant voltage conditions[7] with $V = 12V$. Figures 3-5 show how the steady
state $j_F$, EL and EL/$j_F$ characteristics vs T change after photodegradation. The open
circles refer to the original, undegraded state while the solid circles represent the same
data after degradation. As seen in Figure 3, the forward bias current at $T < 160°K$

decreases by as much as one or two decades due to photodegradation but only a much smaller effect is observed at T > 160°K . These low T results suggest a significant drop in $\tau$ with photodegradation as expected but apparently little decrease at the higher temperatures.  Perhaps figure 5 is most significant because the straight line plot of the log EL/$j_F$ reflects the fact that $\tau_L$ does not depend upon degradation (EL/$j_F \propto (\tau_L)^{-T/T_o}$ ).

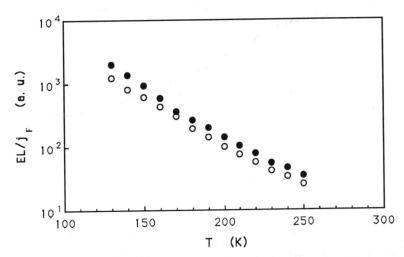

Fig.5  EL/$j_F$ vs. temperature before and after light-soaking. EL and $j_F$ data are from Fig.3 and Fig.4. Open circles before, dark circles after degradation.

The logarithmic slope of the two curves in figure 5 of EL/$j_F$ are both 27°K which yields a value for $T_o$ = 550°K and $\tau_L$ = $10^{-3}$ sec .  Armed with this information one sees in figure 4 that $\tau$ (see Eq. 4) at low T depends upon degradation.  This suggests the possibility that the recombination mechanism[9] at low T is due to tunneling to a defect while the recombination at high T is due to electrons in extended states diffusing to the hole.

Finally, comparing figures 4 and 5, we see that EL at low T is degraded in a similar way as $j_F$.  Since EL $\propto (\tau/\tau_L)^{T/T_o}$ and $j_F \propto \tau^{T/T_o}$ , given that $\tau_L$ is constant, the similarity between the results of figures 3 and 4 are not surprising.

One caution should be given regarding the degradation of $j_F$ and EL and that is that it might be due to a degradation of the contact.  Hopefully, future research will reveal the answer to the question of how much of the degradation effect is due to the bulk and how much is due to the contact.  However, the results shown in figure 5 will not depend upon the contact and we are confident that $\tau_L$ is not affected by degradation.

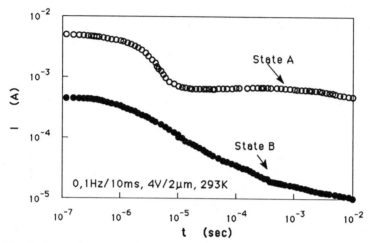

Fig. 6   Curves of the transient space charge limited current for a nin device at annealed state A and light-soaked state B respectively.

Figure 6 shows how the transient space charge limited current changes with degradation. What should be noted is that degradation affects the long-time current. From Eq. 10 one expects a significant drop in the current because the density of deep state $N_R$ is increased. This effect is seen. If this is all that happens under degradation, then one could determine the energy and density of the occupied deep states as a function of the degradation parameters. Unfortunately, we see that at short times (less than the time to trap into the deep states) degradation has also decreased the current. We assume that the quality of the injecting contact has also been degraded. More systematic data is needed before we can analyze in detail the DOS produced by photodegradation.

## CONCLUSIONS

Photodegradation significantly reduces the forward bias current and EL while not affecting the radiative lifetime. Further, transient space charge limited currents are decreased by at least one decade at short times as well as long. These dramatic effects suggest that these experiments can be used to study the Staebler-Wronski effect. Further experiments should enable us to determine the kinetics, density and energy of the photo-induced defects.

We acknowledge the support of SERI under sub-contract and also we are grateful for the samples supplied by Alan Delahoy at Chronar and the amorphous silicon group in SERI. Fruitful discussions were held with R. Crandall, H. Branz and A. Delahoy.

## REFERENCES

1. M. Stutzmann, W. B. Jackson and C. C. Tsai, Phys. Rev. B32, 23 (1985).
2. D. L. Staebler and C. R. Wronski, Appl. Phys. Lett., 31, 292 (1977).
3. J. I. Pankove and J. E. Berkeyheiser, Appl. Phys. Lett., 37, 705 (1980).
4. Keda Wang, Daxing Han, M. E. Zvanut and M. Silver, Phil. Mag. B63, 175 (1991).
5. E. Snow and M. Silver, J. Non-Cryst. Sol., 77/78, 451 (1985).
6. M. A. Lampert and P. Mark, 'Current Injection in Solid' (Academic Press, N.Y. 1970).
7. K. Wang, D. Han, M. Silver and H. M. Branz, to be published in Solar Cells, 1991.
8. M. Silver, E. Snow and D. Adler, Solid State Comm. 53, 637 (1985).
9. M. E. Zvanut, K. Wang, D. Han and M. Silver, MRS/symposium proc. 192, 305 (1990).

# INTERFACE STATES IN a-Si:H AS PROBED BY
# OPTICALLY INDUCED ESR

J. Hautala and P.C. Taylor
Department of Physics, University of Utah, Salt Lake City, UT 84112

J. Ristein
Institut für Technische Physik, Universität Erlangen
Erwin-Rommel-str. 1, 825 Erlangen, FR Germany

## ABSTRACT

Using low temperature above- and below-gap excitation of light induced electron spin resonance (LESR) large densities ($> 10^{17}$ cm$^{-3}$) of charged midgap states are shown to be associated with the interfaces and/or surfaces of a-Si:H films. In all cases (films of 5, 10 and 15 $\mu$m thicknesses) the below-gap LESR produces the same asymmetry in the intensities of the electron and hole lines as we reported earlier, and the induced spin densities scale with the number of interfaces within $\pm 15\%$. When considered as bulk spin densities, the results vary by more than a factor of three.

## INTRODUCTION

Normal LESR on a-Si:H with above-bandgap energy excitation produces two easily distinguishable lines in equal intensities. These lines are attributed to electrons trapped in conduction band-tail states and holes trapped in valence band-tail states. In recent measurements using below-gap light remarkably large densities of LESR were produced, and interestingly the hole resonance was suppressed by a factor of two to three, suggesting the presence of large densities of charged defect states in the gap. We have investigated this phenomenon in much thicker films (5, 10 and 15 $\mu$m) in order to verify the conjecture that the presence of the charged centers is due to the effects of band bending in surface regions of the films and not from charged states distributed throughout the bulk.[2] Low temperature, transient optically-induced electron spin resonance, or LESR, has long been used to probe the nature of electrons and holes trapped in band-tail states of a-Si:H.[3] All previous experiments have employed photo-excitation with sufficient energy to permit band-to-band excitations. In these experiments two LESR lines are observed with equal intensities, a narrow line centered at g = 2.004 that is attributed to electrons trapped in the conduction band-tail states and a broad line centered at g = 2.013 that is attributed to holes trapped in valence band-tail states. With IR excitation one-photon, band-to-band carrier excitation is precluded, and only transitions into and/or out of deep defect states are possible. The distinction between the optically produced transient paramagnetic holes and electrons is straightforward with LESR, and thus provides a critical test of models of defect states in a-Si:H.

In the commonly accepted model for device-quality a-Si:H, neutral dangling bonds, or $D^o$ states, represent the only deep defect level with a concentration of $10^{15}$ to $10^{16}$ cm$^{-3}$. Electrons can be optically excited into $D^o$ states producing doubly occupied dangling bond states (D$^-$). A positive correlation energy, U, of 0.2 to 0.4 eV is estimated for this defect level.[4] Electrons can also be optically excited out of singly occupied dangling bond states leaving positively charged unoccupied states (D$^+$). These transitions have been observed as a sub-band-gap absorption shoulder by photothermal deflection spectroscopy (PDS)[5] and constant photocurrent method (CPM).[6] The difference in the absorption shoulder for a-Si:H using these two techniques indicates the possible presence of surface and/or interface states as important defects.[7] Photoluminescence absorption spectroscopy (PLAS) shows a resolved absorption peak at 1.15 eV, which is not observed by PDS or CPM studies, which also suggests transitions between energetically distinct levels within the gap may be associated with interface states.[8]

Previous LESR experiments on a-Si:H using sub-gap energy light showed a suppression of transient optically produced holes with respect to electrons and neutral silicon dangling bonds and suggested the presence of a large density of filled D$^-$ states near the substrate interface due to band bending. The data were interpreted by Branz[2] as indicative of a bulk property, which was suggested to arise from the existence of large concentrations of charged defect states (D$^+$ and D$^-$) produced by electrostatic potential fluctuations within the material. The results of the experiments presented here, however, clearly support the model that a large density ($>10^{17}$ cm$^{-3}$) of charged defects is associated with the interface and/or surface and not the bulk. There is no evidence, however, that would exclude material inhomogeneities produced in the first $\sim$ 1000 Å of film deposition as the source for the charge fluctuations and charged defects predicted by Branz[2] for the bulk.

## EXPERIMENTAL DETAILS

Samples of a-Si:H with thicknesses of 5, 10 and 15 $\mu$m were deposited by the glow-discharge technique using a silane source. Samples were prepared on 4 mm $\times$ 20 mm $\times$ 0.1 mm ESR grade quartz substrates in consecutive runs by Glasstech Solar Inc. Three 10 $\mu$m and two 5 $\mu$m films were stacked together to yield effective thicknesses of 30 $\mu$m and 10 $\mu$m, respectively. A single 15 $\mu$m film was also used.

The three film samples were directly immersed in a commercial helium flow cryostat which was maintained at 40 K for all experiments. No heating effects were observed with the 2.0 W IR light focused on the films. All ESR measurements were performed at 40 K using an optical transmission cavity in a standard Bruker X-band spectrometer.

Samples were excited directly by a 15 mW 633 nm HeNe laser or by 1.06 $\mu$m light from a cw Nd:YAG laser through a glass fiber. Care was taken to use all the same experimental parameters as in Ref. 1, where proper temperature, microwave power, modulation amplitudes and time constants were carefully determined. It should also be noted that $> 99\%$ of the 1.06 $\mu$m light transmits through even the thickest 30 $\mu$m stack of films because $\alpha \approx 1.5$ cm$^{-1}$ for 1.06 $\mu$m light. Therefore, all films were assumed to be exposed to the same intensity of light.

## EXPERIMENTAL RESULTS

The LESR spectra of Fig. 1 show both the original derivative spectra and numerical integrations of those derivatives, which yield the actual microwave absorption spectra for all three sets of samples. In all spectra the "dark" ESR signal measured before exposure to light has been subtracted off. Figure 1(a) is the LESR signal obtained with the $3 \times 10$ $\mu$m films after exposure to above-bandgap light (633 nm). This spectrum is virtually identical with those which are observed in the 5 and 10 $\mu$m samples and with those commonly observed by others when intrinsic films of a-Si:H have been irradiated with above-bandgap light.[1,3] The important feature to note is the equal intensity of the broad hole line ($n_b$) and narrow electron line ($n_n$) as suggested by a casual inspection of the absorption spectrum in Fig. 1(a).

In contrast to Fig. 1(a), the IR LESR spectra in Figs. 1(b), 1(c) and 1(d) (5, 10 and 15 $\mu$m samples, respectively) exhibit a clear suppression of the broad hole line relative to the narrow line ($n_n/n_b \approx 2$ to 3). This result is in exact accordance with that which was seen by Ristein et al. earlier.[1] Experiments were run by using independently varying microwave and light excitation intensities in order to eliminate any saturation effects. The relative decrease in intensity of the broad LESR line is therefore interpreted as a genuine suppression of light-induced holes relative to light-induced electrons in the conduction band-tail states and neutral dangling bonds (D$^o$ states).

The calculated spin densities for all measured dark ESR and IR LESR spectra are shown in Table I. The numbers were calculated by doubly integrating the derivative spectra, taking into account the effective sample size and the effective volume of the irradiated samples. Surface densities were calculated assuming only one surface per film, and (1000 Å) densities indicate densities of defects if all the measured defects were produced during the first 1000 Å of film deposition. The $n_n$ and $n_b$ densities were calculated assuming a 1000 Å interface layer.

## DISCUSSION

The primary purpose of this research was to determine whether the LESR states produced with sub-bandgap light are associated with the surfaces or with the bulk of the material. The results listed in Table I strongly suggest these defects

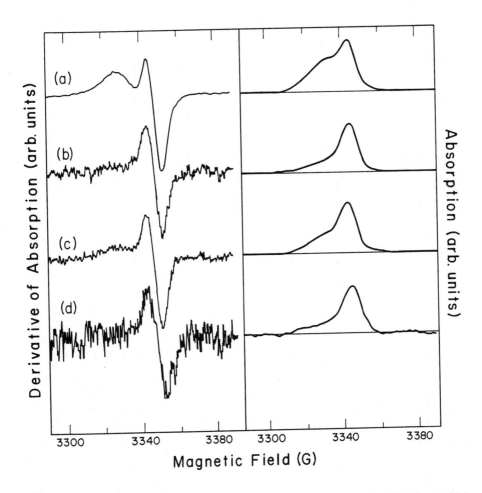

Figure 1. LESR derivative spectra for red- and IR-light excitation (left-hand side) along with the first integrals (right-hand side) of intrinsic a-Si:H 5, 10 and 15 $\mu$m films at 40 K (0.2 W cm$^2$ of $\lambda$ = 633 nm). Spectrum (a) above-bandgap excitation of the 3 × 10 $\mu$m films and exhibits a one-to-one ratio of broad-to-narrow, or holes-to-electrons intensities. Spectra (b), (c) and (d) are the IR-light (below-bandgap) excitation for the 2 × 5 $\mu$m, 3 × 10 $\mu$m and 1 × 15 $\mu$m samples, respectively, using 50 W cm$^{-2}$ of $\lambda$ = 1064 nm light. A clear suppression of the broad, or hole, intensity is observed for all IR LESR spectra. Identical ESR parameters were used for all samples. The spectrum (a) is indistinguishable from above-bandgap excitation spectra of the 5 and 15 $\mu$m samples.

## TABLE I

### ESR and LESR Results on Glasstech Solar Inc. Intrinsic a-Si:H Samples

| | ESR | | | |
|---|---|---|---|---|
| | 3 × 10 $\mu$m | 2 × 5 $\mu$m | 1 × 15 $\mu$m | |
| Spins | 3.2 | 2.4 | 1.3 | × $10^{12}$ |
| Bulk density | 1.8 | 2.0 | 2.4 | × $10^{15}$ cm$^{-3}$ |
| Surface density | 1.8 | 2.0 | 2.2 | × $10^{12}$ cm$^{-2}$ |
| (1000 Å) | (1.8) | (2.0) | (2.2) | × $10^{17}$ cm$^{-3}$ |

| | LESR    $\lambda$ = 1064 nm | | | |
|---|---|---|---|---|
| | 3 × 10 $\mu$m | 2 × 5 $\mu$m | 1 × 15 $\mu$m | |
| Spins | 5.7 | 3.1 | 1.4 | × $10^{12}$ |
| Bulk density | 1.5 | 2.5 | 0.74 | × $10^{16}$ cm$^{-3}$ |
| Surface density | 1.5 | 1.2 | 1.1 | × $10^{13}$ cm$^{-2}$ |
| (1000 Å) | (1.5) | (1.2) | (1.1) | × $10^{18}$ cm$^{-3}$ |
| $n_n$ | (10) | (9) | (8) | × $10^{17}$ cm$^{-3}$ |
| $n_b$ | (5) | (3) | (3) | × $10^{17}$ cm$^{-3}$ |

Calculated spin densities for all measured dark ESR and IR LESR spectra. All numbers were calculated by double integration of the derivative spectra, taking into account the effective sample size and the effective volume of the irradiated samples. Surface densities were calculated using only one surface per film. Assuming a > 1000 Å of band bending in to the bulk from the interface, defect densities extending from 1000 Å from the interface are also calculated. The $n_n$ and $n_b$ densities were calculated assuming a 1000 Å interface layer. Data clearly show the defect densities scale with the surfaces rather than the bulk.

are predominantly contained at or near the interfaces and/or surfaces and are **not** a bulk property. Scaling to a single surface per film the IR LESR data show an average surface density of $1.3 \times 10^{13}$ cm$^{-2}$ with $\pm$ 15% error. If scaled as a bulk density, an average of $1.6 \times 10^{16}$ cm$^{-3}$ is obtained with an error of $\pm 55\%$. If it is assumed that most of these defects are contained in the first 1000 Å of film deposition, then an average defect density of $1.3 \times 10^{18}$ cm$^{-3}$ $\pm 15\%$ is obtained.

As stated earlier, the relative intensities of broad and narrow lines can be separated by analyzing the integrated spectra, and in this manner, the ratio $n_n/n_b \approx 3$ has been determined, which means $n_b \sim 3 \times 10^{17}$ cm$^{-3}$ and $n_n \sim 9 \times 10^{17}$ cm$^{-3}$ if the defects are assumed to be contained in the first 1000 Å. These values exceed those bulk densities in good quality films by two orders of magnitude.

The ESR line-shape parameters for a singly occupied dangling bond ($D^0$) state that comprises the dark ESR signal in a-Si:H are g = 2.0055 and $\Delta \approx$ 5 Gauss for the g value and linewidth (peak-to-peak width of the derivative spectrum), respectively. Electrons that are optically excited into the conduction band rapidly thermalize into localized band-tail states and conversely, the holes are thermalized into localized valence band-tail states. Transition 1 in Fig. 2(a) shows this optical transition which requires above bandgap energy radiation, and always results in a 1:1 ratio of electron-to-hole LESR intensities. The ESR line-shape parameters are g = 2.0048 and $\Delta \approx 4$ Gauss for the electron and g = 2.011 and $\Delta \approx 18$ Gauss for the hole as shown in Fig. 1(a).

Unfortunately, the line-shape parameters for the dangling bond are too close to those of the tail state electrons to distinguish them in a LESR spectrum. Thus only two distinct lines can be resolved. Electrons in tail states and dangling bonds ($D^0$ states) show up as a common narrow line ($n_n$), and holes are easily distinguished as a broader resonance with a higher g value ($n_b$).

As mentioned by Ristein et al.[1] and Branz[2] the IR LESR data in general, and the $n_n/n_b \neq 1$ asymmetry especially, provide crucial tests for defect models in a-Si:H. Figure 2 shows the possible transitions allowed using the 1.17 eV radiation for both the usually accepted model (a) and the potential fluctuation model (b) after Ref. 2.

All possible one- (transition 4) or two-step (transitions 2 followed by 2' or 3 followed by 3') processes in Fig. 2(a) result in $n_n/n_b = 1$, and therefore this standard picture cannot explain the $n_n/n_b \approx 3$ observed by IR LESR. Transition 4 is improbable because of small densities of highly-localized tail states in this energy range.

Several possibilities exist to explain the asymmetry of narrow-to-broad IR LESR intensities. A filling of the $D^-$ states in a region of the films near the substrate interface due to band bending is one plausible explanation. If, for example, the conduction band is "bent" down by $> 0.3$ eV (roughly the coulomb repulsion energy) over a 1000 Å region from the interface with the quartz substrate and $> 10^{17}$ cm$^{-3}$ $D^-$ states exist, then transition 3' in Fig. 2(a) becomes possible producing a dangling bond and a tail-state electron, both of which contribute to $n_n$. Broad or band-tail "hole" lines could be produced by one of the two-step processes

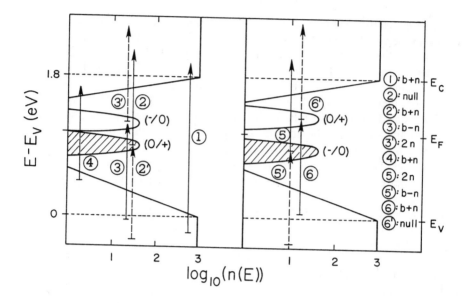

Figure 2. Schematic density of states diagram for (a) the usually accepted model of a-Si:H, and (b) the charge fluctuation model after Ref. 2. Levels filled with an electron are shaded. Single-photon excitations are indicated by solid arrows, and the second steps of two-step excitations are indicated by dashed arrows. Transition 1 in (a) is from 633 nm light, all other transitions are from 1064 nm light. The excitation-induced changes in narrow (n) and broad (b) spin resonances are indicated. Both a dangling bond ($D^0$ state) and an electron trapped in the conduction band-tail states are considered as contributing to a single narrow (n) resonance. Energetic positions of the defect states are determined by the energy required to excite an electron <u>out</u> of that state; thus (0/+) in (a) represents a filled $D^0$ level and (-/0) is an (empty) $D^-$ level, and in (b) (-/0) represents a filled $D^-$ level and (0/+) is an empty $D^+$ level.

discussed earlier assuming a significant number of $D^0$ states remain in the interface region.

A second possible explanation for $n_n/n_b \approx 3$ in the films is presented in Fig. 2(b), assuming that the argument of Branz[2] can be restricted to regions close to the interfaces. The basic argument suggests that inhomogeneities in the a-Si:H produce charge fluctuations that locally shift the band edges such that even though a positive U is assumed, large numbers of $D^-$ states exist lower in energy than nearby $D^0$ states, and conversely, $D^+$ states exist which are lower in energy in other regions. Under these conditions transitions 5 and 6 in Fig. 2(b) are possible producing a combined $n_n/n_b = 3$. Transitions 5' and 6' are predicted to be weak because they depend on transitions 5 and 6 taking place first and the matrix elements are smaller for transitions involving neutral defects.

A third possibility consists of oxygen-induced surface defect states that have been measured in large densities ($> 10^{12}$ cm$^{-2}$) when a-Si:H is exposed to $< 10^{-3}$ Torr of oxygen at 293 K.[9] These authors[9] suggest the created defects consist of singly and doubly occupied dangling Si bonds originating from three-fold coordinated oxygen.

## CONCLUSIONS

The IR LESR results on 5, 10 and 15 $\mu$m a-Si:H samples confirm that large densities ($> 10^{17}$ cm$^{-3}$) of charged defect states are associated with the interfaces and/or surfaces of a-Si:H films.

## ACKNOWLEDGEMENTS

H. Chatham, P. Bhat and GlassTech Solar are gratefully acknowledged for supplying the samples.

## REFERENCES

1.  J. Ristein, J. Hautala and P.C Taylor, Phys Rev. **B40**, 88 (1989).
2.  H.M. Branz, Phys. Rev. **B39**, 5107 (1989).
3.  R.A. Street and D.K. Biegelsen, J. Non-Cryst. Solids **35+36**, 651 (1980); A. Friederich and D. Kaplan, J. Electron. Mater. **8**, 79 (1979).
4.  H. Dersch, J. Stuke and J. Biechler, Phys. Status Solidi **B105**, 265 (1981).
5.  W.B. Jackson, N.M. Amer, A.C. Boccara and D. Fournier, Appl. Opt. **20**, 1333 (1981).
6.  M. Vanecek, J. Kocka, J. Stuchlik, Z. Koziseh, O. Strick and A. Triska, Sol. Energy Mater. **8**, 41 (1983).
7.  K. Pierz, B. Higenberg, H. Mell and G. Weiser, J. Non-Cryst. Solids **97+98**, 63 (1987).
8.  F. Ranganathan and P.C. Taylor, J. Non-Cryst. Solids **97+98**, 707 (1987).
9.  K. Winer and L. Ley, Phys. Rev. **B37**, 8363 (1988).

# DEPENDENCE OF THE SATURATION BEHAVIOUR OF THE METASTABLE DEFECT CREATION ON a–Si:H MATERIAL PROPERTIES MEASURED BY keV ELECTRON IRRADIATION

A. Scholz and B. Schröder

Fachbereich Physik, Universität Kaiserslautern, Erwin–Schrödinger–Str.,
D–6750 Kaiserslautern, FRG

## A B S T R A C T

The electronic stability of a–Si:H films deposited by different methods and a wide range of preparation parameters has been investigated by keV–electron irradiation. Employing an electron dose of about $70 J/cm^2$ a metastable defect density near its saturation level was created. The defect density of the intial and the irradiated state has been determined by CPM and PDS. It was found that the saturation level of the creation of metastable defects $\Delta N_{sat}$ is independent on the previous (stable) defect density $N_0$ within the experimental inaccuracy. Meanwhile an evident increase of $\Delta N_{sat}$ with the hydrogen content $c_H$, the $c_H$–fraction bonded at inner surfaces, and the optical gap was observed. Obviously, there exists a dependence of $\Delta N_{sat}$ on the preparation temperature. The saturation behaviour of the films does not always correlate with that of pin solar cells prepared with the same i–material.

## INTRODUCTION

Metastable defects are created in hydrogenated amorphous silicon (a–Si:H) by different treatments of the material such as light illumination[1,2], keV–electron irradiation[3,4], inducing excess charges (e.g. by doping)[5], and thermal quenching[6]. Since the formation of metastable defects in the intrinsic or doped material, respectively, is of fundamental scientific interest and also strongly influences the device applications of a–Si:H material, numerous investigations were carried out to obtain a better understanding of this primary so–called Staebler–Wronski effect.

Lee and co–worker[7] first reported on saturation of the metastable, optically induced ESR dangling bond signal after prolonged exposure to band gap light ($3W/cm^2$). A clear indication of the saturation behaviour was also found by keV–electron irradiation[8,9]. Although there are other statements[10], not only keV–electron irradiation[11] but also light soaking experiments[12] show that the saturation behaviour or level, respectively, is dependent on the irradiation intensity and the temperature at which the perturbation takes place. This fact indicates that the saturation is due to a balance between defect generation and annealing.

In this paper we investigate the saturation level of the metastable defect creation of a–Si:H films deposited by different methods and a wide range of preparation parameters using the method of keV–electron irradiation. We are convinced that the saturation behaviour or level, respectively, is one of the most reliable quantities to characterize the stability of amorphous semiconductor materials. Of course, it must be taken into account that the saturation level is dependent on the defect creation method, the creation rate, and the temperature at which the creation takes place.

## EXPERIMENTAL

The keV–electron irradiation was carried out in a scanning electron micro-scope. The experimental arrangement and the detailed procedure are published elsewhere[4]. The electron energy was 20keV. At this energy a quite homo-geneous energy dissipation of the irradiating electrons is guaranteed within the a–Si:H films of thicknesses around $1\mu m$. Therefore, a homogeneous distribution of the defect creation can be expected. The irradiation was performed at room temperature. The heating of films due to the electron bombardment was negli-gible since the irradiation was performed with defocussed electron beam ($j_0 \approx 100\mu A/cm^2$) in the scanning mode. An electron dose of $70J/cm^2$ was employed into the films resulting in a defect density near its saturation level[4]. The a–Si:H films investigated have been prepared by dc magnetron sputtering technique and rf glow discharge decomposition of silane. The deposition para-meters were varied in a wide range to obtain a–Si:H films with a wide spread of properties. Nevertheless, the preparation was not a random one but some optimization was performed to minimize the initial defect density for the different deposition conditions. The defect density of the initial and the irradi-ated films has been determined by photoconductivity spectroscopy in the constant photocurrent mode (CPM) and by photothermal deflection spectro-scopy (PDS).

## RESULTS

In fig. 1 the dependence of the metastable saturation defect density $\Delta N_{sat}$ on the initial (stable) defect density $N_0$ is depicted. The squares represent sput-tered, the circles glow discharge samples. The solid line represents the depen-dence found by Stutzmann[13] according to his defect creation model for small

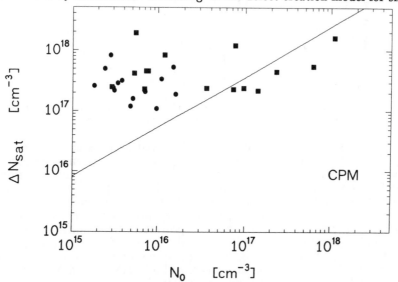

Fig. 1:    *Dependence of the metastable saturation defect density $\Delta N_{sat}$ on the initial defect density $N_0$. The solid line represents the dependence found by Stutzmann[13] for light soaking.*

intensity light soaking. The CPM results indicate that $\Delta N_{sat}$ is obviously independent on $N_0$ within the limit of the statistical fluctuations of the results $10^{17} \leq \Delta N_{sat} \leq 2 \cdot 10^{18}cm^{-3}$. The evaluation of the PDS results leads to the same result. Certainly the PDS results are not so powerful as the $N_0$ obtained in this case vary only between $2 \cdot 10^{16}$ and $5 \cdot 10^{17}cm^{-3}$. In fig. 2 the dependence of $\Delta N_{sat}$ and $N_0$ (both determined by CPM) on the total hydrogen content is shown. As marked by the dashed lines sputtered samples obviously need $c_H > 10at\%$ to reach their minimum $N_0$ while gd–a–Si:H can be prepared with $N_0 < 10^{16}cm^3$ for $c_H \geq 3.5at\%$. Independent on the preparation method the $\Delta N_{sat}$ values weekly increase with $c_H$. In contrast to the CPM results the evaluation of the PDS measurements indicates a stronger increase of $\Delta N_{sat}$ with $c_H$, but the $N_0$–values also increase with $c_H$. Fig. 3 exhibits the variation of $\Delta N_{sat}$ and $N_0$ with the optical gap $\Delta E_{opt}$. These results can also be attributed to the dependence of $\Delta E_{opt}$ on $c_H$ (see discussion). The increase of $\Delta N_{sat}$

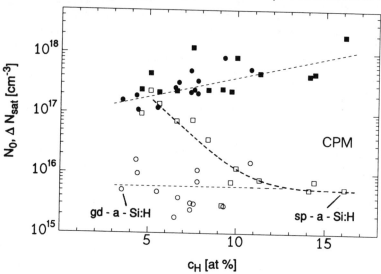

Fig. 2:    *Dependence of the metastable saturation defect density $\Delta N_{sat}$ and the initial defect density $N_0$ on the total hydrogen content.*

with $\Delta E_{opt}$ (dashed line) is similar to that observed by Park et al.[14] for light soaking (solid line). In fig. 4 the change of $\Delta N_{sat}$ and $N_0$ is shown which can be attributed to the fraction of $c_H$ bonded in surface–like bonding configurations. This fraction was determined by evaluation of IR absorption measurements. Due to detailed investigations of Beyer and Wagner[15] the integral absorption of the $2090cm^{-1}$ stretching vibrational band can be correlated with the hydrogen fraction which is bonded in clustered monohydride (SiH) or dihydride (SiH$_2$) configurations at inner surfaces (voids). Due to the large fluctuations of $N_0$ the statement given by the result in fig. 4 is not very powerful. Also the evaluation of the PDS data which result in an increase of both, $\Delta N_{sat}$ and $N_0$, with the ratio $c_H (2090)/c_H$ (total) does not give more clearness.

To investigate the influence of the deposition rate R and the deposition temperature $T_S$ on the saturation defect density $\Delta N_{sat}$ special sets of glow discharge samples have been prepared by the PST company[16]. The preparation was carried out in such a manner that as far as possible only $T_S$ and R are varied but all other parameters are kept constant. Parallel to the separate film deposition a standard solar cell was deposited with the same i–layer material.

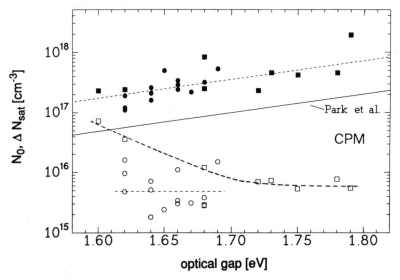

Fig. 3: *Dependence of the metastable saturation defect density $\Delta N_{sat}$ and the initial defect density $N_0$ on the optical gap $\Delta E_{opt}$.*

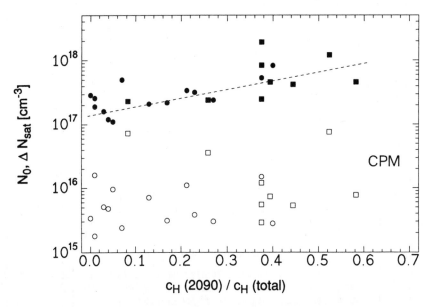

Fig. 4: *Dependence of the metastable saturation defect density $\Delta N_{sat}$ and the initial defect density $N_0$ on the ratio of the hydrogen content in surface–like bonding configurations $c_H(2090cm^{-1})$ to the total hydrogen content $c_H$.*

As shown in fig. 5a, $\Delta N_{sat}$ decreases with increasing $_{TS}$ and goes through a flat minimum near $T_S \approx 250^{\circ}C$. In the same temperature range $N_o$ increases (CPM) or stays constant (PDS). On the other hand $\Delta N_{sat}$ and $N_o$ (both from CPM and PDS measurements) are independent on the deposition rate within $1.5 \leq R \leq 6$ Å/s. The PDS and the CPM results are very similar except that the $N_o$ level obtained with PDS is higher and there is no minimum in the temperature

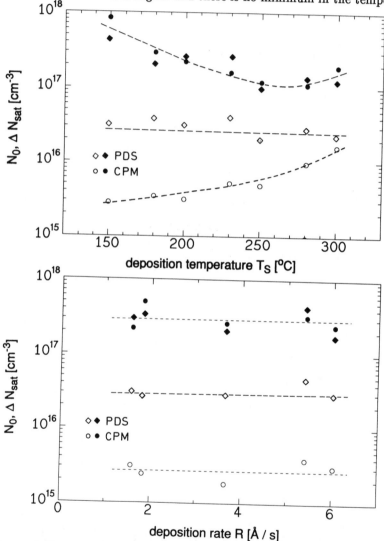

Fig. 5:    *Influence of the deposition temperature $T_S$ (a) and the deposition rate R (b) on the metastable saturation defect density $\Delta N_{sat}$ and the initial defect density $N_o$ (see text for details)*

dependence of $\Delta N_{sat}$ in the PDS measurement. The relative change of the conversion efficiency $\eta$ due to 1000h or 500h, respectively, AM1 light soaking of solar cells are depicted in fig. 6. The i–layers of the solar cells are simultaneously prepared to the films, whose results are shown in fig. 5. Concerning the temperature dependence, a good correlation between the $\Delta N_{sat}$ and $\eta$ changes can be stated. As expected from the decrease of $\Delta N_{sat}$ with increasing deposition temperature the stability of the appropriate solar cells is enhanced. Of course, the maximum stability of the solar cells has been found for $T_S \approx 200^\circ C$ which lies about 50 degrees below the minimum of $\Delta N_{sat}$. Since unfortunately the deposition temperature of the solar cells could not be raised above $220^\circ C$ (due to incipient thermal instability of the $TCO/p^t$ interface) the

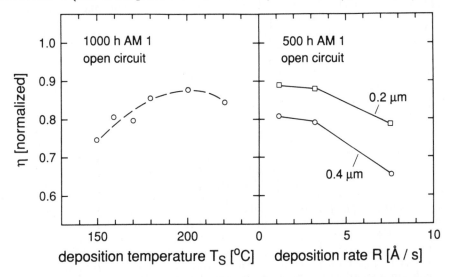

Fig. 6:    *Reduction of the conversion efficiency $\eta$ of pin solar cells after long term light soaking as a function of deposition temperature $T_S$ (left) and deposition rate R (right)*[16]

flat maximum observed should not be overinterpreted. In contrast to the deposition temperature dependence, the results of films and solar cells do not fit together in regard to the deposition rate dependence. While $\Delta N_{sat}$ is surprisingly independent on R, the stability of the solar cells clearly decreases with increasing R.

## DISCUSSION

As already mentioned in the introduction, the metastable saturation defect density is dependent on the defect creation method, the creation rate, and the temperature at which the creation takes place. Therefore the results reported in the chapter above should be only qualitatively compared to results obtained by other degradation methods. The statement just mentioned answers already the very important question concerning the origin of the metastability. Due to the dependence of $\Delta N_{sat}$ on the irradiation temperature, the annealing behaviour of $\Delta N_{sat}$ or $\eta$ after keV–irradiation [11,17] and the femto to second laser soaking behaviour observed by Stutzmann[18] (high pulse power, AM1 average) we conclude that the metastable saturation defect density represents a balance between defect generation and annealing. This statement is in agreement with new light soaking results of Santos and co—workers[12]. The dependence of the metastable saturation defect density on temperature and illumination intensity can be explained in the corresponding paper by a chemical equilibrium model for defect density under illumination. The experimental data mentioned exclude other creation models (see e.g. ref. 19) which propose that saturation is caused be depletion of defect creation sites. Taking the foregoing discussion into account, the independence of $\Delta N_{sat}$ on $N_0$ (see fig. 1) is a direct consequence. The correlation reported by Stutzmann[13] (see solid line in fig. 1) only represents the very special case of defect creation at low light intensity.

Hydrogen content and hydrogen bonding configuration were often discussed in connection with the metastability of the a-Si:H material. The correlation observed (s. fig 2 and 3) between $\Delta N_{sat}$ and $c_H$ (total) or the ratio $c_H$ $(2090cm^{-1})/c_H$ (total), respectively, is not very strong. Additionally it has to be taken into account that the material properties $c_H$, the ratio $c_H$ $(2090cm^{-1})/c_H$ (total) and $\Delta E_{opt}$ are completely interconnected to each other. Therefore it cannot be clearly decided which of the quanitities is responsible for the increase of $\Delta N_{sat}$. Since the concentration of hydrogen in fig. 2 only varies between $2 \cdot 10^{20} \lesssim c_H \lesssim 8 \cdot 10^{20} cm^{-3}$ and these concentrations are more than two orders of magnitude larger than $\Delta N_{sat}$ it is hard to believe that this always dominating quantity is responsible for the $\Delta N_{sat}$ increase. From this point of view we would assume that the change of the energy gap due to $c_H$ variations is the reason for different generation and annealing processes. On the other hand experimental deposition methods like hot wire CVD[20] and chemical annealing[21] enable the preparation of good quality a—Si:H material with $c_H \approx 2 \cdot 10^{19} cm^{-3}$ ($c_H \lesssim 0.5at\%$). The first results given concerning these low—$c_H$ materials certify a much higher stability although $c_H$ is still much larger than $\Delta N_{sat}$. Another fact may be taken into account. a—Ge:H material of high stability was reported to contain a significantly reduced amount of small voids[22]. Therefore the reduction of small voids may improve also the a—Si:H stability. It should be mentioned that for a given volume small voids contain much more hydrogen than large voids. The avoidance of small voids by a special preparation method should enable a drastical reduction of the hydrogen concentration.

In good agreement between film (see fig. 5) and solar cell degradation results (see fig. 6) a deposition temperature $T_S \approx 200^oC$ seems the optimal one for best stability. From the dependence $\Delta N_{sat} = f(T_S)$ and assuming a general annealing improvement at higher $T_S$ we would expect that temperatures $T_S >$ 250 are favorable to obtain the highest stability not only for films but also for solar cell devices. Due to the limitation of the device stability which is not connected to the a—Si:H metastability this assumption could not yet be proved in a solar cell device.

In contrast to the temperature dependence, the deposition rate dependence of $\Delta N_{sat}$ and the solar cell degradation are not correlated. The independence of $\Delta N_{sat}$ (film) on R must be attributed to the fact that $\Delta N_{sat}$ is mainly determined by the bulk properties which do not significantly change with R. In contrast to the film the device quality and stability are determined by interface regions, too. The increase of R strongly influences especially a thin i—layer near the p—i—interface[23]. These changes may heavily influence the degradation of the fill factor or open circuit voltage but stays within the measurement accuracy of $\Delta N_{sat}$ films.

## CONCLUSION

The independence of the metastable saturation defect density $\Delta N_{sat}$ on the initial defect density $N_o$ measured by keV–electron irradiation is shown to be in agreement with origin of the metastability as a balance between defect generation and annealing. The observed dependence of $\Delta N_{sat}$ on the hydrogen concentration $c_H$ is interpreted in connection with the recently reported improved stability of low–$c_H$ a–Si:H material. Correlations found or missing between film ($\Delta N_{sat}$) and device degradation behaviour are discussed.

This research work was supported by the German Ministry of Research and Technology (BMFT)

## REFERENCES

1.   D.L. Staebler and C.R. Wronski: Appl. Phys. Lett. **31**, 292 (1977)
2.   H. Dersch, J. Stuke, and J. Beichler: Appl. Phys. Lett. **38**, 456 (1980)
3.   C. Wagner, S. Gangopadhyay, B. Schröder, and J. Geiger: AIP Conf. Proc. **157**, 46 (1987)
4.   U. Schneider and B. Schröder: in Amorphous Silicon and Related Materials, ed. H. Fritsche, Vol. 1A (World Scientific Publ., Singapore 1989) p. 687
5.   R.A. Street: Phys. Rev. Lett. **49**, 1187 (1982)
6.   R.A. Street, J. Kakalios, C.C. Tsai, and T.M. Hayes: Phys. Rev. **B35**, 1316 (1987)
7.   C.Lee, W.D. Ohlsen, P. Taylor, H.S. Ullal, and G.P. Ceasar: Phys. Rev. **B31**, 100 (1985)
8.   U. Schneider, B. Schröder, and F. Finger: J.Non–Cryst. Solids **97&98**, 795 (1987)
9.   U. Schneider and B. Schröder: Solid State Commun. **69**, 895 (1989)
10.  H.R. Park, J.Z. Lui, and S. Wagner: Appl. Phys. Lett. **55**, 2658 (1989)
11.  U. Schneider, B. Schröder, and F. Finger: J. Non–Cryst. Solids **114**, 633 (1989)
12.  P.V. Santos, W.B. Jackson, and R.A. Street: this Proceeding Volume
13.  M. Stutzmann: Phil. Mag. **B60**, 531 (1989)
14.  H.R. Park, J.Z. Liu, P. Roca i Cabarrocas, A. Maruyama, M. Isomura, S. Wagner, J.R. Abelson, and F. Finger: MRS Conf. Proc. Vol. **192**, p. 751 (1990)
15.  W. Beyer and H. Wagner: J. Non–Cryst. Solids **59/60**, 161 (1983)
16.  The film and solar cell preparation and the solar cell degradation experiments were carried out by Phototronics Solar Technology Corporation, Putzbrunn, FRG
17.  U. Schneider, B. Schröder, and P. Lechner: J. Non–Crystal. Solids **115**, 63 (1989)
18.  M. Stutzmann, priv. Communication
19.  D. Redfield: Appl. Phys. Lett. **48**, 846 (1986)
20.  A. H. Mahan and M. Vanacek: see this Proceeding Volume.
21.  H. Shirai, D. Das, J. Hanna, and I. Shimizu: see this Proceeding Volume
22.  R. Plättner, E. Günzel, G. Scheinbacher, and B. Schröder: see this Proceeding Volume
23.  J. Sopka, U. Schneider, B. Schröder, and H. Oechsner: IEEE Transactions on Electron Devices **36**, 2848 (1989)

# Accuracy of Defect Densities Measured by the Constant Photocurrent Method

N.W. Wang, X. Xu and S. Wagner
Department of Electrical Engineering, Princeton University
Princeton, NJ, 08544

## ABSTRACT

We report a study of the accuracy with which the Urbach energy and defect density of a-Si:H and a-Si,Ge:H alloys may be extracted from optical absorption spectra determined by the constant photocurrent method. Surprisingly, a great part of the discrepancy results from evaluation of the CPM spectra, rather than from the normalization of the spectra to an absolute scale of the optical absorption coefficient. We will discuss the sources of error in several existing methods of data analysis and the adaptability of these methods for studying alloys. As well, we propose a new method of analysis, one in which we look at the excess subgap absorption over a range of energies.

## INTRODUCTION

As an optoelectronic technique which measures the bulk defect density, CPM provides a convenient means of characterizing a-Si:H and, in particular, is a useful tool in investigating the Staebler-Wronski effect. Studying the kinetics of this effect requires considerable accuracy, as one wishes to detect small changes in defect density. In a round-robin test conducted in our laboratory we have found variations by a factor of up to three in the defect density determined by different operators on the same sample.

In the constant photocurrent method, one shines light on a thin film of a-Si:H and varies its intensity such that a constant photocurrent is maintained while the energy of the light is swept. When the absorbance, $\alpha d$, where $\alpha$ is the absorption coefficient and d is the thickness of the sample, is much less than 1, the absorption coefficient of the material is inversely proportional to the intensity required for keeping the photocurrent constant. The data from this measurement are, however, not on an absolute scale and must be normalized to known absorption information for the sample. Commonly, they are normalized to absorption data from a transmission measurement.

For a-Si:H, the resulting absorption spectrum can be broken into two regions--an exponential region typically between 1.3 eV and 1.7eV and a shoulder occurring below 1.3 eV. For a-Si,Ge:H alloys with their lower bandgaps, the spectra also have this form but are scaled to lower energy. The absorption coefficient is exponential over the range of photon energies for which valence band tail to conduction band transitions can take place and is given by the Urbach relation, $\alpha(h\nu)=\alpha_o \exp(h\nu/E_u)$. At low energies the absorption is dominated by transitions between occupied defect states and conduction band states. From the slope of log $\alpha$ vs. energy in the exponential region one gets the Urbach energy and from the intensity of the shoulder region one infers the defect density. We will discuss the sources of error in the procedures for normalization and for extracting Urbach energy and defect density from the data.

## NORMALIZATION

The difficulty of normalizing CPM data by matching them to the absorption spectrum from a transmission measurement lies in that often the two spectra do not overlap at all (fig.1). The transmission data are bounded at low energies for two reasons: when the absorbance of the film is less than ~1%, the change in intensity between the incident and transmitted light is too small to measure; when the wavelength of the light is similar to the film's thickness, interference fringes distort the data. The range of energies over which CPM is useful, on the other hand, is bounded above by the optical gap of the material and by the breakdown of the assumption that the photocurrent is simply proportional to the absorption coefficient. To avoid error, it is necessary to realize where the useful regions of transmission data begin and CPM data end and not to use spurious data points in matching the two curves. In using solely the transmission data as the normalization reference, one is often forced to extrapolate from the CPM and transmission curves and use some guesswork to connect them. We have found it useful, therefore, to refer as well to the absorption spectrum measured by photothermal deflection spectroscopy (PDS) which overlaps with the regions of both the other techniques. Normalization based on comparing CPM to transmission and PDS data has proved very reproducible. When we use this method, most of the error in finding the defect density stems not from normalization but from analysis of the spectrum.

## ANALYSIS OF SPECTRUM

Though there are, as will be shown, numerous uncertainties in analyzing CPM data, the idea behind using optical absorption spectra for determination of defect density is simple. Were the material defect-free and to possess only strained bond states, the absorption coefficient would be expected to behave according to the Urbach relation for all energies less than the optical gap. That the measured absorption coefficient is larger than that predicted by the Urbach expression implies transitions from extra states near the center of the gap and reflects the presence of dangling bond defects in actual amorphous materials. The amount of excess absorption at subgap energies is related to the amount of defects in the sample. It should be noted that this relationship is particularly useful to studying stability. In stability experiments one is concerned with changes in defect density brought about by illumination or annealing. Such changes can be detected as relative shifts in the amplitude of the excess subgap absorption. Between samples of similar bandgap, it is also easy to tell which is relatively higher in defect density, simply by superimposing their spectra.

We do, however, need to extract the defect density more quantitatively. Several methods are employed for correlating the shape and magnitude of the absorption curve to the defect density. It is possible to deconvolve the absorption spectrum or to calibrate the defect density to either the so-called integrated excess subgap absorption or to the value of the absorption coefficient measured at a particular energy in the shoulder region.

**Deconvolution of the spectrum**: The absorption coefficient, $\alpha(E)$, is the integral over all possible transitions of energy E,

$$\alpha(E) = \frac{\text{const.}}{h\nu} \int N(\varepsilon)g(\varepsilon+E)d\varepsilon \qquad (1)$$

where $N(\varepsilon)$ is the density of initial states, that is, the density of occupied valence band, bandtail and defect states, and $g(\varepsilon)$ is the density of the final, conduction band states. In this expression the transition matrix element and refractive index have been assumed constant.  For photon energies less than the optical gap, only transitions from tail to conduction band and defect to conduction band states will contribute to $\alpha(E)$:

$$\alpha = \alpha_D + \alpha_T. \tag{2}$$

Here $\alpha_D$ is the absorption due to defect state transitions and $\alpha_T$ is the absorption due to tail state transitions.  If it is assumed that the conduction band density of states has the parabolic free electron form, that the valence bandtail density of states falls off exponentially with distance from the band edge, and that the defect distribution is Gaussian, then we have concrete functions with which to evaluate equation (1) numerically.  The absorption from defect transitions has the form

$$\alpha_D(E) = N_c \frac{A}{\sqrt{2\pi W^2}} \frac{const.}{E} \int exp\,[-\,(\varepsilon - E + E_i)^2 / 2W^2]\,d\varepsilon \tag{3}$$

where $N_c$ is the free electron value of the density of states at the conduction band edge, $E_i$ is the position of the Gaussian defect distribution relative to the conduction band edge, and $W$ is the halfwidth and $A$ the area of the Gaussian.  By making the area of the Gaussian defect distribution the variable parameter in a numerical fit of the actual absorption data to equation (3), one obtains the defect density.  Vanecek et al.[1] and Curtins and Favre [2] describe this method of deconvolution and list values they have chosen for parameters for doped and undoped a-Si:H.  Alloys, with their varying bandgaps, require different distribution widths and positions.  If these parameters are not known with great certainty,  there is considerable latitude in fitting the convolution integral to experiment.  The parameters need to be set or calibrated carefully, so this method is more complicated  and subject to assumption than is desirable in a routine measurement.

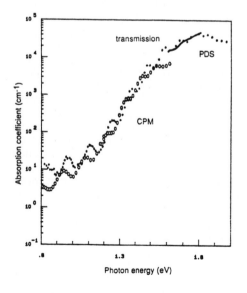

<u>Fig. 1</u>
$\alpha(h\nu)$ for an a-Si,Ge:H sample. The ranges of transmission and CPM data do not overlap, so normalizing by comparing to PDS <u>and</u> transmission is more accurate. The dotted line is PDS, the solid line is transmission, and the open circles are CPM data.

**Subgap absorption:** Another method of analysis looks at the excess, defect-related absorption, $\alpha_D$, which is the difference between measured absorption and the absorption extrapolated from the exponential portion of the spectrum. In the subgap absorption method, one assumes that $\alpha_D$, at a particular energy, is proportional to the defect density. The energy is chosen to be the photon energy high enough to induce all transitions from defects to the conduction band but too low to excite many valence tail to conduction band transitions. For a-Si:H, Vanecek chose 1.2 eV as the energy at which to calibrate.[3] This method works well for spectra that are flat at low energies, that is, for samples whose defects lie primarily outside the tail region. The spectra of a-Si,Ge:H alloys are often not of this type: they exhibit larger Urbach energies and slope steeply in the defect absorption regime (fig.3). As well, it is clearly not possible to choose a single energy at which to calibrate the excess absorption for evaluating a-Si,Ge:H alloys. This analysis procedure could work for alloys only if the position of the defect distribution were constant among alloys of the same optical gap and if that position could be predicted.

This method of using the subgap absorption at one specific photon energy depends on how accurately the line through the linear region of the log $\alpha$ vs. E plot, representing the tail state absorption, can be drawn. With the interference fringes that appear for ~1um thick films and the especially limited range of the exponential Urbach region for high defect density samples and alloys, it becomes difficult to determine the slope in the Urbach region of the plot with precision. Though the shape and the normalization of data measured repeatedly on the same sample change negligibly, the small changes that do result then cause variation in extrapolated $E_u$ that cannot be neglected. The necessity for precision in reading the tail state absorption will be less critical, however, in samples with smaller Urbach energies and for which the calibration energy, $h\nu_0$, is not very near the intersection of $\alpha(h\nu)$ and $\alpha_T(h\nu)$. For such samples, $\alpha_T(h\nu_0)$ will be much smaller than $\alpha(h\nu_0)$, so that $\alpha_D(h\nu_0) = \alpha(h\nu_0)$. Since a-Si:H samples tend to have smaller Urbach energies than a-Si,Ge:H alloys, this method is further seen to be better suited to the evaluation of a-Si:H than to alloys.

**Integrated subgap absorption:** The integrated excess subgap absorption method refers to the calibration of defect density to $\int \alpha_D dE$.[4] To integrate $\alpha_D$ is, in effect, to set the defect density proportional to its average over a range of subgap energies. This method encounters a few difficulties, especially when one applies it to evaluate alloys. A logical problem is that, since the absorption coefficient is itself the integral in equation (1), to then integrate $\alpha_D$ again is physically incorrect. A practical difficulty is that, as with the subgap absorption method, the analysis depends crucially on the way in which the line is drawn in the Urbach region of the log $\alpha$ vs. h$\nu$ plot, which is often imprecise. The result also depends on the energy range over which the integration is performed. In conducting a round-robin test in which five people measured the spectrum of a particular sample and evaluated the Urbach energy, we saw a discrepancy of up to 10 meV in Urbach energy. This discrepancy resulted in a spread in defect density of almost a factor of three and highlighted the loss of accuracy that occurs when several operators instead of one evaluate samples (fig. 2).

For the integrated excess subgap absorption approach to predict defect density with more precision than a factor of three, then, the critical task is to make the slope of the Urbach region in the log $\alpha$ vs. h$\nu$ plot more apparent. The removal of interference fringes helps and several procedures for removing interference fringes have been reported. It has been seen that using the ratio of absorption to transmittance [5] or the ratio of transmittance to reflectance [6] can correct the measured absorption

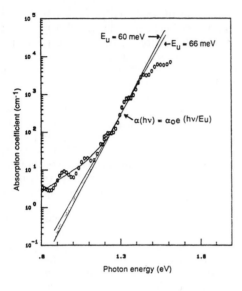

**Fig. 2**
When the Urbach tail is drawn by hand, quite large discrepancies in $E_u$ may result. The shaded area is the discrepancy generated in
$\int \alpha_D dE$.

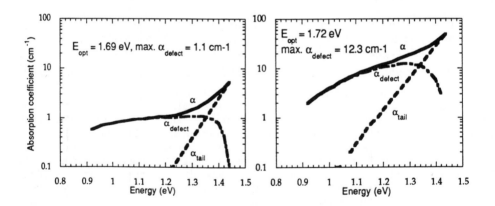

**Fig. 3**
$\alpha_D(h\nu)$ for two types of CPM spectrum:  a) nearly constant in the subgap regime,
b) increasing in the subgap regime.
The maximum value of $\alpha_D$ can be calibrated to defect density.

spectrum of distortion from interference fringes. It is also evidently possible to remove the fringes in the data by filtering them.

Extending the region over which the absorption appears exponential will make the Urbach slope easier to interpret. When the absorption coefficient is high enough to make the absorbance nearly equal 1, which is often the case at energies near the optical gap, the assumption fails that the incident intensity required for keeping $\sigma_{ph}$ constant is inversely proportional to the absorption coefficient. If the actual absorption coefficient needs to be calculated from such data, it is necessary to calculate it from the exact expression for the fraction of light absorbed, $1-\exp(-\alpha d)$. For some samples, performing this calculation will lengthen the range over which the $\log \alpha$ vs. $h\nu$ plot is linear.

**Subgap absorption technique for alloys:** To use again the measure of excess, defect-related absorption and yet to overcome some of the problems with applying the above method to alloys, we propose another means of analysis. As Wronski et al.[7] have mentioned, the plot of $\alpha_D(h\nu)$ contains information on the shape of the defect distribution. Instead of reading $\alpha_D(h\nu)$ at a single given energy, one can look at all of $\alpha_D(h\nu)$. In figure (3), as $h\nu$ increases, $\alpha_D(h\nu)$ increases and then flattens somewhat until it falls to zero at the intersection $\alpha(h\nu) = \alpha_T(h\nu)$. The highest value of $\alpha_D(h\nu)$, which represents the total contribution to absorption by all defects at energies $h\nu$ or less away from the conduction band edge, can then be calibrated to the defect density. This method is based on the assumption that the peak of the defect distribution will lie below the energy at which the intersection $\alpha(h\nu) = \alpha_T(h\nu)$ takes place. That is, it is assumed that the majority of defects will lie outside the valence bandtail so that their contribution to absorption will not be masked by absorption from tail states.

This procedure is independent of bandgap and can be applied both to a-Si:H and a-Si,Ge:H alloys. It is still hampered, though, by its reliance on the precision with which the exponential tail absorption is drawn. As discussed above, there is heavier reliance when the Urbach slope is large and the defect regime absorption curve is steeply sloping, because in this case $\alpha_T(h\nu)$ is comparable in magnitude to $\alpha(h\nu)$ over a longer energy range. This type of spectrum is seen more commonly for alloys than for a-Si:H.

## CONCLUSIONS

To conclude, we have found that, by referring to optical absorption data from PDS and transmission measurements, we can normalize CPM data reliably. The inaccuracy in obtaining defect density lies in the evaluation of the spectrum. The various analysis techniques have their individual advantages and weaknesses. The method of fitting data to the analytic, convolution form of the absorption has the advantages of being physically accurate and allowing different material parameters such as the size and position of the defect distribution to be taken into account. That there are multiple parameters which must be known or estimated, is, on the other hand, a drawback, for the result of the fit is only as trustworthy as the chosen parameters.

The excess subgap absorption technique is simple and consistent with the physics controlling the absorption. Its weakness is that it cannot be applied to a-Si,Ge:H alloys. Its accuracy depends on how precisely the exponential tail absorption can be extrapolated to lower energies, and also on how well the assumption holds that the defect distribution position does not vary between samples.

The integrated excess subgap absorption is also a simple measure of defect density. It is prone to uncertainty, though, for it depends very heavily on the precision of the extrapolated tail absorption and depends also on the the length of the energy interval over which it is calculated.

The maximum of excess subgap absorption can be used to analyze alloys if the defect density is assumed to be proportional to the largest value of $\alpha_D(h\nu)$. The accuracy of this procedure also depends on the reliability of the extrapolated tail absorption. As the extrapolation in thin alloy samples is often subjective, evaluation of CPM spectra by a single operator, rather than several, leads to more consistent results.

## ACKNOWLEDGMENTS

We thank Dr. Milan Vanecek for the highly instructive conversations that led to this study. This work is supported by the Thin-Film Solar Cell Program of the Electric Power Research Institute.

## REFERENCES

1. M. Vanecek, J. Kocka, J. Stuchlik, Z. Kozisek, O. Stika, and A. Triska, Solar Energy Mat. **8**, 411 (1983).
2. H. Curtins and M. Favre, in Amorphous Silicon and Related Materials, ed. H. Fritzsche (World Scientific, Singapore, 1988) p.329.
3. Personal communication with M. Vanecek.
4. W.B. Jackson, N.M. Amer, Phys. Rev. B **25**, 5559 (1982).
5. D. Ritter, K. Weiser, Optics Comm. **57**, 336 (1986).
6. Y. Hishikawa, S. Okamoto, K. Wakisaka, N. Nakamura, S. Tsuda, S. Nakano, M. Ohnishi, and Y. Kuwano, Proc. $9^{th}$ EC-PVSEC, Kluwer, Dordrecht (1989) p.37.
7. C.R. Wronski, Z E. Smith, S. Aljishi, V. Chu, K. Shepard, D-S. Shen, R. Schwarz, D.Slobodin, S. Wagner, AIP Conf. Proc. **157**, 70 (1987).

# PART IV

# MATERIAL
# STUDIES

# A REDUCTION IN THE STAEBLER-WRONSKI EFFECT OBSERVED IN LOW H CONTENT a-Si:H FILMS DEPOSITED BY THE HOT WIRE TECHNIQUE

A. H. Mahan

Solar Energy Research Institute, Golden, CO, USA 80401-3393

Milan Vanecek

Institute of Physics, CSAV, Prague 8, CS-18040, Czechoslovakia

## ABSTRACT

Constant photocurrent (CPM) and steady state photograting (SSPG) measurements have been performed on a series of hot wire (HW) and glow discharge (GD) hydrogenated amorphous silicon (a-Si:H) films, where the substrate temperature was varied in each case to affect the bonded H content. A reduction in the magnitude of the Staebler-Wronski effect, as observed by CPM after saturation light soaking, is seen when the H content is reduced in both sets of samples, but the lowest saturated CPM values and the highest SSPG diffusion lengths measured after saturation occur for HW films having H contents in the range 1-4 at.%. Although correlations exist between the number of excess defects produced by saturation light soaking and microscopic parameters such as the H content and the optical bandgap, there is not a simple correspondence for both sets of samples. This suggests that the Staebler-Wronski saturation may be influenced in part by the film microstructure.

## INTRODUCTION

Since Staebler and Wronski showed that reversible changes occur in the electronic properties of hydrogenated amorphous silicon (a-Si:H) when the films are exposed to light,[1] this phenomenon has been studied extensively.[2] Many early studies utilized relatively short illumination times, because the light induced change in material properties during this time scale is typically rather large. However, since subsequent studies have observed[3] that films with different electronic properties may degrade at different rates, some of the conclusions drawn from these earlier studies may be inappropriate. Of more relevance is the saturation value, N(B), that the defect density reaches. This saturation number is important for industrial applications, because it is this value that ultimately determines the long term electronic properties and/or device performance of the material. On the other hand, the number of excess defects that are generated, that is, the saturated value minus the initial value, N(B)-N(A), becomes important when attempting to probe the underlying factors governing the Staebler-Wronski effect. Park et al.[4,5] introduced the concept of saturation, using the constant photocurrent method (CPM) to examine the defect density, and went on to measure a large number of samples deposited at different laboratories under a wide range of deposition conditions with the hope of identifying the Staebler-Wronski mechanism. Unfortunately, only weak correlations were found between the saturated CPM values and microscopic parameters, and the major factors governing the light induced effect were not identified. A more important question, perhaps, is whether the saturated gap state density can be reduced below that exhibited by device quality glow discharge (GD) a-Si:H containing the standard 10 at.% H.

We have previously reported [6] the electronic properties of a series of a-Si:H films deposited by the HW technique, and found that device quality a-Si:H which contains as little as 1 at.% bonded H can be deposited for the first time. In this technique, silane gas is decomposed on a heated tungsten filament, which is followed by evaporation of atomic silicon and hydrogen, enabling film growth on a heated substrate. In the previous study,[6] we varied the substrate temperature to change the H content in the samples. We now present a detailed systematic study of the effects of saturation light soaking on these samples, using the CPM and steady state photograting (SSPG) techniques to probe changes in the material . Data for a series of GD films, also deposited at different substrate temperatures, have been included for comparison purposes. We find that the saturation defect density as observed by CPM is reduced as the H content is lowered for both sets of samples, but the HW saturation values for H contents less than 10 at.% are consistently lower than the values for the corresponding GD samples, and are lowest for H contents in the range 1-4 at.%. A similar type of behavior is noted for the SSPG diffusion lengths. The present results thus demonstrate that the saturation defect density after light soaking can be reduced below that exhibited for device quality GD a-Si:H. Further, by examining the excess defect density as a function of selected macroscopic material parameters, we comment that H bonded in the monohydride mode may also contribute to the Staebler-Wronski effect, and that differences in the behavior between the HW and GD results suggest that the film microstructure may also play a role in the light induced metastability.

## EXPERIMENT

The deposition conditions used to deposit the HW and GD films have been reported previously.[6] For this study we varied the deposition temperature from 80-610 °C (135-550°C) for the HW(GD) samples, producing H contents in the range 26.6-0.3 (15-2.5) at.%, as measured by infrared (IR) spectroscopy. We were unable to deposit GD films with lower H contents, as such films require substrate temperatures >600°C, and the resultant films contain a sizeable microcrystalline Si component. On the other hand, the HW films remained amorphous over the entire range of substrate temperatures. 0.4 mm spacing Cr contacts were evaporated onto the top surfaces of the films after deposition. CPM and SSPG measurements were then made on the films in the as grown state, whereupon they were alternatively light soaked from the front and back sides[7] for 20-30 hr. at a time using focused and filtered 650 nm radiation (370 mW/cm$^2$) from a 1000 W tungsten halogen lamp. During light soaking, the sample temperature was kept below 30°C. The CPM spectrum was typically measured after each 20-30 hr. light soak interval, and saturation was determined when no detectable increase in absorption was observed after an additional 20-30 hr. light soak interval. Typically, 100 hours were needed to reach saturation. In addition, selected samples were saturation light soaked using a Kr ion laser.[8] SSPG measurements were then made with the material in the saturated state. Finally, the CPM absorption was calibrated by measuring both electron spin resonance and CPM on a device quality high frequency GD sample from GSI that was light soaked to saturation, enabling an alpha at 1.25 eV (2 cm$^{-1}$) to correspond to $8 \times 10^{16}$ cm$^{-3}$ spins.

## RESULTS

In Fig. 1 we show the defect density as measured by CPM for the HW and GD films, both in the as grown and saturated light soaked states, plotted as a function of the bonded H

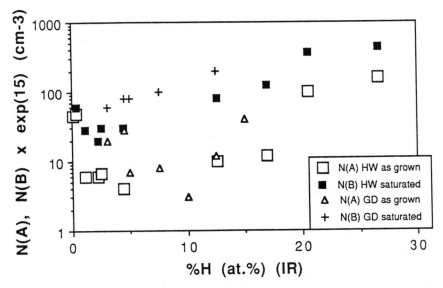

Figure 1. Defect densities, as measured by the CPM technique, for both HW and GD films in the as grown ,N(A), and saturated light soaked ,N(B), states, plotted versus the H content.

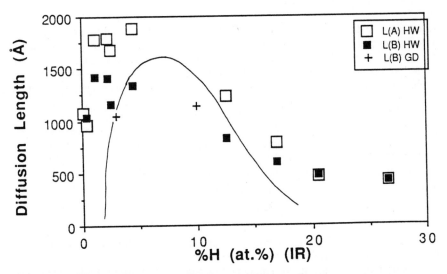

Figure 2. Ambipolar diffusion lengths for both HW and GD films in the as grown, L(A), and saturation light soaked ,L(B), states, plotted versus the H content. The solid line, representing the GD as grown data, is taken from Fig. 4 of Ref. 6.

content. Representative sample deposition temperatures are shown. Note that while the as grown CPM values for the GD samples exhibit a minimum for H contents around 10 at.%, i.e., the device quality deposition temperature (250°C), in agreement with previous measurements,[9,10] the HW samples continue to exhibit low CPM as grown values for H contents down to about 1 at.%, before they finally increase above the value of $1 \times 10^{16}$ cm$^{(-3)}$. At the same time, in this region of low H content, the saturated CPM defect densities for the HW samples are also considerably lower than those for the GD films, and also occur at low H contents (1-4 at.%). In Fig. 2 we show measurements of the ambipolar diffusion length for the same films, also as a function of H content. For the GD samples, the curve representing the as grown data was taken from Fig. 4 of ref. 6. Again, the same trends are evident. While the GD diffusion lengths for the samples in the as grown state peak around the H content contained in device quality a-Si:H, and decrease on either side of this peak, the HW samples exhibit their highest as grown values at the lower H contents mentioned previously. Similarly, the highest saturated SSPG values for the HW samples occur in a similar range, and are higher than any of the values for the GD samples. Several of the saturated values result from Kr ion laser saturation and were saturated and measured elsewhere.[8]

In Fig. 3 we plot the initial spin density, N(A), versus the width of the Urbach edge, Eo. As can be seen, there is a direct correlation between the width of Eo and the defect density, as predicted by thermodynamic models.[9,10] In this case, there is no discernible difference between data for the HW and GD samples. Note, however, that narrow Urbach edges are observed, not only for the high temperature HW films,[6] but also for the GD samples as well. This will be discussed elsewhere.[11]

To further illustrate differences between the HW and GD samples, we now examine the excess defect density obtained by light soaking. In Fig. 4, we plot N(B)-N(A) versus the H content. As can be seen, this excess spin density decreases dramatically as the H content is reduced for both series of samples, even in the (very) low H content regime where the initial spin density, N(A), is seen to increase (see Fig.1). However, once again the data for the HW and GD samples fall on different curves. That is, for a given bonded H content, the HW samples exhibit a smaller excess defect density than the GD samples. In Fig. 5, we plot N(B)-N(A) versus the optical bandgap of the samples. Excellent correlations are also observed in this case. In particular, as the bandgap is reduced, the excess defect density is reduced sharply, but again, the data for the two sets of samples fall on different curves, with the curve for the GD films being higher in magnitude than the curve for the HW samples.

## DISCUSSION

Before discussing what we can learn about the Staebler-Wronski effect from the present results, we caution once again that we varied the substrate temperature in both sets of samples to vary the H content, so we cannot distinguish between, for example, the effects of the H content and the compacting of the a-Si lattice[12] on the Staebler-Wronski metastability. The interplay between these two parameters, and their affect on other macroscopic parameters such as the optical bandgap[6,13] and the Urbach edge[14,15] must be considered when attempting to fit these data to any models involving the metastability.

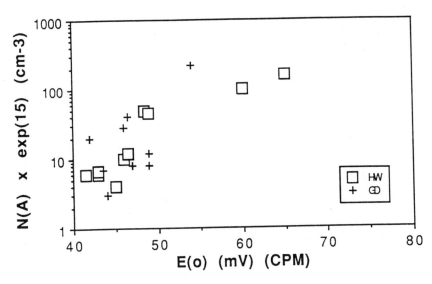

Figure 3.  As grown defect density, N(A), plotted versus the Urbach edge, as measured by CPM.

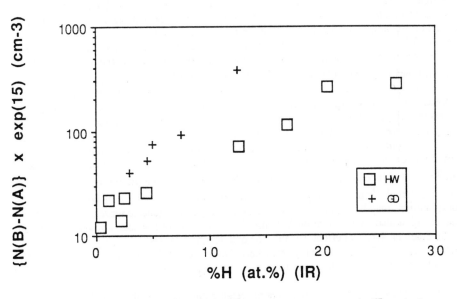

Figure 4.  The excess defect density, N(B)-N(A), plotted versus the H content.

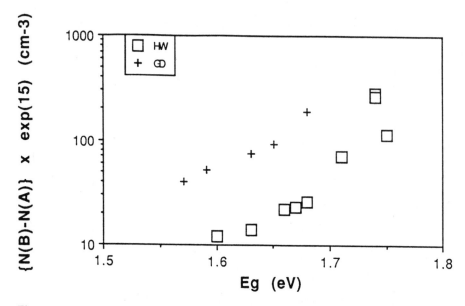

Figure 5. The excess defect density, N(B)-N(A), plotted versus the optical Tauc's bandgap.

Nevertheless, we can make several points concerning the present data. First of all, since this is the first time that device quality material has been made with such small H contents, we can thus examine the Staebler-Wronski effect in this region. If we assume that the Staebler-Wronski effect is reduced purely because of the reduced number of H atoms, then the present results suggest, in contrast to earlier measurements linking the light induced effect primarily to dihydride bonding,[16-18] that H bonded in the monohydride state may contribute to the metastability as well. That is, since our samples, both HW and GD, deposited at substrate temperatures >250°C (containing less that 10 at.% H) contain no discernible dihydride bonding from infrared measurements and still exhibit significant changes in defect density upon light soaking, then H bonded in the dihydride mode cannot be the sole contributor. To put this another way, if these low H content films do contain dihydride bonding, then the present results show that the lower dihydride films exhibit a smaller light induced effect.[19] These agruments suggest that the microscopic models of Stutzmann et al.[20] and of Ching et al.,[21] involving H in the monohydride bonding configuration, may be applicable in a specific as well as in a generic sense. However, the data in Fig. 4 also suggest that the amount of H by itself does not solely govern the amount of the Staebler-Wronski effect, because the data for the HW and GD samples fall on different curves. That is, in the regions where the two sets of films have identical bonded H contents, the HW films exhibit a consistently lower metastability.

Secondly, it has been argued that the energy released upon recombination plays a role in the Staebler-Wronski effect. Since this recombination energy is thought to be proportional to the optical bandgap of the material, Park et al.[5] plotted the saturated defect density versus the bandgap, and found a correlation. Since the initial state density is found to vary considerably for our samples (see Fig.3), we prefer, again, to plot N(B)-N(A) versus the bandgap. Our results, seen in Fig. 5, also show a correlation between the number of excess defects and the bandgap, except that, again, the data for the HW and GD samples fall on different curves. That is, for films having equivalent bandgaps, i.e., available recombination energies, the GD films exhibit significantly larger light induced changes. This is further evidence of a difference in light soaking behavior between the two types of a-Si:H samples.

We previously suggested[6] that the HW films are better ordered than the GD films. That is, we explored the bandgaps for both HW and GD films as a function of H content and found that, for H contents less than 10 at.%, the bandgaps of the HW films were consistently higher than those for the GD films containing comparable H contents. From the model of Maley,[13] we divided the changes in bandgap from the value at 0 at.% H (no alloying) into two components, the first due to the amount of H alloying, and the second due to the amount of ordering produced by the incorporated H. From this we concluded that the HW films were better ordered than GD films containing comparable H contents because their bandgaps were higher. Further support for this claim of ordering come from preliminary Raman measurements of the full width at half maximum of the Si-Si transverse optical (TO) mode, which are consistently narrower for the HW films. Assuming that this narrower mode translates into a smaller bond angle deviation,[22] this also suggests better ordering for the HW samples. Additional evidence supporting differences in the microstructure between the GD and HW samples was alluded to earlier when we commented that we were unable to lower the H content in the GD samples below approximately 3 at.% (deposition temperature on the order of 500°C) without the material turning microcrystalline, while the HW samples remained amorphous for all deposition temperatures and H contents explored here.

We now suggest that the differences in ordering of the respective HW and GD lattices can also affect the values of the saturated Staebler-Wronski effect. We have already shown in Fig. 3 that the initial state densities of the HW and GD films, when plotted against the Urbach edge, are quite similar, so, following thermodynamic models relating the number of weak bonds to the width of the Urbach edge,[5,9,10] the number of weak bonds susceptible to being broken by recombination processes should also be similar for the two sets of films. If the weak bond model[5] is correct, then the light induced saturation changes, i.e., N(B)-N(A), for the two sets of samples should also exhibit identical behavior when plotted against either the H content or the optical bandgap, and clearly they do not. Thus, we assert that the microscopic nature of the lattice may play a role in the Staebler-Wronski effect, and that the HW films exhibit a lower saturation because the ordering of the HW a-Si lattice is better. These structural differences are being further explored by Small Angle X-Ray scattering measurements on high temperature HW and GD films.

## SUMMARY

Constant photocurrent and steady state photograting measurements have been performed on a series of HW and GD hydrogenated amorphous silicon films, where the substrate temperature was varied in each case to affect the bonded H content. A reduction in the magnitude of the Staebler-Wronski effect, as observed by CPM, is seen when the H content is reduced in both sets of samples, but the lowest saturated CPM values and the highest SSPG diffusion lengths occur for HW films having H contents in the range 1-4 at.%. Although correlations exist between selected macroscopic parameters and Staebler-Wronski saturation, there is not a simple correspondence for both sets of samples. This suggests that Staebler-Wronski saturation may be influenced in part by the film microstructure.

## ACKNOWLEDGEMENTS

The authors thank B. P. Nelson for deposition of the GD films used in this study, I. Balberg for diffusion length measurements, T. J. McMahon for the ESR measurements, N. Wada for the Raman measurements, and A. C. Gallagher and R. S. Crandall for many helpful discussions. This work is supported by the U. S. Department of Energy under Contract No. DE-AC02-83CH10093.

## REFERENCES

(1). D. L. Staebler and C. R. Wronski, Appl. Phys. Lett. 31, 292 (1985).
(2). See, i.e., AIP Conf. Proc. 157, ed. by B. L. Stafford and E. Sabisky (AIP, New York, 1987) and references therein.
(3). M. Pinarbasi, M. J. Kushner and J. R. Abelson, Phys. Rev. B 56, 1685 (1990).
(4). H. R. Park, J. Z. Liu and S. Wagner, Appl. Phys. Lett. 55, 2658 (1989).
(5). H. R. Park, J. Z. Liu, P. Roca i Cabarrocas, A. Maruyama, M. Isomura, S. Wagner, J. R. Abelson and F. Finger, MRS Symp. Proc. 192, 751 (1990).
(6). A. H. Mahan, B. P. Nelson, R. S. Crandall and I. Balberg, J. Appl. Phys. (May 1991), in press.
(7). For correct CPM measurements of saturated samples, it is important to light soak the samples homogeneously.
(8). The authors thank S. Wagner for the saturation light soaking and saturation measurements (CPM, SSPG) performed on selected samples at Princeton University.
(9). M. Stutzmann, Phil. Mag. B 60, 531 (1989).
(10). Z Smith and S. Wagner, Phys. Rev. Lett. 59, 688 (1987).
(11). M. Vanecek, B. P. Nelson, A. H. Mahan, T. J. McMahon and R. S. Crandall, to be published.
(12). A. H. Mahan, D. L. Williamson, B. P. Nelson and R. S. Crandall, Solar Cells 27, 465 (1989).
(13). N. Maley and J. S. Lannin, Phys. Rev. B 36, 1146 (1987).
(14). A. H. Mahan, P. Menna and R. Tsu, Appl. Phys. Lett. 51, 1167 (1987).
(15). G. D. Cody, T. Tiedje, B. Abeles, B. Brooks and Y. Goldstein, Phys. Rev. Lett. 47, 1480 (1981).
(16). E. Bhattacharya and A. H. Mahan, Appl. Phys. Lett. 52, 1587 (1988).
(17). M. Ohsawa, T. Hama, T. Akasawa, T. Ichimura, H. Sakai, S. Ishida and U. Uchida, Japan. J. Appl. Phys. 24, L838, (1985).
(18). D. E. Carlson, in "Disordered Semiconductors," edited by M. A. Kastner, G. A. N. Thomas, and S. R. Ovshinsky (Plenum Press, New York, 1987), pg. 613.
(19). S. Nakano, S. Okamoto, T. Takahama, M. Nishikuni, K. Ninomiya, N. Nakamura, S. Tsuda, M. Ohnishi, Y. Kishi and Y. Kuwano, IEEE PV Spec. Conf. 21, 1656 (1990).
(20). M. Stutzmann, W. B. Jackson and C. C. Tsai, Phys. Rev. B 32, 23 (1985).
(21). W. Y. Ching, D. J. Lam and C. C. Lin, Phys. Rev. B 21, 2378 (1980).
(22). D. Beeman, R. Tsu, and M. L. Thorpe, Phys. Rev. B 32, 874 (1985).

# VERY STABLE A-Si:H PREPARED BY "CHEMICAL ANNEALING"

Hajime SHIRAI, Jun-ichi HANNA and Isamu SHIMIZU

The Graduate School at Nagatsuta, Tokyo Institute of Technology
4259, Nagatsuta, Midori-ku, Yokohama, 227, Japan

## ABSTRACT

Highly stable a-Si:H was prepared by a novel technique termed "chemical annealing". The light induced metastable defects were reduced dramatically in the films prepared by this method at 300C or higher. The structural relaxation in the vicinity of the growing surface was enhanced by the treatments with atomic hydrogen or excited noble gases, He* and Ar*.

## 1. INTRODUCTION

Hydrogenated amorphous silicon (a-Si:H) has been widely used for various electronic devices. Now, it is important to improve its fundamental properties, such as transport of holes and the stability of photoconductivity under light illumination. Especially, the instability found in a-Si:H is one of the most serious obstacles which must be eliminated. So far, several plausible models have been proposed to explain the metastable defects caused in high quality a-Si:H films by light illumination.[1,2] In addition, the kinetics of the formation and annealing of the metastable defects have been interpreted by the models involving H-motions in Si-network.[3,4] Attempts have been made to eliminate the precursors responsible for the metastable defects from Si-network by modifying preparation of a-Si:H films.[5,6] Recently, the authors proposed a novel preparation technique termed "chemical annealing" (CA) in which the deposition of very thin layer of a-Si:H and the treatment of its surface with active species such as atomic hydrogen or excited inert gases (Ar* or He*) were alternately repeated.[7,8] One of the most prominent features of this technique is that the hydrogen content, $C_H$ (at%), is continuously reduced down to 1 at% at the substrate temperatures ($T_s$) of 350°C or less. Amorphous structure is maintained as far as the films are deposited under the CVD-like condition[9] with $SiH_3$ as the precursor. The optical gap, $E_g$ (eV), obtained from the Tauc plot is linearly correlated to the $C_H$ in the films prepared by this technique. This result suggests us that Si-network in the growing surface is reconstructed by the release of hydrogen during the CA treatment, resulting in more rigid structure. On the other hand, the density of the dangling bonds is in a range of $10^{15}$-$10^{16}$ spins/cm$^3$, which was obtained by ESR of the films prepared by the CA process. This low density of defects in the films is supported by the other optoelectric properties: high photoconductivity and high efficiency in the low temperature photoluminescence. In addition, the drift mobility of 0.2 cm$^2$/Vs (300 °K) was obtained for the hole-transport in a-Si:H films with $C_H$ of 5 at% and $E_g$ of 1.60 eV prepared by the CA technique.[10] This evidence implies that the localized states in the vicinity of the valence band are markedly reduced by the CA treatment.

In this study, chemical reactions in the growing surface responsible for the structural relaxation during the CA treatment are investigated to reveal the roles of these active species played in the growing surface. In addition, the authors concentrate their attention on the stability of these films under light illumination.

## 2. EXPERIMENTAL

The preparation apparatus used for CA is described in ref.8. The reactor is composed of an RF (13.56 MHz) diode for the decomposition of SiH$_4$ and a microwave plasma (2.45 GHz) tube for generation of long-lived active species, atomic hydrogen and excited states of inert gases such as Ar* and He*, carried through a quartz tube. Atomic deuterium was used in place of atomic hydrogen to distinguish hydrogen in the film from that brought by the precursors, SiH$_n$ (n=3). a-Si:H of about 1 μm in thickness was deposited on c-Si(111) and glass substrate (Corning 7059) by varying the exposure time to atomic hydrogen, $T_2$ and $T_S$, while other conditions, the flows of gases, powers and pressure were kept constant. The growth rate and deposition time ($T_1$) of a-Si:H layer were kept constant at 2 A/s and for 10 sec. The flow of atomic hydrogen was measured by coloration in a-WO$_3$ film, which resulted in a range of $10^{14}$-$10^{15}$ atoms/cm$^2$ s. The gases in our system were replaced within 1 second.

The $C_H$ in the film was estimated from the measurements of IR-absorption and secondary ion mass spectrometry (SIMS, CAMECA IMS-4f). The depth profiles of deuterium diffusion in the films were measured by SIMS for a-Si:H after exposing their surface to atomic deuterium for one hour to reveal the behavior of atomic deuteriums.

The changes in the photoconductivities ( $\sigma_p$) were measured under illumination with light of 100 mW/cm$^2$ from a tungsten-lamp fitted with IR-cut filter.

## 3. RESULTS AND DISCUSSION

### 3-1 Surface reactions in CA a-Si:H film growth

In our system, the thin a-Si:H layer is deposited and the surface is then treated with active species such as atomic hydrogen, He* or Ar* during a certain period, $T_2$. In other words, in CA process, the deposition and the annealing are repeated alternately. Accordingly, the annealing effects are observed as a function of $T_2$. Figure 1 show the changes in $C_H$ plotted as a function of $T_2$ for a-Si:H prepared at various $T_S$. $C_H$ decreases monotonically with an increase in $T_2$ and the reduction rate increases with a rise in $T_S$ when the surface is treated by H or He*. On the other hand, $C_H$ is rapidly reduced to 4-5 at% by annealing with Ar*, which is almost independent of $T_2$. Excited species, Ar* or He*, are responsible for the decrease in $C_H$ because no obvious changes were observed in $C_H$ of films prepared under condition that the microwave power was turned off. In addition, it is certain that modulation of the growing surfaces with these active species is the important process in this chemical annealing. This is because the influences of these active species become prominent when flow of SiH$_4$ is terminated periodically to avoid deactivation of these species by interaction with silanes.

Preliminary kinetic analysis was made for the results shown in Fig.1 according to a stretched exponential function given by the following equations[4]:

$$\Delta C_H = \Delta C_{HO} \exp (T_2/\tau)^\beta$$
$$\tau = \tau_o \exp E_b/kT \qquad (1)$$
$$\Delta C_H = C_H(T_2) - C_H(\infty)$$
$$\Delta C_{HO} = C_H(0) - C_H(\infty) \qquad (2)$$

where, $\tau$, $\tau_o$, $E_b$, k, T and $\beta$ indicate relaxation time, activation energy for the relaxation, Boltzmann constant, absolute temperature and dispersive parameter, respectively. It is assumed in this analysis that the structural relaxation or the formation of Si-Si bonds, accompanied by the release of hydrogen, takes place in the growing surface. $E_b$=0.3 eV and $\beta$=0.8 were obtained by fitting the experimental data to Eqs. (1) and (2). This activation energy in the CA process is larger than that in the conventional RF-GD process[11], suggesting that the $C_H$ is efficiently reduced at higher $T_S$ in the former compared with the latter.

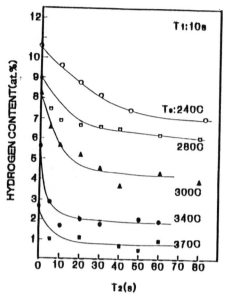

Fig.1 $C_H$ versus $T_2$

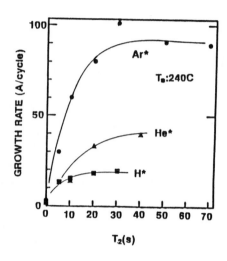

Fig.2 Growth rates plotted as a function of $T_2$

Atomic hydrogen and excited noble gases exhibit distinctly different behavior when used as the active species in the CA process. In Fig.2, the growth rate in each cycle is plotted as a function of $T_2$ for films prepared with different active species, H , He* and Ar*. A slight change in growth rate with $T_2$ were observed when H was used for the annealing, whereas the marked increases were seen with the excited inert gases. It is considered that the sticking probability of precursor, $SiH_3$, on the surface is increased by the treatment with Ar* or He*. Furthermore, deposition is unlikely during $T_2$, because the generation of precursors was negligible for $T_2$, as was established by the OES analysis. The direct deposition from $SiH_4$ on the annealed surface is also ruled out since the increases in the growth rate were observed only when RF power was turned on. Our interpretation is supported by the fact that the growth rate was saturated at a certain value corresponding to the amount of precursors to cover the whole surface.

All films annealed with these active species exhibited a strong correlation between the $C_H$ and the $E_g$ in addition to a high photoconductivity. Consequently, the structural relaxation is considered to be promoted with atomic hydrogen, He* and Ar* treatment. Motion of dangling bonds through hydrogen on the surface is assumed to be a plausible cause for enhancement of the structural relaxation.

Accordingly, $T_S$ is considered to be the parameter most strongly related to the stability, in contrast to structural relaxation which is related to both $T_S$ and $T_2$. The origin of the instability is still unclear, but it probably arises during the construction of Si-network in the vicinity of the growing surface. Impurities such as O, N and C in the films are unlikely to be the origin of degradation because there are no marked differences in these contents. In addition, there are no strong relationships between $C_H$ and the degradation of the photoconductivity. These observations lead us to conclude that the photo-induced degradation takes place at localized inhomogeneous sites remaining in Si-network after its construction in the growing surface.[12] In addition, the origin of the degradation is considered to be eliminated by the CA treatments at $T_S > 300$ °C. The structural relaxation of Si-network in the growing surface is important in order to avoid crystallization or formation of other inhomogeneous structures.

### 3-3 Surface reaction and diffusion of atomic hydrogen

The chemical reactions in the vicinity of surface are not simple when the surface of a-Si:H is exposed to the flow of atomic hydrogen. There are the several plausible processes[13,14]: abbtraction of hydrogen bonded to Si, breakage of weak Si-Si bonds, chemical etching by forming $SiH_4$, recombination, formation of Si-H bonds with the surface dangling bonds, and diffusion. These reactions are competetive on the surface when the film is exposed to atomic hydrogen.

To qualitatively reveal the chemical processes caused by atomic hydrogen, measurements of the depth profiles with SIMS were made together with the observation of the surfaces with SEM for a-Si:H after exposing the surface to the flow of atomic deuterium for one hour at various temperature, $T_a$. The surfaces were treated by the same flux of atomic deuterium as that used in the CA process for an hour corresponding to the annealing period.

Figure 4 show the depth profiles of D and H in the films exposed to atomic deuterium for 50 min. at various $T_a$. Chemical etching is dominant at 200°C or lower. The diffusion of deuterium is considered to be suppressed at low $T_a$ by the chemical etching and due to its large activation energy of around 1.5 eV.[15] Deuterium diffusion is enhanced greatly by raising $T_a$ up to about 300C, but tends to be reduced again at 340C. While, chemical etching was negligible at $T_a$ higher than 200 °C. This is supported by the SEM observations where no marked changes are seen on the surfaces treated at $T_a > 200$°C. These results suggest that the accumulation of deuterium in the surface causing to the chemical etching is hard at higher $T_a$, because the contributuion of surface reactions, i.e, subtraction of hydrogen and recombination become dominant comparing with

### 3-2 Stability of CA a-Si:H

It has been established that a-Si:H prepared by the CA-process exhibited high photoconductivity indicating that the density of states in the gap being as low as that in the device-quality a-Si:H. Films with $E_g$ ranginging from 1.5 to 1.65 eV corresponding to $C_H$ from 1.0 to 5 at% had $\sigma_p > 10^{-4}$ S/cm under light illumination of white light of 100 mW/cm$^2$.

Such high photoconductivity tends to be degraded gradually by continuous light illumination for a-Si:H films prepared by the conventional RF-GD, resulting from a gradual increase in the metastable defects. The degradation in $\sigma_p$ is expressed as a function of illumination time (t) given by following equation.[3]

$$p(t) = \sigma_{po} \, t^{-1/3} \qquad (3)$$

Where, po indicates the initial value of $\sigma_p$. In Fig. 3, the changes in $\sigma_p$ are plotted as a function of illumination time in log-log plots for CA a-Si:H prepared under various conditions. The slope of curve (1) is close to 1/3, which is consistent with the relation given in Eq.3. On the other hand, the degradation in $\sigma_p$ is fairly slow in the films prepared by the CA-process at $T_S > 280°C$. On the other hand, the p of the film prepared at 240 °C shows a similar degradation rate to that indicated by curve (1). The numbers in the figure indicate $T_S$ in °C and $C_H$ in at%. It is noteworthy that after light soaking 50 hrs or more, samples prepared by CA process at 300°C or higher retained more than 80 % of the initial $\sigma_p$ was retained.

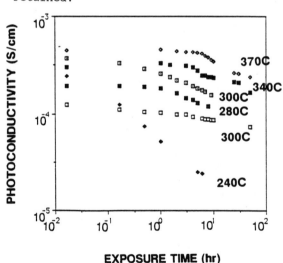

Fig.3 Photo-induced degradation of films prepared by the CA process

white light of 100 mW/cm$^2$

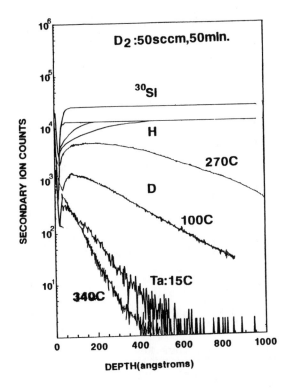

*Fig.4 The depth profiles of H and H of a-Si:H after exposure to atomic deuterium for 50 min at various temperature, $T_a$.*

diffusion process. In addition, the diffusion of deuterium is consequently suppressed at $T_a$ higher than 300 °C owing to a preferential surface reactions for elimination of hydrogen. According to the stability measurements, very stable a-Si:H films were obtained by the CA treatment at 300°C or higher, at which the surface reactions for elimination of hydrogen are dominant. In other words, some localized inhomogeneities are considered to be efficiently eliminated during the construction of Si-network in the surface by promotion of surface reactions with the aid of active species at $T_S > 300°C$.

According to the measurements with SIMS, a very small amount of deuterium remains in Si-network prepared by the CA process with atomic deuterium. This result gives a strong support to the idea that atomic deuterium is mainly the hydrogen abstraction from the surface and consu motion of surface dangling bonds. The activation energy for relaxation of about 0.3 eV in the CA treatment shows a good agreement with that in the diffusion[16] of hydrogen through the dangling bonds in a-Si films.

## 4. CONCLUSION

The authors propose a novel preparation technique, called "chemical annealing", for making high quality a-Si:H with rigid network. In this technique, the deposition of thin a-Si:H layer and the treatment with active species such as atomic hydrogen and excited inert gases are alternately repeated. The stability under light illumination is markedly improved in the films prepared by the CA process at $T_S$ higher than 300 ˚C. Structural relaxation in the vicinity of growing surface is efficiently promoted by the surface treatment with the active species. Consequently, Si-network with more stable and rigid structure is obtained by this technique.

## Acknowledgements

The authors wish to thank to Mr. T. Ariyoshi and K.Nakamura for their assistance in measurements of photoconductivity. This work is supported in part by Grant-in-Aid for Scientific Research, Ministry of Education, Science and Culture No 02402021 and in part by NEDO as part of Sunshine Project under MITI.

## References

1.  H. Schade,Semicond.& Semimetals,21B 359 (1984)
2.  W.B.Jacknon, J.M.Marshall and M.D.Moyer, Phys.Rev.B39 3609 (1989)
3. M.Stutzmann, W.B.Jackson and C.C.Tsai,Phys.Rev.B32 23 (1985)
4. W.B.Jackson,J.Non-cryst.Solids,114 591 (1989)
5. G.Ganguly, A.Suzuki, S.Yamazaki, K.Nomoto, and A.Matsuda, J.Appl.Phys.,68 3738 (1990)
6. M.Pinarbasi, J.R.Abelson and M.J.Kushner, Appl.Phys.Lett.,1685 (1990)
7. H.Shirai,D.Das,J.Hanna and I.Shimizu, Tech. Digest Int, PVSEC-5 59 (1990)
8. H.Shirai, D.Das, J.Hanna,and I.Shimizu, Appl.Phys.Lett. (to be submitted)
9. C.C.Tsai,J.C.Knights,G.Chang and B.Wacker, J.Appl.Phys.,59 2998 (1986)
10. D.Das,H.Shirai,J.Hanna and I.Shimizu;Jpn.J.Appl.Phys., 364 (1991)
11. K.Tanaka and A.Matsuda,Mat.Sci.Report,2 139 (1987)
12 D.L.Williamson,A.H.Mahan, P.B.Nelson and R.S.Crandall,J.Non-cryst.Solids,114 226 (1990)
13.R.Robertson and A.Gallagher,J.Chem.Phys.,85 3623 (1986)
14.S.Veprek and F.A.Sarott, Phys.Rev.B 36 3344 (1987)
15. M.Nakamura.T.Ohno,K.Miyata,K.Konishi, Appl.Phys.Lett.,65 3061 (1989)
16. D.E.Carlson and C.M.Magee, Appl.Phys.Lett.,33 81 (1978)

# A COMPARATIVE STUDY OF THE LIGHT-INDUCED DEFECTS IN INTRINSIC AMORPHOUS AND MICROCRYSTALLINE SILICON DEPOSITED BY REMOTE PLASMA ENHANCED CHEMICAL VAPOR DEPOSITION

M.J. Williams, Cheng Wang and G. Lucovsky
Departments of Physics, and Materials Science and Engineering
North Carolina State University, Raleigh, NC 27695-8202

## ABSTRACT

This paper discusses the deposition of microcrystalline silicon, $\mu$c-Si, by the remote plasma enhanced chemical vapor deposition process. We discuss the deposition process, and the properties of undoped and doped $\mu$c-Si thin films. We emphasize the properties of an "intrinsic" $\mu$c-Si thin film material that is obtained by light boron doping. The properties of this material, in particular its effective band-gap of 1.44 eV, its relatively high photoconductivity and its undetectable Staebler-Wronski degradation make it a candidate material for the i-layer photo-active constituent of p-i-n PV devices.

## INTRODUCTION

The deposition of microcrystalline Si ($\mu$c-Si) thin films using the glow discharge deposition (GD) method has been investigated and the process parameters for depositing heavily-doped $\mu$c-Si films are well established[1,2]. We have deposited doped and intrinsic hydrogenated amorphous Si (a-Si:H) thin films using the Remote Plasma Enhanced Chemical Vapor Deposition (RPECVD) method, and have recently extended this process to the deposition of intrinsic $\mu$c-Si thin films. Once undoped $\mu$c-Si films were successfully deposited by the RPECVD method, we initiated studies to determine whether they could be doped by the downstream injection of the appropriate dopant gases: phosphine for n-type doping and diborane for p-type doping. We observed that the $\mu$c-Si thin films could be doped in much the same manner as RPECVD a-Si:H films. Dark conductivities as high as 40 S/cm for phosphorous doped $\mu$c-Si films and 6 S/cm for boron doped films have been obtained; the respective activation energies are 0.02 eV and 0.04 eV. During these doping studies we varied the ratios of the dopant gases to silane and thereby identified an "intrinsic" $\mu$c-Si material with an unusually low dark conductivity, $\sim 10^{-8}$ S/cm, and a relatively high dark conductivity activation energy, 0.66 eV. This film was produced by light boron doping using a $B_2H_6/SiH_4$ ratio of $10^{-5}$. Undoped $\mu$c-Si films have a dark conductivity of about $10^{-3}$ S/cm with an activation energy of approximately 0.3 eV. Further investigation showed that these boron-doped "intrinsic" $\mu$c-Si film had a high photoresponse approaching that of PV-grade a-Si:H. We have also observed that these films showed no detectable decrease in the photoconductivity after exposure to >100 mW/cm$^2$ of "white light" for periods of time exceeding 40 hours. These electronic and photoelectronic properties make these "intrinsic" $\mu$c-Si films candidate materials for the photo-active i layers in p-i-n PV devices. In this paper we contrast the RPECVD method with the GD method for deposition of both amorphous and $\mu$c-Si thin films. Doping of the $\mu$c-Si films will be described, and lastly the unique optoelectronic properties of the lightly boron-doped "intrinsic" $\mu$c-Si films will be discussed.

## RPECVD DEPOSITION OF AMORPHOUS SILICON

In the RPECVD process for a-Si:H, the Si-containing and dopant-atom source gases are not subjected to direct plasma excitation. Rather, they are injected into the deposition chamber downstream from the plasma. In addition the sample substrate is located outside of the plasma region. Downstream injection results in fewer excited species in the gas phase[4]. That is, rather than directly plasma-exciting and fragmenting the SiH_4 reactant to generate a multiplicity of excited, ion and radical species such as H, Si, SiH, SiH_2, SiH_3 etc. as in the GD method, plasma-excited He interacts with downstream injected SiH_4 yielding an excited state of SiH_4, in which the molecule is not fragmented. There are therefore fewer possible precursor species and fewer possible reaction pathways available for film deposition in the RPECVD process than in the GD process. The deposition chamber also has an option for direct plasma excitation of process gases and this is used in the deposition of silicon oxides and nitrides. For these depositions the oxygen and nitrogen-containing source gases can be directly excited, but as in the RPECVD process of a-Si:H, the SiH_4 reactant is always introduced downstream from the plasma and is never subjected to direct plasma excitation.

The presence of a number of hydrogen-containing precursor species in the GD process for a-Si:H depositions means that hydrogen is available from a number of different sources, as in H, SiH, SiH_2, SiH_3, etc., so that it can be readily incorporated into the growing film through surface reactions involving these active precursors. Hydrogen incorporation in the RPECVD process also occurs through a surface reactions; however these are qualitatively different than the GD reactions since the deposition precursor is the excited SiH_4 molecule rather than the multiplicity of fragmentation products that derive from the break-up of that molecule. The specific deposition reactions that lead to these differences are discussed in Ref. 5, and at this juncture, we point out that the first step in this process involves the release of molecular hydrogen. This means that less hydrogen is then available for incorporation into the

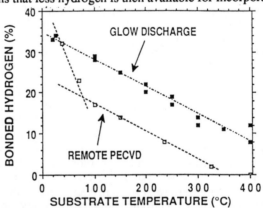

Figure 1. **Hydrogen content in films as a function of temperature with GD and RPECVD. Lower H concentration in RPECVD films reflects lower availability of hydrogen during deposition.**

growing film. As indicated in Figure 1, the amount of hydrogen incorporated into the a-Si:H films is lower in RPECVD films for all deposition temperatures above about 100°C than in GD films. Previous work has established that the amount of hydrogen which gets incorporated into an a-Si:H film is a function of the substrate temperature, and is proportional to the product of an availability factor, that is specifically related to the molecular and/or atomic sources of the hydrogen, and a surface rejection factor which is determined by the substrate temperature and related to the energy required for surface desorption [5,6]. For temperatures greater than about 100°C, the data in Fig. 1 indicate substantially less hydrogen in the RPECVD films.

We have established optimal conditions for the deposition of device quality intrinsic and doped a-Si:H thin films using RPECVD. Device quality films are typically deposited at 250°C under a total system pressure of 300 mTorr. A 200 sccm He flow plasma is remotely excited using an RF power of approximately 50 W. $SiH_4$ gas is delivered at a rate of approximately 1 sccm; the $SiH_4$ is diluted 10:1 with He so that the total flow of the $SiH_4$,He mixture is ~10 sccm. The typical electronic properties of intrinsic a-Si:H films deposited with these process parameters are a dark conductivity of less than $10^{-10}$ S/cm, an activation energy for the dark conductivity of about 0.8 eV and a photoconductivity of approximately $10^{-4}$ S/cm under 50 mW/cm$^2$ illumination. We have also found that it is also possible to deposit films at significantly lower deposition temperatures, 150 to 200°C. At these temperatures the total hydrogen concentration is in the range of 10 - 15 at. %, and the hydrogen is incorporated into the film in monohydride bonding configurations[6]. Films deposited under these conditions frequently require a post-deposition anneal to obtain optimized photoelectronic properties.

### RPECVD DEPOSITION OF MICROCRYSTALLINE SILICON

We have also investigated the deposition of $\mu$c-Si thin films using the RPECVD process. For the glow discharge deposition of $\mu$c-Si thin films an excess of hydrogen must be present in the gas stream. Typical $H_2/SiH_4$ ratios are in the range of 50 to 100:1. The hydrogen slows the deposition down so that to nucleation of microcrystallites can occur. It may also assist in the removal of an undesired amorphous constituent; in any event, the deposition process reactions are not completely understood[7]. In RPECVD there are two options for the injection of molecular hydrogen. The first option is to directly excite the hydrogen in the plasma with the helium. The second option is to inject the hydrogen downstream from the plasma along with the $SiH_4$ and with the dopant gases. Both methods were used and it was found that microcrystalline growth occurred in either case. In order to compare the quality of the films produced Raman spectroscopy and TEM were used to determine the degree of microcrystallinity and film microstructure (Figs. 2,3). It was found that the films with the highest degree of crystallinity were produced using downstream $H_2$ injection with a substrate temperature of 250°C, and a $H_2/SiH_4$ ratio of 30:1. Under these conditions the 520 cm$^{-1}$ peak associated with the microcrystallite component of the film had the smallest line-width and the highest intensity. There were other non-crystalline phases present in the film as evidenced by additional features in the Raman spectra below about 500 cm$^{-1}$. These non-crystalline features were minimized for the deposition conditions identified above[8].

When studying the films by TEM we did not observe any induction period in the films before the onset of microcrystalline growth. This type of behavior is frequently observable for $\mu$c-Si films deposited by the GD process. If there is an

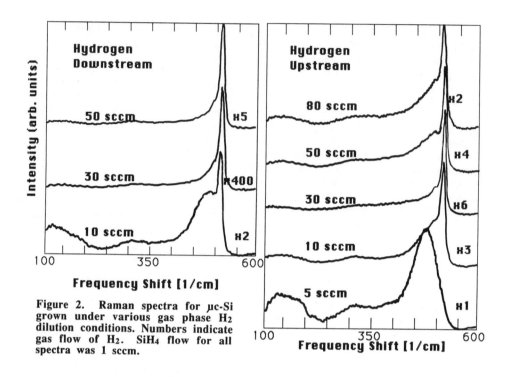

Figure 2. Raman spectra for μc-Si grown under various gas phase H₂ dilution conditions. Numbers indicate gas flow of H₂. SiH₄ flow for all spectra was 1 sccm.

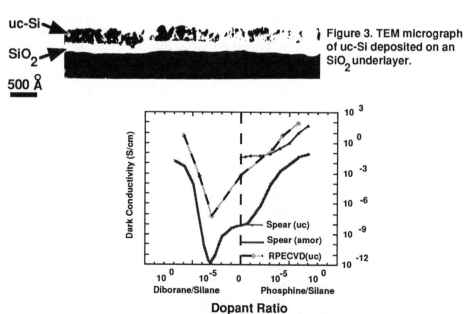

Figure 3. TEM micrograph of uc-Si deposited on an SiO₂ underlayer.

Figure 4. Dark conductivity as a function of dopant level for mc-si. Other curves are doping profiles for amorphous and microcrystalline films as measured by Spear (Ref. 7).

induction period it produces less than  30 Å of a non-crystalline phase; this is the minimum thickness observable by TEM.

## DOPING OF MICROCRYSTALLINE SILICON

Once the conditions for producing quality microcrystalline films were established, doping of the RPECVD films was undertaken. The dopant gases used were $PH_3$ for n-type doping and $B_2H_6$ diluted in He or $SiH_4$ for p-type doping. Doping was successful and the heaviest doping levels for which microcrystallinity could be maintained obtained were a 0.01 $PH_3/SiH_4$ ratio and a 0.001 $B_2H_6/SiH_4$ ratio. The n-type films had  a dark conductivity of 40 S/cm and a dark conductivity activation energy of 0.018 eV, while the p-type films had a dark conductivity of 6 S/cm and an activation energy of  0.040 eV.  The doping level for p-type films was limited to 0.001 because greater $B_2H_6$ concentrations films caused a reversion back to amorphous growth.

Fig. 4 shows the dark conductivities for a number of μc-Si films grown from different ratios of the dopant gases to silane[9].  This doping profile parallels that observed for GD deposition of doped a-Si films.  In particular, a minimum in dark conductivity occurs for a boron dopant gas ratio of $B_2H_6/SiH_4 \sim 10^{-5}$.  We believe that this reduced dark conductivity with doping level is due to a  compensation of donor-like, native defects by the boron doping making the film "intrinsic" in character. This film will be discussed in greater detail below.

The  low conductivity of  $6\times10^{-8}$ S/cm for the "intrinsic" material mentioned above  warranted more investigation so its photoresponse and its dark conductivity activation energy were determined.  The activation energy was 0.66 eV and its photoconductivity was $1\times10^{-4}$ S/cm.  The effective band-gap of this material was 1.4 eV, as determined by the E04 approach (energy at which $\alpha=10^4$ cm$^{-1}$), so that the activation energy corresponded to about half the band-gap.These photoelectronic characteristics are similar to those of device quality intrinsic a-Si:H and prompted us to investigate the possibility of suggesting the use of this material for the photo-active i layer in a p-i-n device structure. We first studied the optical stability to determine whether or not this material displayed a Staebler-Wronski effect.  The optical stability was first evaluated for relatively short exposure times, and it was found that the photoconductivity of the films remained stable for 6 hours under 50 mW/cm$^2$. The absence of any detectable Staebler-Wronski degradation was pursued further and we have now found these films to be stable for up to 40 hours under illumination levels of 50  mW/cm$^2$ (Fig. 5).  This is in direct contrast to PV-quality intrinsic a-Si:H films for which the photoconductivity is generally reduced by about two  orders of magnitude over the course of 6 hours of illumination at levels of about 50 mW/cm$^2$ [10]. The spectral response of these films in the visible is quite flat as indicated by Fig. 6. This is consistent with an effective optical bandgap of 1.4 eV.

We have since attempted to reproduce the growth of these lightly-boron doped intrinsic films and have found that film quality is dependent on chamber history and chamber geometry . Further work is underway to investigate repeatability and to obtain low dark conductivities consistently. For example, Figure 7 shows the variation in dark and photoconductivity for several of the films deposited using a diborane/silane doping gas level of $10^{-5}$. While the photoconductivity is high and does not vary significantly for any of the films, the dark conductivity and activation energy vary greatly from film to film.  The slope of the dark conductivity  as a function of activation energy is not 1/kT, indicating that the dark-conductivity conduction process may be modulated by a combination of internal

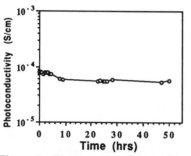

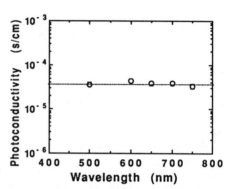

**Figure 5.** Photoconductivity of "intrinsic" uc-Si over 50 hours at 50 mW/cm2. Photoconductivity shows no Staebler-Wronski effect. Small fluctuations are due to temperature variations.

**Figure 6.** Photoconductivity as a function of wavelength for G=1E19 1/(cm^3*s) for "intrinsic" uc-Si.

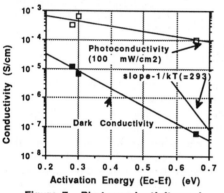

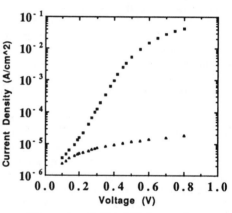

**Figure 7.** Photoconductivity and dark conductivity of "intrinsic" uc-Si.

**Figure 8.** Current-Voltage characteristics for an all uc-Si solar cell.

barriers and band-bending at the crystallite boundaries, as well as the position of the Fermi level in the the microcrystallites.

We have evaluated the suitability of the "intrinsic" μc-Si as a photo-active i layer by incorporating the material in an all microcrystalline p-i-n diode. The I-V characteristics of this device are shown in Figure 8. The dimensions of the diode were 300 Å for the P and N layers and 2800 Å for the i layer. The device was deposited on P+ c-Si with Al and Ag back and front contacts, respectively . The diode showed a rectification ratio of $\sim 10^4$ for $|V| \sim 0.5$ V. Jsc≈ 4 mA/cm2 under 50 mW/cm2 and Voc =0.3 V. This was a preliminary evaluation of the material in a device and the device geometry was not optimized. Its stability as a photodiode has yet to be determined.

## SUMMARY

In conclusion, we have deposited both undoped and doped a-Si:H and μc-Si using the remote PECVD process. Both types of films have photoelectronic and electrical properties which make them suitable candidates for device applications. The dopability of the amorphous and microcrystalline films has been demonstrated, and the highest levels of dark conductivity obtained are comparable to what can be obtained by conventional GD deposition processes. We have further identified and investigated an "intrinsic" lightly boron-doped μc-Si which shows high conductivity and excellent stability under prolonged optical exposure. Future work will concentrate on determining the deposition conditions needed for reproducibility of "intrinsic" μc film qualities. Additional studies will be undertaken to determine in a quantitative way the relationship between the degree of microcrystallinity and the electrical properties in these "intrinsic" μc-Si films. We will also determine and compare carrier transport mobilities using time-of-flight mobility, Hall and photo-Hall effects. These approaches should allow us to better understand the transport mechanism of these "intrinsic" μc-Si thin films.

## ACKNOWLEDGEMENTS

This work was done with the support of the Solar Energy Research Institute under subcontract XM-9-18142-2. We would like to thank E. Buehler and Dr. R.J. Nemanich for Raman spectroscopy and C. Jahncke for electrode deposition and profilometry.

## REFERENCES

1. A. Matsuda, S. Yamasaki, K. Nakagawa, H. Okushi, K. Tanaka, S. Iizima, M. Matsumara and H. Yamamoto, Japan J. Appl. Phys. 19, 1305 (1980).
2. W.E. Spear, G. Willeke and P.G. LeComber, Physica 117B & 118B, 908 (1983).
3. C. Wang and G. Lucovsky, Proceedings of the 21st IEEE Photovoltaic Specialists Conference 2, 1614 (1990).
4. D.V. Tsu, G.N. Parsons, M.W. Watkins and G. Lucovsky, J. Vac. Sci. Technol. A7, 1115 (1989).
5. G. Lucovsky and G.N. Parsons, Optoelectronics, Devices and Technology 4, 119 1989).
6. D.V. Tsu, G.N. Parsons, M.W. Watkins and G. Lucovsky, J. Vac. Sci. Technol. A7, 1115 (1989).
7. C.C. Tsai, R. Thompson, C. Doland, F.A. Ponce, G.B. Anderson, and B. Wacker, MRS Symp Proc. 118, 49 (1988).
8. C. Wang, M.J. Williams, and G. Lucovsky, J. Vac. Sci. Technol. A9 (1991), in press.
9. W.E. Spear and P.G. LeComber, Philos. Mag. 33, 935 (1976).

# LIGHT STABILITY OF AMORPHOUS GERMANIUM

R.Plättner,E.Günzel,G.Scheinbacher,B.Schröder*
Siemens AG, Corporate Research and Development
Otto-Hahn-Ring 6, D-8000 München 83,FRG,Fax 08963649164
*Universität Kaiserslautern,Fachbereich Physik

## ABSTRACT

The conditions for the preparation of high quality and light stable a-Ge:H material were investigated. By "hard" deposition, i.e. by strong ion bombardment during the rf-plasma deposition, it is possible to obtain a material with a photoconductivity of $\eta\mu\tau = 10^{-6}cm^2/V$ (at 950nm). This value does not change after 2500 hours AM1 illumination. Also a stress test under keV-electron irradiation shows the material to be stable. By small angle X-ray scattering this stable material was characterized as free from small voids. Larger voids are present in both "hard" and "soft" deposited a-Ge:H material and therefore seem not to be mainly responsible for light degradation. The high photoconductivity and the good stability are a precondition for the use of a-Ge:H in tandem solar cells.

## INTRODUCTION

In the last few years, the extension of the spectral sensitivity range of amorphous silicon solar cells has been the subject of many investigations around the world. The initial aim of these developments was to increase the cell efficiency. Ways for this are offered by tandem cells made from a-Si:H and a-Ge:H, which can also utilize the spectral range of solar light between 650 and 1000 nm. By dividing the necessary optical absorption length of the solar cells into two subcells, the light stability of the entire tandem cell is also improved due to the higher field strength in each subcell. It would be an advantage for the cells if the material of the thicker back-side cell also had a high stability to light degradation.

We therefore concentrated on the development of amorphous germanium as the narrow-gap cell material with a bandgap of 1.0 eV. Such a material has not been available hitherto with good electrical properties.

We found that the properties of the a-Ge:H material manufactured with low-pressure plasma CVD depend strongly on the plasma conditions used. Investigation of the various materials with respect to their light stability showed that the electronically superior materials also exhibit better stability.

## EXPERIMENTAL

Starting from the consideration that an electronically superior a-Ge:H material would have a denser structure, we have attempted to produce such a material by more intensive ion bombardment. This material should contain fewer voids, i.e. fewer internal surfaces, which are normally saturated with hydrogen .

We distinguish "hard" and "soft" deposition conditions in manufacturing the a-Ge:H in a low-pressure CVD diode reactor. "Hard" deposition firstly means the injection of higher RF powers into the plasma. This means a higher ion bombardment of the growing film surface and, as we believe, the removal of initially weakly-bonded germanium atoms during the growth of the film. Superior a-Ge:H films were obtained at higher powers. Another approach to increase the ion bombardment is to increase the bias voltage. However the ion bombardment is little affected by the application of an external bias voltage, as this is not transferred to the growing layer due to the nonconductive glass substrate. In contrast, the bias voltage of a sample on a cathode $U_1$ or anode $U_2$ is affected by the geometry of the reactor. According to W. Kasper et al.[1], the functional

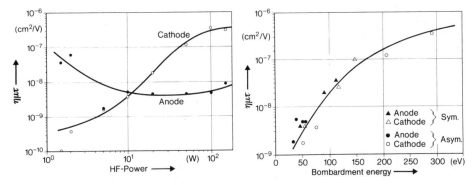

Fig.1: Normalized photoconductivity of a-Ge:H as a function of discharge power (asymmetrical electrodes, 5% dilution in hydrogen)

Fig.2: Photoconductivity vs.bombardment energy for variously deposited a-Ge:H films

dependence is obtained from $U_1:U_2 = (F_2:F_1)^2$. The bias voltage is proportional to the square of the surface ratio of the RF-powered electrode $F_1$ and the grounded electrode or reactor area $F_2$ respectively which confines the plasma. We therefore performed investigations with a higher anode/cathode ratio, i.e. a higher ion impact. The a-Ge:H films with superior electronic properties[2] really were formed when using higher powers and cathodic deposition simultaneously (Fig.1). The process gas $GeH_4$ was diluted (5% in hydrogen). The pressure was held at about 100 Pa. We identified the bias voltage of the substrate surface relative to the bulk plasma as the fundamental parameter influencing the growth process.

## RESULTS AND DISCUSSION

Characterization with the normalized photoconductivity (at 950nm) showed values up to $\eta\mu\tau = 10^{-6} cm^2/V$. It was also possible to demonstrate (Fig.2) a correlation between all the different deposition types and uniquely the parameter of substrate bias voltage (ion bombardment). This thus represents the critical plasma parameter for the quality of the a-Ge:H material[3]. For good a-Ge:H material, it should have a value of at least 250 eV.

The light stability of the a-Ge:H manufactured in this way was also surprisingly good. Even after more than 2500 hours AM1 illumination, no decline of the photoconductivity could be observed (Fig.3a). The dark conductivity also remains unchanged (Fig.3b). In comparison to this, the Staebler-Wronski effect of hitherto existing a-$Si_{1-x}Ge_x$:H materials already appeared after a illumination time of a few hours.

However, the lower bandgap in these materials must be taken into account, especially for a narrow-gap material as a-Ge:H. The excess carriers responsible for the degradation are present only in about 10 times higher concentration at AM1 compared to the dark. This is because of the narrow gap of a-Ge:H, which gives rise to its higher dark conductivity. In contrast to the large increase of $10^5$ for illumination of a-Si:H there results a factor of $10^4$ between both materials. Assuming in a worst case a linear proportional defect generation by the excess carrier concentration, the illumination time for a-Ge:H should be extended by the same factor ($10^4$) resulting in the same degradation effect as for a-Si:H. The illumination time of 2500 hours (Fig.3) should therefore be sufficient to see a distinct effect.

A longer degradation test of the samples is impractical for reasons of time. Therefore we performed a degradation stress test with 20 keV-electrons according to a procedure developed at the University of Kaiserslautern[4,5].In this technique, a degradation up to saturation is achieved in one hour with a radiation dose of 60 Joule/cm$^2$.It is then possible to anneal the samples after keV

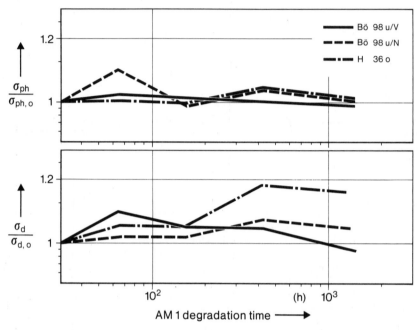

Fig.3: Light stability of a-Ge:H (a: photoconductivity, b: dark conductivity; set of similar samples)

radiation thermally up to the initial value during 6 hours (standard procedure) at 160 °C. Although it is not yet certain that the same defect generation mechanism is at work here as in photodegradation, the course of degradation and annealing of this process indicates an probably identical mechanism.

In Fig.4 it can be seen that the degradation defined as photoconductivity $\sigma_p$ before(0) and after(1) the irradiation $\sigma_p(0)/\sigma_p(1)$, is significantly smaller in samples of higher germanium content. Pure a-Ge:H (x = 1) shows up to three orders of magnitude smaller degradation than pure a-Si:H. We can also see that our samples show a higher stability by a factor of 10 compared with those produced by different international research groups. An influence of the dark conductivity will be excluded by giving the values $\sigma_p/\sigma_d$ before(0) and after(1) the irradiation, $\sigma_{p/d}(0)/\sigma_{p/d}(1)$. Only a marginal change is observed compared to $\sigma_p(0)/\sigma_p(1)$. Germanium-Silicon alloys already show a significantly improved stability at 5% Ge content.

Matsuda et al.[6] were able to produce a good a-Si:H material of improved light stability when they applied a heavy ion bombardment to the growing layer during the plasma deposition. This is not in contrast to our ion bombardment condition. In the case of a-Si:H, a high energy bombardment is clearly not advantageous, but a bombardment with heavy nobel gas particles such as krypton or xenon, that are added to the process gas, is. Obviously there is clearly a significant difference in the behavior of GeH$_4$ and SiH$_4$. Our results show that a-Ge:H originates with improved electronic properties and stability when a high energy bombardment is used. So we could also prepare good a-Ge:H samples by a wave resonance plasma deposition[7],combined with a capacitive superposed discharge, which is characterized by a higher ionization density compared to normal diode plasma excitation. With a new reactor the hitherto best a-Ge:H of $\eta\mu\tau = 1.12\cdot10^{-6}cm^2/V$ was prepared, on what we will report in a next paper.

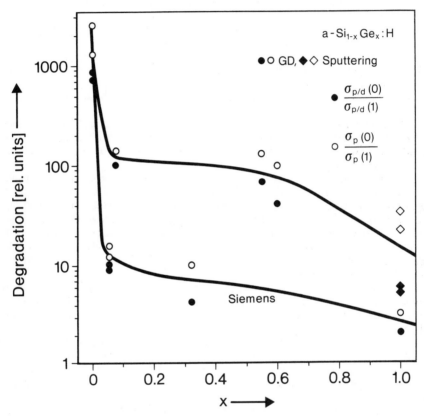

Fig.4: Degradation of a-Si$_{1-x}$Ge$_x$:H by keV electron impact

In contrast, it has not been possible to date to produce good a-Si:H material under high-energy ion bombardment. For this material a nobel gas bombardment, as used by Matsuda, is obviously more favorable. This conclusion is also supported by the fact that the noble gases of higher mass, such as krypton and xenon, that simultaneously have excitation levels lower than helium or argon, are better suited for a-Si:H fabrication as they do not ionize the SiH$_4$ species by penning impacts.

To clarify the content of voids in the a-Ge:H layers, we investigated the "hard" and "soft" deposited material with SAXS (Small Angle X-ray Scattering). The samples were deposited with a thickness of 0.5 to 1 μm on 10 μm thick Al foils of high purity. The irradiation was performed with monochromatic synchrotron radiation of 8.040 keV at DESY in Hamburg. For the evaluation, the sample was mounted together with a central beam stop in front of a twodimensional positionally-dependent gas counter and tilted by 0°, 15°, 30°, 45° and 50° with respect to the direction of radiation. A detailed paper on these measurements will be published. The result in the form log scattering intensity I vs. square of scattering vector Q (with Q = 2π·scattering angle/wavelength) was evaluated. From a Guinier plot , a Porod plot and a contour plot we found different sizes and by calibration with a glassy carbon standard a different content of voids in "hard" and "soft" samples. A $Q^{-4}$ vs.I behavior in the Porod plot indicates relatively smooth surfaces of the voids.

Table 1: SAXS results on void content in "hard" and "soft" deposited a-Ge:H films.

| | void diameter | shape | "hard" | "soft" |
|---|---|---|---|---|
| small voids | ≈ 40 - 300Å | cigar | <0.05 vol% | < 1 vol% |
| large voids | ≈300 - 1000Å | disc | 1 vol% | 0.7vol% |

The large diameter voids have the shape of flat discs and a large axis diameter of mostly 500Å (with a 20% part of 1000Å). These large voids exist in both "hard" as well as in "soft" deposited samples with a content of 1 or 0.7 vol% respectively. The orientation of these discs is flat to the substrate surface.

A striking difference between both materials is seen in the content of small diameter voids, which have the shape of a cigar, a diameter of 40 to 300Å and a long axis orientation perpendicular to the substrate. While in "soft" samples a content of 1 vol% was found, in "hard" samples no small voids exist. We conclude that a connection exists between electronically good and stable a-Ge:H materials and a low content of small voids in the bulk of these materials. This result agrees well with the results of Williamson et al.[8] who found rod like microvoids only in nondevice quality a-Si:H, an overall void concentration of about 1% in device quality a-Si:H and 25% in a-SiGe:H (obviously nondevice quality).

To confirm the electronic properties of our samples, we performed investigations by PDS and optical transmittance. In these, we found the materials exhibiting a very low tail-state density, depending on the preparation conditions[3]. In "hard" samples of a-Ge:H, an Urbach tail width $E_o$ of only 50 meV was found, comparable to device quality a-Si:H material. The $\eta\mu\tau$ values for this type of material were $10^{-6}$ cm$^2$/V, while "soft" samples decreased in $\eta\mu\tau$ down to $10^{-8}$ cm$^2$/V. In Fig.5 results are shown for the Urbach tail width $E_o$ obtained by various fabrication processes. The deep states in our material were evaluated from the defect absorption range of PDS measurements. They give a result of $6·10^{16}$ cm$^{-3}$ in "hard" deposited samples, assuming[2] a proportionality factor of $3·10^{15}$ cm$^{-2}$.

The "hard" and "soft" materials also differ considerably with respect to their activity in solar cells. The red response is clearly higher for the $\eta\mu\tau = 10^{-6}$cm$^2$/V material (see Fig.6) than for the $5·10^{-8}$ cm$^2$/V material. The measurements were performed with reverse voltage of 1 V and show almost the same maximum efficiency as for a-Si:H cells, but with an additional large scope for an efficiency improvement from the red response in the 650 to 1100 nm range. It can also be seen in the figure from the zero bias measurement that the cells do not extract the carriers sufficiently because the electrical field is to low. Corresponding to this, the hitherto manufactured a-Ge:H solar cells have reached an efficiency of 2.5%, which is the best value published to date. Due to the narrow-gap materials, the cell structure is somewhat different than that known for a-Si:H cells. We believe that a great deal still remains to be learnt here, so that ultimately an efficiency of 10% for a single cell or more than 15% for the tandem cell can be attained.

Further measurements on a-Ge:H material relate to their hydrogen and oxygen content and their effect on stability. The hydrogen content in our electronically good a-Ge:H samples is about 14 at% which is somewhat higher than in a-Si:H (10 At%). According to IR measurements[3] by far the largest share of this hydrogen is  contained in GeH bonds (1880cm$^{-1}$) and thus corresponds to the SAXS results of a low content of voids in high quality materials. We would expect only a small share of GeH$_2$ bonds to occur at a higher proportion of voids. This also agrees with the measured IR spectrum.

Evolution measurements of the hydrogen show a small low-temperature peak (about 280°C) in some of the "hard" deposited a-Ge:H samples. There are strong indications that in these "hard" samples seeds of microcrystalline material cause a crystallisation which drives the hydrogen

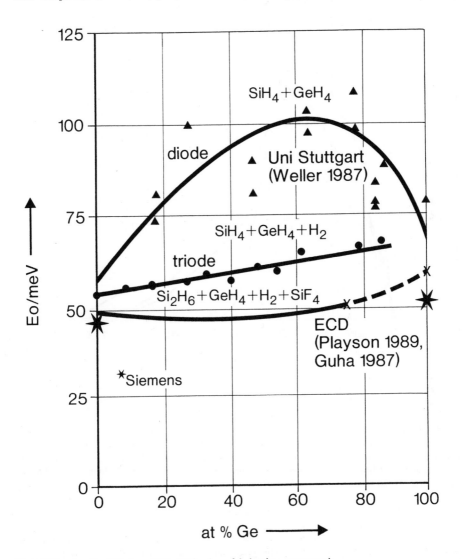

Fig.5: Urbach tail width for a-SiGe:H (various fabrication processes)

out during the evolution process. We could not detect such seeds by Raman measurements probably because the seed volume is too small. There seems to be a limit of bombardment for the preparation process which has to be avoided for device quality material.

The oxygen content of "soft"-deposited a-Ge:H samples also implies other drawbacks under specific deposition conditions allowing reasonably good electronic properties. The relatively high number of voids causes a fast absorption of oxygen, which diffuses from the air into the bulk. SIMS measurements on our samples showed that "soft" samples already have a 20 times higher oxygen content than "hard" samples after 5 minutes contact with air, and after 72 hours a

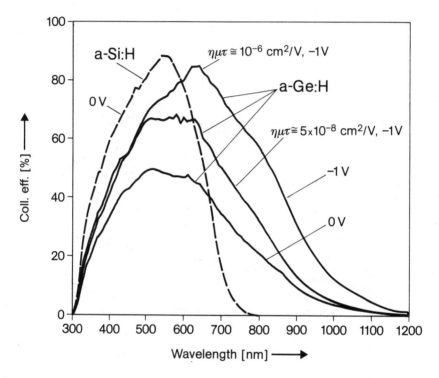

Fig.6: Spectral sensitivity of a-Ge:H and a-Si:H solar cells

saturation is almost attained. In the case of saturation, even a 50 times higher content is observed. In contrast, "hard" samples show no absorption of oxygen at all even after 200 hours. Since the electrical properties of the "soft" samples show a correlation with the oxygen absorption, the photoconductivity is impaired, e.g. decreasing from a best value of $10^{-6} cm^2/V$ by a factor of about 10 due to the oxygen absorption.

## CONCLUSION

It was shown that a-Ge:H can be manufactured exhibiting good electronic properties and good stability. A "hard" deposition is suitable in conjunction with a high ion bombardment. With this kind of deposition, voids in a-Ge:H material can be almost completely avoided as could be shown by SAXS measurements. There is no decrease in photoconductivity of a-Ge:H during illumination for 2500 hours AM1. A stress test with keV-electrons shows the stability of the material to be up to 1000 times better than that of conventional good a-Si:H. Strong bonding of hydrogen as GeH was shown by IR and evolution in "hard" samples. "Soft" samples are unstable in photo and dark conductivity for light soaking and oxygen incorporation from air.

## ACKNOWLEDGEMENT

The authors would like to thank H.G.Haubold for taking and evaluating the SAXS measurements. We also appreciate the stimulating discussions with F.Karg, W.Kusian and W.Kasper. The measurements of the oxygen and hydrogen effects by W.Juergens and P.Zeitler are gratefully acknowledged. This work was financially supported by the Federal Ministry of Research and Technology.

## REFERENCES

1. W.Kasper, H.Böhm, B.Hirschauer, submitted to J.Appl.Phys.

2. F.Karg, H.Böhm, K.Pierz, J.Noncryst.Sol.114,477(1989)

3. F.Karg, B.Hirschauer, W.Kasper, K.Pierz, Solar Energie Mat,(1991) to be published

4. U.Schneider, Thesis, University Kaiserslautern (1989)

5. U.Schneider, A.Scholz, B.Schröder, F.Karg, H.Kausche, Jap.J.Appl.Phys.Lett. 30(1991) in press

6. A.Matsuda, S.Mashima, K.Hasezaki, A.Suzuki, S.Yamasaki, P.McElheny, submitted to Appl. Phys. Lett.

7. H.Kausche, K.Prasad, R.Plättner, Proc.9th. E.C.PVSEC p.595(1989)

8. R.S. Crandall, Y.S. Tsuo, Y. Xu, and A.H. Mahan, Solar Cells, to be published

# CHARACTERIZATION AND STUDY OF LIGHT DEGRADATION EFFECTS IN ECR a-Si:H,Cl FILMS

C.P.Palsule, S.Gangopadhyay, C.Young, T.Trost, and M.Kristiansen
Texas Tech University, Lubbock, Tx 79401

## ABSTRACT

We have studied the electrical and optical properties along with light induced degradation of a-Si:H,Cl films prepared by electron cyclotron resonance(ECR) plasma. We find that there is an irreversible decrease in dark conductivity of these films after vacuum anneal at 200 C. The degradation in photoconductivity of these films due to the Staebler-Wronski effect is smaller than that observed in glow discharge prepared a-Si:H films.

## INTRODUCTION

Recently, investigations of the reversible light induced changes in hydrogenated amorphous silicon(a-Si:H) have sparked increasing interest in the kinetic behavior and microscopic origin of the metastable defect generation. Staebler and Wronski[1] were the first to discover that both doped and undoped glow discharge deposited a-Si:H films show a decrease in dark and photoconductivity when exposed to light for a prolonged time. These changes were found to be reversible by annealing at elevated temperatures (>150 C). This effect has been attributed to a reversible increase of density of gap states. The light induced defects have been shown to be silicon dangling bonds[2] which act as recombination centers for photoexcited carriers and lead to a shift of the dark Fermi level towards the midgap. Different techniques[3-6] have been used to study the light induced increase of the density of states and several microscopic models have been proposed to explain the effect.[6-10]

The growing attention devoted to halogenated and hydrogenated glow discharge amorphous silicon films (GD:a-Si:H,Cl and a-Si:H,F) is due to the good possibility of utilizing them as materials for low cost photovoltaics.[11] Madan et al[12] have proposed that fluorine acts as a dangling bond terminator and its incorporation in amorphous silicon films leads to a midgap density of states comparable or lower than those of a-Si:H films without fluorine. There are some studies performed on a-Si:H,Cl [13-16] where it has been shown that chlorine could also be acting as a dangling bond compensator in an amorphous silicon matrix. We have prepared a-Si:H,Cl films using a silicon tetrachloride and hydrogen gas mixture in an electron cyclotron resonance plasma. We have previously reported our preliminary investigations on the ECR plasma and the characterization of the films.[17]

Despite the number of investigations on the role of chlorine on the optical and electronic properties of a-Si:H,Cl, the light degradation effects on these films have not been studied so far. This work is the first examination of the light induced degradation on a-Si:H,Cl films.

## EXPERIMENTAL DETAILS

In our ECR system[17] microwave power was introduced into the plasma chamber at a frequency of 2.45 GHz via a microwave waveguide and a vacuum window. A magnetic mirror-like static magnetic field with a peak to minimum ratio of approximately 1.25 was applied axially to the plasma chamber by a pair of coils. The films were deposited with absorbed microwave power of 20 W, $H_2/SiCl_4$ ratio of 9:1 and at a substrate temperature of 285 C. For comparison purposes two device quality glow discharge deposited a-Si:H films were obtained from Solar Energy Research Institute and University of Arkansas.

The steady state photoconductivity and dark conductivity measurements were made using coplanar colloidal silver contacts deposited on top of the film with a gap of 1 mm. White light from a solar simulator was used for illumination of the samples. We repeated some of the measurements with monochromatic light (575 nm) and found that the results were same irrespective of the source of illumination. Annealing of the samples was performed in a closed cycle refrigeration system in which the temperature can be varied from 15 K to 475 K. Hydrogen and chlorine contents of the samples were analyzed by infrared absorption using a Perkin-Elmer spectrometer.

## RESULTS

### OPTICAL AND ELECTRICAL PROPERTIES OF THE FILMS

Infrared spectra of these films were recorded in the range of 400-3000 $cm^{-1}$. Figure 1 shows the spectrum of a typical a-Si:H,Cl film as prepared and after vacuum annealing at 200 C for 2 hours. The origin of the bands at 2000, 2090, 890, 840 and 630 $cm^{-1}$ is well known. Referring to the vibrational properties of a-Si:H, these bands are attributed to the stretching modes of Si-H(2000 $cm^{-1}$) and Si-$H_2$(2090 $cm^{-1}$), the bending modes of Si-$H_2$ groups(890 $cm^{-1}$) and Si-$H_n$ chains (890 and 840 $cm^{-1}$), and the wagging rocking mode of Si-H and Si-$H_2$ species(630 $cm^{-1}$), respectively.[11] However, in this case 2090 $cm^{-1}$ band has some contribution from SiHCl mixed groupings as proposed by Kalem et al[18,19]. The three dominant bands which are not present in a-Si:H samples but are observed here are at 523, 800 and 2250 $cm^{-1}$. The band at 523 $cm^{-1}$ is attributed to the stretching mode of Si-Cl[18]. Kalem et al[18] observed a band at 795 $cm^{-1}$ and assigned it to the wagging mode of Si-$H_2$ in Si-$H_2$Cl configuration. They have suggested that electronegativity increase due to the replacement of one silicon atom by one chlorine atom would induce an increase of 50 $cm^{-1}$ in the wagging mode frequency of Si-H radicals. For this reason we suggest that the band at

800 cm$^{-1}$ is due to the wagging mode of Si-H in SiHCl$_3$. The band at 2250 cm$^{-1}$ then is the stretching mode of Si-H in SiHCl$_3$. The correlation of these two peaks can be verified from the fact that vacuum annealing at 200 C almost doubled the intensity of these two bands while other bands remained the same. The hydrogen and chlorine contents of these films are found to be about 4% and 3.5%, respectively.

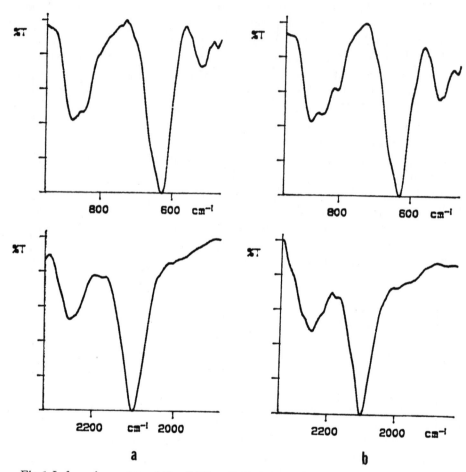

Fig.1 Infrared spectra of the ECR a-Si:H,Cl film (a) as deposited (b) after annealing in vacuum at 200 C.

A drastic irreversible change in dark conductivity was observed when a-Si:H,Cl samples were annealed at 200 C in vacuum. A typical sample showed a dark conductivity change from as deposited value of $1.1 \times 10^{-8}\Omega^{-1}cm^{-1}$ to $7.5 \times 10^{-13}\Omega^{-1}cm^{-1}$ after annealing. Photoconductivity of the sample with AM1 illumination (100mW/cm$^2$) of white light changed from $8 \times 10^{-6}\Omega^{-1}cm^{-1}$ to $3 \times 10^{-7}\Omega^{-1}cm^{-1}$ after annealing.

We measured the room temperature photocurrent of a-Si:H,Cl films as a function of photon energy to obtain the band gap. As can be seen from Figure 2 there is a shoulder at low energies followed by a sharp rise in the photocurrent. The photocurrent $I_p$ can be expressed by the following expression:

$$I_p = eN(1 - R)[1 - \exp(-\alpha d)]\eta\mu\tau EL, \tag{1}$$

where N(1-R) is the incident flux corrected for surface reflection, $\alpha$ is the absorption coefficient, d is the film thickness, $\eta$ is the generation efficiency of free carriers, $\tau$ is the electron life time, $\mu$ is the electron mobility, E is the electric field, and L is the length of the electrodes. In the weakly absorbing region, the above equation can be approximated as:

$$\frac{I_p}{eN(1-R)} \sim \alpha dEL\eta\mu\tau \tag{2}$$

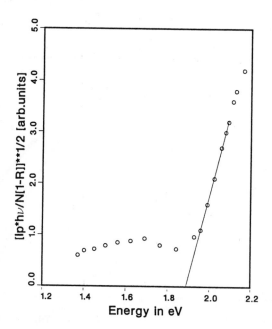

Fig.2 $[I_p h\nu/eN(1-R)]^{1/2}$ plotted as a function of energy.

So by plotting $[I_p h\nu/eN(1-R)]^{1/2}$ as a function of $h\nu$, we obtained an optical bandgap of 1.9 eV(Figure 2).[14] This value is comparable to those obtained before by photoconductivity measurements in a-Si:H,F[12] and a-Si:H,Cl[14] samples.

## LIGHT INDUCED DEGRADATION

The photoconductivity of a-Si:H,Cl samples decreased with illumination time similar to a-Si:H samples. Figures 3a and 3b show the plots of log of normalized photoconductivity versus log of illumination time for a-Si:H and a-Si:H,Cl samples at 100 mW/$cm^2$ and 500 mW/$cm^2$, respectively. For GD a-Si:H film, at short illumination times (<5minutes), there is a slow but monotonic decrease of photoconductivity with time but at longer times the curve exhibits a $t^{-1/3}$ dependence on illumination time as suggested by Stutzmann et al.[6] For the chlorinated film there are two regions with slopes of 0.14(t<5 min.) and 0.24(5 min<t<60 min) separated by a kink. We have observed this behavior in all our a-Si:H,Cl samples, but at present we are unable to explain it. As can be seen from Figure 3, the final value of normalized photoconductivity of ECR a-Si:H,Cl films is higher than that of GD a-Si:H for the same illumination intensity and time of illumination. This smaller decrease in normalized photoconductivity indicates smaller light degradation of a-Si:H,Cl films than a-Si:H films.

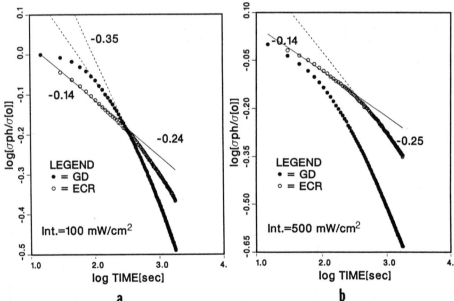

Fig.3 Normalized photoconductivity versus illumination time (a) 100 mW/$cm^2$ (b) 500 mW/$cm^2$.

The intensity dependence of the GD a-Si:H and ECR a-Si:H,Cl films is shown in Figure 4. In case of GD a-Si:H film there are two regions with $\sigma \sim I^{0.89}$ (<50 mW/$cm^2$) and $\sigma \sim I^{0.40}$(>50 mW/$cm^2$). In case of a-Si:H,Cl $\sigma \sim I^{0.74}$ holds for the entire intensity range used for measurement. This difference in $\gamma$ (in the relation of $\sigma \sim I^\gamma$) for GD a-Si:H film reflects a difference in the recombination kinetics of the two regions. At low intensities the recombination has the monomolecular form ($\gamma \sim 1.0$) whereas it is bimolecular in the high intensity region($\gamma \sim 0.5$). For a-Si:H,Cl films the value of $\gamma$ is in between the two limits- $0.5 < \gamma < 1.0$. Madan et al[12] have observed $\gamma = 0.7$ for a-Si:H,F films. It does not seem possible to explain this on the basis of a mixture of monomolecular and bimolecular recombination. But Kagawa et al[20] have reported that such an intermediate value of $\gamma$ can be explained if the recombination rate is controlled by the capture rate of free holes by donor like recombination centers. So we think that this can justify $\gamma = 0.74$ for the a-Si:H,Cl films.

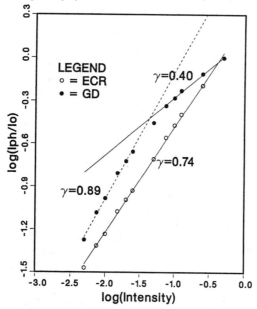

Fig.4 Intensity dependence of photocurrent for GD a-Si:H and ECR a-Si:H,Cl films.

Stutzmann et al[6] have suggested that the variation of the photoconductivity with illumination time and intensity assuming a monomolecular recombination is given by

$$\left(\frac{1}{\sigma_{ph}(t)}\right)^3 - \left(\frac{1}{\sigma_{ph}(0)}\right)^3 = C_1 \cdot \frac{t_{ill}}{G} \tag{3}$$

in the long time limit, and by

$$\left(\frac{1}{\sigma_{ph}(0)}\right)\left[\frac{1}{\sigma_{ph}(t)} - \frac{1}{\sigma_{ph}(0)}\right] = C_2 t_{ill} \tag{4}$$

for short times and/or small generation rates, where $C_1$ and $C_2$ are constants. Figures 5a and 5b show the plot of $\left(\frac{1}{\sigma_{ph}(t)}\right)^3 - \left(\frac{1}{\sigma_{ph}(0)}\right)^3$ versus time for t> 5 min.

at different intensities for GD a-Si:H and ECR a-Si:H,Cl samples, respectively. For GD a-Si:H a linear dependence on illumination time is only obtained for the monomolecular recombination region($<50$ mW/cm$^2$). In case of ECR a-Si:H,Cl films a linear dependence is obtained for intensities up to 500 mW/cm$^2$. This seems to support the claim of Kagawa et al[20] about the recombination mechanism. For short time regimes($<5$ min.) according to Equation 4 we obtain linear dependence on time for GD a-Si:H but we do not see the linear dependence in ECR a-Si:H,Cl films.

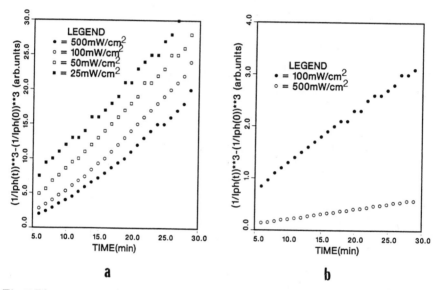

a    b

Fig.5 Photocurrent plotted according to Equation 3 (a) for GD a-Si:H film (b) for ECR a-Si:H,Cl film.

## CONCLUSIONS

In our ECR deposited a-Si:H,Cl films, a vacuum anneal at 200 C shows an irreversible decrease in dark current which is accompanied by a doubling of the intensity of infrared absorption bands at 800 cm$^{-1}$ and 2250 cm$^{-1}$ We claim that these bands can be assigned to wagging and stretching modes of Si-H in SiHCl$_3$, respectively. We suggest that there might be a correlation between increase in the intensity of the abovementioned infrared absorption bands and the reduction of dark conductivity. The optical energy gap of these films is around 1.9 eV which is quite a high number for a material containing only 4% hydrogen. So we feel that the chlorine atoms do contribute to this increase of the bandgap. The decrease in photoconductivity of these films at a particular intensity is smaller than glow discharge prepared a-Si:H films, so we think that these films undergo smaller light degradation than a-Si:H films. The variation of photocurrent with intensity seems to be consistent with the explanation of Kagawa et al[20] about the recombination taking place through the capture of free holes by donor like recombination centers.

## ACKNOWLEDGEMENTS

We are grateful for the support provided by the Texas Advanced Technology Project, the National Science Foundation and J.C.Schumacher Company. We would also like to thank Dr.R.Crandall from Solar Energy Research Institute and Dr.H.Naseem from University of Arkansas for providing us with the glow discharge prepared a-Si:H films.

## REFERENCES

1. D.L.Staebler, and C.R.Wronski, Appl.Phys.Lett.31, 292 (1977)
2. R.A.Street, Appl.Phys.Lett.42, 507 (1983)
3. H.Dersch, J.Stuke, and J.Bleichler, Appl.Phys.Lett.38, 456 (1980)
4. J.Kocka, M.Vanecek, H.Stuchlik, O.Stika, I.Kubelik, and A.Triska, Proceedings of the Int.Conf.on Amorphous Semiconductors,vol.3 (CIP-AP, Bucharest,1982), p.150
5. N.M.Amer, A.Skumanich, and W.B.Jackson, Physica.B 117-118, 897 (1983)
6. M.Stutzman, W.B.Jackson, and C.C.Tsai Phys.Rev.B, 32, 23 (1985)
7. F.Schauer, and J.Kocka, Phil.Mag.B, 52, L25 (1985)
8. N.Ishii, M.Kumeda, and T.Shinizu, Jap.J.appl.phys, 24, L244 (1985)
9. R.S.Crandall, Phys.Rev.B, 36, 2645 (1987)
10. E.Bhattacharya, and A.H.Mahan, Appl.Phys.Lett.52, 1587 (1988)
11. M.H.Brodsky, M.Cardona, and J.J.Cuomo, Phys.Rev.B, 16, 3556 (1977)
12. A.Madan, S.R.Ovshinsky, and E.Benn, Phil.Mag.B, 40, 259 (1979)
13. W.W.Kruehler, R.D.Plaettner, M.Moeller, B.Rauscher, and W.Stetter, J.Non-crystalline Solids, 35-36, 333 (1980)
14. J.Chevalier, S.Kalem, S.Al Dallal, and J.Bourneix, J.Non-crystalline Solids, 51, 277 (1982)
15. P.Danish, S.Georgiev, and U.Jahn, Sol.Energy Mater., 9, 405 (1984)
16. S.Kalem, J.Chevalier, S.Al Dallal, and J.Bourneix, J.Physics Paris, 42, C4-361 (1981)
17. S.Gangopadhyay, T.Trost, M.Kristiansen, C.Young, P.Zheng, C.Palsule, and M.Pleil, MRS Symposium Proceedings,192(San Francisco, 1990), p.627
18. S.Kalem, R.Mostefaoui, J.Bourneix, and J.Chevalier, Phil.Mag.B, 53, 509 (1986)
19. G.Luckovsky, Solid State Commun., 29, 571 (1979)
20. T.Kagawa, N.Matsumoto, and K.Kumabe, Phys.Rev.B, 28, 4570 (1983)

Preparation and Properties of Amorphous Silicon Films Produced
Using Electron Cyclotron Resonance Plasma

Vikram L. Dalal,* Ralph D. Knox,† B. Moradi,* A. Beckel,* and S. VanZante*
*Department of Electrical and Computer Engineering
†Microelectronics Research Center
Iowa State University
Ames, IA 50011

## ABSTRACT

Electron-cyclotron-resonance (ECR) plasma offers a potentially better way of controlling the growth chemistry of a-Si:H. Such control can be expected to improve the microstructure of a–Si:H, and hence its stability. In this paper, we report on the preparation and properties of a-Si:H films produced using a remote ECR plasma at low pressures. It is shown that the ECR- a-Si:H films have electronic properties comparable to glow-discharge produced films. The stability of ECR-films appears to be superior to glow-discharge-films.

## INTRODUCTION

Light-induced degradation of a-Si:H films (Staebler-Wronski effect) [1] is a major technological problem. Much work has been done to identify the mechanisms responsible for the effect. Recent evidence from both theoretical simulations [2] and experimental data [3,4,5] suggests that localized structure, impurities, and H bonding may play a role in light-induced degradation. It is well known that a-Si:H is not a perfect continuous-random-network (CRN). Rather, it is full of voids, [6] poor H bonding ($SiH_2$, polysilanes) and impurities such as c,o and N. In particular, the poor bonding arrangement surrounding internal voids, such as $SiH_2$ and other poorly bonded H, are believed to play a major role in degradation, since low-energy Si-H bonds can be broken easily by carrier recombination.

In this paper, we examine a growth technique which may be useful in reducing voids and improve microstructure and H bonding. This technique, electron-cyclotron-resonance (ECR) remote plasma deposition, produces a-Si:H films of acceptable quality and stability.

## GROWTH CHEMISTRY OF a-Si:H

It is well known that growth of a-Si:H is a multi-step process [6]. The steps in the process are depicted in Fig. 1. First, the bonded H has to be removed, which can be done with an H ion or radical. Next, $SiH_3$ has to insert itself into the surface bond, and finally, the $H_2$ has to be eliminated from neighboring bonded $SiH_3$ radicals, giving rise to Si-Si bonds. Each of these steps must occur perfectly to get a CRN. Such perfection, of course, is impossible. In particular, multiple $SiH_3$ radicals (hills) and no radical attachments (valleys or voids) are statistically likely to occur. A consequence of hill-and-void attachments is that to get a void-free network, we have to provide energy to the growing surface to facilitate the migration of multiply-attached radicals to neighboring voids. Increasing growth temperature is one method of providing surface energy. However, increasing the temperature of the entire lattice leads to breaking of the few statistically inevitable imperfect (weak) bonds, leading to increased defect states.

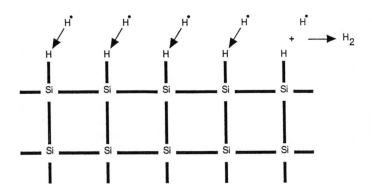

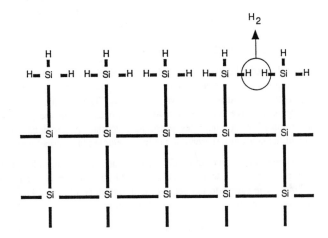

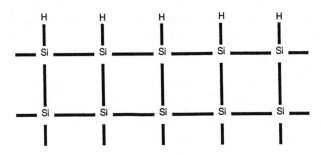

Figure 1
The Critical Steps in Formation of a-Si:H Films

A much better way to improve bonding during growth would be to provide excess energy only to the surface. One way is ion bombardment, which was used by Knights et al., and recently by Paul and co-workers, [8] to improve the structure of a-Si:H and a-(Si,be):H. Another way is to provide surface energy by using etching-during-growth, by using reactive H or F ions during growth [9]. Dalal et al. showed [9] 10 years ago that etching by F during growth improved the microstructure of a-Si:H films, leading to microcrystallinity, a result that can also be achieved by using H as an etchant [10]. In this process, poorly bonded Si (e.g., distorted bonds, Si hills) is etched away by H or F, leaving behind a vacant site for a more perfect bonding. Such etching, in combination with ion bombardment, can be expected to produce films with better order and microstructure. This conclusion is validated by the observation that carried to its logical conclusion, such an etching-plus-bombardment leads to crystallinity at low temperatures, not only for Si, but for C (diamond films) as well.

ECR deposition, in principle, is one technique which provides a high density of H (or F) ions, and can be used to provide bombardment of the surface [11]. In this paper, we report on our results with low-pressure ECR remote plasma deposition.

## DESCRIPTION OF THE REACTOR

The reactor is a load-locked UHV system, fitted with an ECR plasma source. The system is pumped by a turbo-pump, even during processing, to keep impurity levels low. A schematic diagram of the apparatus is shown in Fig. 2. The sample can be biased independently negative or positive with respect to the plasma, though the present results report only on films produced using substrates at ground potential. A screen can be inserted between the plasma and the substrate, thereby achieving a remote plasma operation.

In operation, $H_2$ is introduced into the ECR source and $SiH_4$ is introduced near the sample. Typical flow rates are 4-4.5 sccm for $SiH_4$, and 25-30 sccm for $H_2$. The operating pressure is low ($\sim 10$ m$\tau$). The substrate temperature is in the range of 275–295° C. Typical growth rate is 0.5-1Å/sec. Film thicknesses are in $\sim 1$ μm range.

## RESULTS

1. H-bonding.

The H-bonding, as determined by FTIR spectroscopy, is shown in Fig. 3. There is minimal $SiH_2$ bonding, with primary bonding being Si-H. The concentration of H, determined by measuring Si-H peak at 2000 cm$^{-1}$, is 7-8%.

2. Dark conductivity and Activation Energy.

The conductivity activation energy, measured between 30° C and 200° c, is in the range of $\sim 0.8$ eV. The corresponding dark conductivities are $1-4 \times 10^{-10}$ s cm$^{-1}$, and photo-conductivities under AM1.5 spectrum are $5-7 \times 10^{-5}$ s cm$^{-1}$. The typical photo-to-dark conductivity ratios are $3-10 \times 10^5$. The γ factor of photo-conductivity, $\sigma_{pc} \sim$ (Intensity)$^\gamma$, are 0.87-0.9. The electron (μτ) product is typically $5 \times 10^{-7}$ cm$^2$/V.

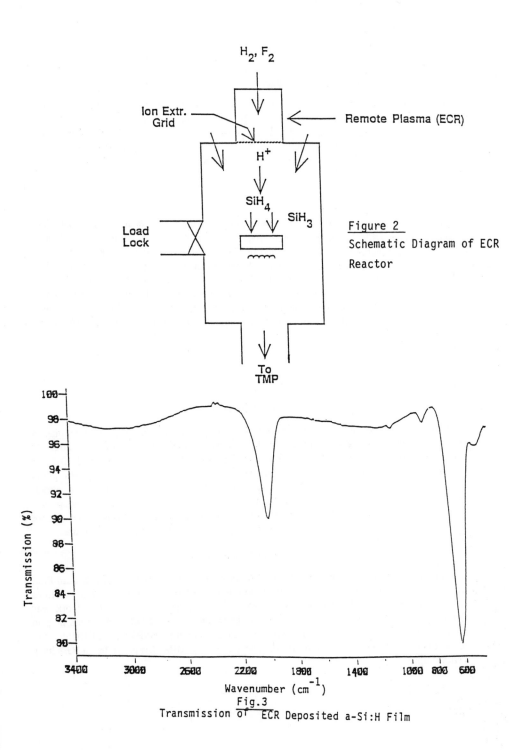

Figure 2
Schematic Diagram of ECR
Reactor

Fig.3
Transmission of ECR Deposited a-Si:H Film

### 3. Mid-gap Defect Density and Urbach Edge.

The mid-gap defect density and Urbach energy were determined by sub-gap photo-conductivity measurements. The results for a typical film, prepared at 275° C, are shown in Fig. 4. The Urbach energy for valence band tails is 44 meV, and the mid-gap absorption is in the 0.3-0.5 cm$^{-1}$ range. The Iauc bandgap is 1.75 eV. The results for films prepared at 295° C are similar, except that mid-gap absorption is in the range of 0.5-1.0 cm$^{-1}$, indicating still high quality.

### 4. Stability

The light-induced instability in photo-conductivity was studied using an ELH lamp as the excitation source with intensity of 100 mW/cm$^2$. The results of decrease in photo-conductivity as a function of time are plotted in Fig. 5, and compared with the results for a standard slow-discharge a-Si:H film from which devices of > 8% efficiency had been made. It appears that the ECR-film has somewhat better stability than comparable glow-discharge film. However, these results are preliminary, and indicate only the changes in electron ($\mu\tau$) product. What we need are complementary data on hole ($\mu\tau$) products. Such experiments are now being implemented, and more definitive conclusions can be made only after such data is available.

### CONCLUSIONS

In conclusion, we have shown that a-Si:H films produced using ECR-remote plasma deposition appear to have a quality comparable to that of standard glow-discharge films. In particular, electron ($\mu\tau$) products, Urbach energy for valence band tails, and mid-gap absorption activation energy and dark conductivity are comparable to device quality glow-discharge films. The decay in electron ($\mu\tau$) product, as indicated by decay of photo-conductivity, is less than for a glow-discharge film. However, the lack of data on properties of holes in ECR films prevents us from drawing any definitive conclusions at the present time. Such experiments, which require the use of p-i-n cells, are planned for the future.

### ACKNOWLEDGMENTS

Part of this work was supported by NASA, under a SBIR contract with Iowa Thin Film Technologies. Many interesting discussions with Frank Jeffrey and Derick Grimmer of Iowa Thin Films are gratefully acknowledged. Part of the equipment used in this work was provided as a gift by Polaroid Corp., and we thank Dr. Bieber of Polaroid for his help in securing that gift. A. Becket and S. VanZante were supported by NSF under a REU grant to Iowa State University. B. Moradi was supported by IBM Corp. under a Graduate Fellowship.

Figure 4

**SUB-BANDGAP PHOTOCONDUCTIVITY OF
SAMPLE 2-116**

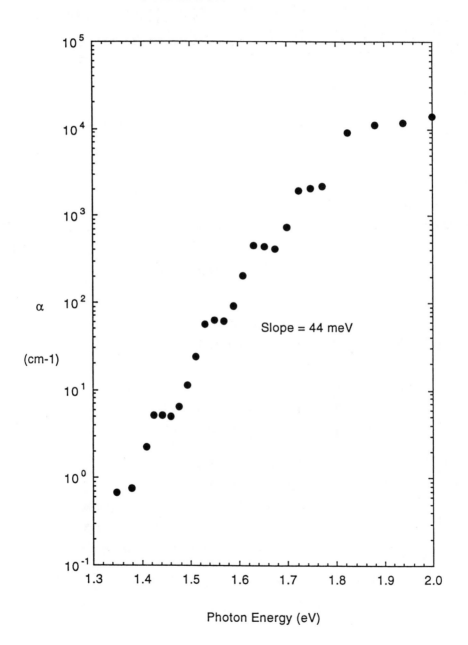

Slope = 44 meV

Photon Energy (eV)

Figure 5

# PHOTOCONDUCTIVITY RATIO VS TIME OF ECR AND GD DEPOSITED a-Si:H

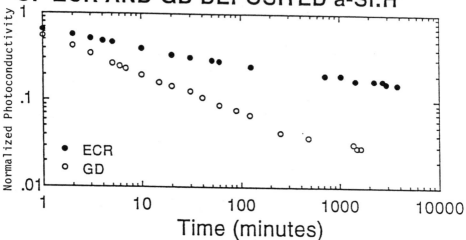

REFERENCES

[1]   D. Staebler and C. Wronski, *Appl. Phys. Lett.* 31, 292 (1977).

[2]   R. Biswas, Proc. of this Conference.

[3]   T. Unold and J. D. Cohen, *Proc. Material Res. Society* 149, 719 (1990).

[4]   V. L. Dalal and C. Fuleihan, *Proc. Material Res. Society* 149, 601 (1989).

[5]   C. M. Fertmann, J. O'Dowd, J. Newton, and J. Fischer, "Stability of a-Si alloy materials and devices," *Proc. AIT Conf.* 157, 103 (1987).

[6]   D. L. Williamson et al., *J. Non-Cryst. Solids* 114, 226 (1989).

[7]   J. C. Knights, *J. Non-Cryst. Solids* 35-36, 159 (1980).

[8]   S. J. Jones, W. A. turner, and W. Paul, *J. Non-Cryst. Solids* 114, 525 (1989).

[9]   V. L. Dalal, C. M. Fertmann, and E. Eser, *Proc. AIP Conf.* Vol. 73, 15 (1981).

[10]  C. C. Tsai, G. B. Anderson, R. Thompson, and B. Wacker, *J. Non-Cryst. Solids* 114, 151 (1989).

[11]  Y. A. Shing, *Proc. IEEE Photovolt. Spec. Conf.* 20, 202 (1988).

# Studies of Light Soaking Stability in r f Sputter-Deposited a-Si:H

A. Wynveen,[1] J. Fan,[1] J. Kakalios[1] and J. Shinar[2]

[1]School of Physics and Astronomy,The University of Minnesota,Minneapolis, MN 55455
[2]Ames Laboratory - U.S. DOE and Physics Dept., Iowa State University, Ames IA 50011

## Abstract

The light-induced degradation of the dark conductivity and photoconductivity (the Staebler-Wronski effect) is reduced in r.f. sputter deposited hydrogenated amorphous silicon (a-Si:H) when the r.f. power employed during deposition is increased. The high r.f. power samples display a photo to dark conductivity ratio of $10^3$ when illuminated with 100 mW/cm$^2$ of heat-filtered white light, with a measured photoconductivity $> 7 \times 10^{-5}$ (ohm cm)$^{-1}$.

## Introduction

There has been great progress in the development of thin film, large area solar cells employing hydrogenated amorphous silicon (a-Si:H) as the photovoltaic material.[1] A major limitation of a-Si:H based solar cells is the light-induced degradation of the dark conductivity and photoconductivity, also known as the Staebler-Wronski effect.[2] Despite considerable investigative effort, the microscopic mechanism responsible for the light induced defect generation remains undecided. Studies of the creation and annealing kinetics of the SWE indicate that hydrogen motion is directly involved in the formation of the light-induced defects.[3-8] As the hydrogen diffuses, it can change the bonding configurations of the silicon network. The hydrogen diffusion coefficient, $D_H$, is found to have a power-law time dependence, decreasing at longer times as

$$D_H(t) = D_{OO}\,(\omega t)^{-\alpha} \tag{1}$$

where $D_{OO}$ is the microscopic diffusion coefficient, $\omega$ is the attempt-to-hop frequency, and $\alpha$ is the power law exponent.[9,10] The time dependent $D_H$ is analogous to dispersive electronic transport, and an exponential distribution in energy of Si-H trapping states, $\exp[-E/kT_O]$, has consequently been proposed to account for the time decay of $D_H$.[9] Following this analogy, $\alpha$ was concluded to depend on temperature as $(1 - T/T_O)$ where $kT_O$ is the width of the trap distribution.

Recently there has been growing interest in employing deposition techniques other than glow discharge in order to improve the light-soaking stability of a-Si:H.[1] Shinar and co-workers have found that the time decay of $D_H$ is much stronger ($\alpha$ closer to unity) in r.f. sputter deposited a-Si:H containing a significant microvoid content.[10] Similar results for glow-discharge grown a-Si:H were recently reported by Tang et. al.[11] This stronger decay is suspected to involve a broader distribution of Si-H trapping states, believed to result from a higher microvoid concentration in these films. This higher microvoid content is reflected in a larger $T_O$, leading to an $\alpha$ in eqn. (1) closer to unity. Shinar and co-workers have also observed significantly slower hydrogen diffusion in r.f. sputter

241

deposited a-Si:H films which have a low hydrogen content.[12]  Such films should have a reduced SWE, if hydrogen motion is indeed responsible for the light-induced defects.[19]  To test this proposal, we report here measurements of the photoconductivity and the light-soaking stability of the electronic properties of these r.f. sputter deposited a-Si:H films.

## Experimental    Procedure

All samples studied were prepared by 13.56 MHz radio frequency sputter deposition from an undoped polycrystalline 15 cm diameter silicon target.  The amorphous silicon films are deposited onto Corning 7059 substrates, held at a distance of approximately 2.5 cm from the sputtering target.  The samples were grown in an atmosphere of 10 mTorr argon and 0.5 mTorr hydrogen.  The substrates were not intentionally heated, and were placed on a water cooled pedastal; nevertheless previous studies have indicated that the electronic and ionic bombardment during film growth raises the effective temperature at the growing film surface to ~ 150C.  The total transmitted r.f. power was systematically varied from 50 to 600 W (0.27 to 3.3 W/cm$^2$).  Film thicknesses were in the range of 1-2 μm.  Full characterization of the optical and microstructural properties of these films has been reported previously.[13]

Co-planar conductance measurements were performed using silver paint electrodes (length 0.4 cm, separation 0.1 cm) applied to the a-Si:H samples.  The conductivity data described here was obtained using voltages of 90V, which yielded linear current-voltage characteristics.  To remove any effects of prior light exposure, the sample was annealed in air at 150C for two hours, defined as state A, and then slowly cooled (cooling rate 2-3 C/min) back to room temperature.  Photoconductivity measurements were performed with a heat-filtered tungsten-halogen lamp, with an intensity of 100 mW/cm$^2$.  The samples were illuminated for 30 min, which then defined the Staebler-Wronski state B.  It was confirmed that the photo-induced changes in the dark and photoconductivity were reversible by reannealing the sample back into state A.

## Results

The ratio of the dark current after light soaking to the annealed state A current for the r.f. sputtered a-Si:H samples as a function of the incident r.f. power is shown in fig. 1.  Samples grown with an r.f. power of less than 200 W show a decrease in the dark current of approximately an order of magnitude after light soaking.  In contrast, samples grown with an r.f. power > 400 W shown no change in dark current after extended illumination.  This improved stability is also reflected in the time dependence of the photoconductivity during light soaking, as shown in fig. 2.  Here the photocurrent, normalized to its value at the start of illumination, is plotted against the exposure time.  For samples deposited at higher r.f. power levels, the decay of the photocurrent becomes slower, and is essentially time-independent for a 30 min exposure for a sample grown with 600 W of r.f. power.  A factor of two reduction in $\sigma_{ph}$ is observed in the 600 W sample after exposures longer than 40 hours.  A complete study of the influence of the r.f. power on the $\sigma_{ph}$ time dependence will be reported separately.

A reduced Staebler-Wronski effect is typically observed in glow discharge deposited films which have a high initial defect density in state A.[14]  If the initial

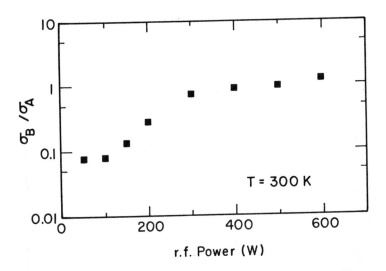

Fig. 1: Ratio of the dark conductivity after 30 min exposure to 100 mW/cm$^2$ heat filtered white light (defined as state B), to the annealed state conductivity, for a series of r.f. sputtered  a-Si:H samples, as a function of the incident r.f. power.

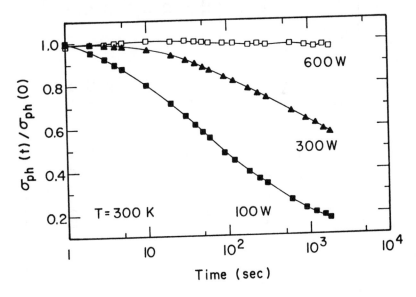

Fig. 2: Time dependence of the photoconductivity (normalized to its value at the start of illumination) during light exposure for increasing r.f. power. The solid lines are guides to the eye.

Table 1

Hydrogen Content and Photoconductivity of r.f. Sputtered a-Si:H

| r.f. power(w) | sample thickness ($\mu$m) | deposition rate (Å/sec) | hydrogen content (at. %) | $\sigma_{ph}(\Omega^{-1}cm^{-1})$ | $\sigma_{ph}/\sigma_A$ | $\gamma$ |
|---|---|---|---|---|---|---|
| 50  | 1.45 | 0.81 | 18   | $2.7 \times 10^{-5}$ | $2 \times 10^{4}$   | 0.93 |
| 100 | 1.90 | 1.76 | 15.4 | $2.9 \times 10^{-6}$ | $1.1 \times 10^{5}$ | 0.88 |
| 150 | 1.65 | 1.53 | 13.7 | $2.1 \times 10^{-5}$ | $3.0 \times 10^{4}$ | 0.88 |
| 200 | 1.35 | 1.43 | 14.3 | $4.3 \times 10^{-6}$ | $1.5 \times 10^{4}$ | 0.87 |
| 300 | 2.06 | 1.91 | 8.5  | $4.3 \times 10^{-6}$ | $2.8 \times 10^{3}$ | 0.86 |
| 400 | 0.73 | 3.92 | 16.9 | $9.7 \times 10^{-7}$ | $9.2 \times 10^{2}$ | 0.79 |
| 500 | 0.95 | 1.68 | 11.4 | $7.4 \times 10^{-5}$ | $1.2 \times 10^{3}$ | 0.79 |
| 600 | 0.95 | 2.64 | 10.9 | $1.2 \times 10^{-5}$ | $1.4 \times 10^{3}$ | 0.74 |

dangling bond density is greater than $10^{17}$ cm$^{-3}$, which is the order of magnitude of the light-induced defect density, then there is minimal change in σ with light soaking. The midgap defects serve as recombination centers which decrease the photoconductivity and pin the Fermi level near midgap, decreasing the dark conductivity. Samples with a high initial defect density therefore typically have low photoconductivities and small $\sigma_{ph}/\sigma_d$ ratios. As shown in Table 1, the r.f. sputtered samples studied here have respectable photo and dark conductivity values, even for the samples grown at high r.f. power which exhibit an enhanced light soaking stability. The photoconductivity of the 500 W sample is $> 7 \times 10^{-5}$ Ω$^{-1}$ cm$^{-1}$ when measured with 100 mW/cm$^2$ of heat-filtered white light. Also listed in Table 1 is the ratio $\sigma_{ph}/\sigma_d$, which for the 500 and 600 W samples is $10^3$. The photocurrent displayed a power law dependence on the light intensity F used, measured using neutral density filters, that is $\sigma_{ph} \propto F^\gamma$, as commonly observed in amorphous semiconductors. A γ value near unity has been interpreted as indicating that the density of photo-excited carriers is much less than the available recombination centers, while a γ ~ 0.5 indicates that the photocarrier density is comparable or exceeds the recombination center density.[15] The observed decrease in γ with increasing r.f. power in Table 1 would is consistent with the dangling bond density is lower in samples grown at higher r.f. power, consistent with an earlier study by Albers and co-workers.[13] The concentration of bonded hydrogen in the a-Si:H films listed in Table 1 was obtained by infra-red absorption measurements from the intensity of the 640 cm$^{-1}$ wag mode.[13,16]

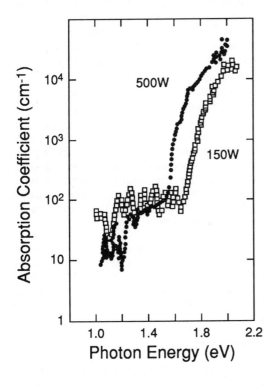

Fig. 3: Plot of the optical absorption coefficient obtained by PPES for r.f. sputtered a-Si:H against photon energy, for samples deposited at 150 and 500 W r.f. power.

The optical absorption of the films was measured using photo-pyroelectric spectroscopy (PPES).[17] This technique is similar to photothermal deflection spectroscopy; in PPES the thermal gradient induced by the weakly absorbed light is detected by a pyroelectric polymer (PVDF) in contact with the sample. By combining measurements when the sample is in thermal contact and not in thermal contact, this technique provides measurements of the optical absorption coefficient $\alpha$ from 0.6 to 2.5 eV, with a sensitivity $\alpha d > 10^{-4}$, where d is the sample thickness. Figure 3 shows measurements of $\alpha$ against photon energy h$\nu$ for samples grown with varying r.f. power. The absorption for the 150 W sample (1.65 µm thick) is blue-shifted compared to the 500 W (0.95 µm) a-Si:H film. Both samples have comparable low Urbach slopes ($E_0 \sim 54$ - 58 meV) and midgap defect densities (and comparable $\sigma_{ph}$ from Table 1). The reduced SWE in the 500 W sample therefore cannot be ascribed to increases in the dangling bond or band tail states with r.f. power. As the r.f. power increases from 200 to 600 W, the Tauc optical gap obtained from plots of $(\alpha h\nu)^{1/2}$ against h$\nu$ for the high photon energy data decreases from 1.7 to 1.45 eV. The photoconductivity data in Table 1 has not been normalized for this decrease of bandgap. A fuller description of the PPES optical absorption data will be published separately.[18]

## Discussion and Conclusion

The amorphous silicon samples grown via r.f. sputtering at a high r.f. power have respectable $\sigma_{ph}$ and $\sigma_{ph}/\sigma_d$ values, and display a reduced Staebler-Wronski effect. It is not clear what aspect of the deposition process is responsible for the superior stability against light-induced degradation in these samples. As mentioned earlier, studies of the hydrogen diffusion coefficient $D_H$ in r.f. sputtered samples have found that the microvoids increase the power law exponent $\alpha$ (eq. 1), that is, they cause a rapid decay of $D_H$.[10] However, the microvoid content, as monitored by the $SiH_2$ and $SiH_3$ bending-scissors mode infra-red absorption at 840-890 cm$^{-1}$, generally decreases with increasing r.f. power. The studies of $D_H$ in r.f. sputter deposited films also found that $D_H$ is significantly slower in films with a low ($\sim 2$ at. %) hydrogen content.[12] As shown in Table 1, there is no systematic decrease in the hydrogen content with increasing r.f. power for the samples studied here. The value of $D_H$ for these films is not known, and may be large. It is suspected that the higher r.f. power employed leads to an unintentional heating of the substrates during deposition, which would result in a better structural relaxation of the a-Si:H. This idea can be tested by examining samples deposited at lower r.f. power levels onto intentionally heated substrates and correlating their light-soaking stability with $D_H$. These studies are presently underway.

This work was supported by the University of Minnesota and the Ames Laboratory, which is operated by Iowa State University for the U.S. Dept. of Energy under Contract No. W-7405-Eng-82. One of us (J.K.) was supported by a McKnight-Land Grant Professorship, and A.W. was supported by a Harry and Viola St. Cyr Scholarship and an UROP award. We are grateful to M. L. Albers (deceased) for sample preparation.

## References

1.  See, for example, P. G. LeComber, Jour. Non-Cryst. Solids, **115**, 1 (1989) and A.I.P. Conf. Proc. no. 157 (1987).

2. D. L. Staebler and C. R. Wronski, Appl. Phys. Lett. **31**, 292 (1977); J. Appl. Phys. **51**, 3262 (1980).
3. J. I. Pankove and J. E. Berkeyheiser, Appl. Phys. Lett. **37**, 705 (1980).
4. H. Dersch, J. Stuke and J. Beichler, Appl. Phys. Lett. **38**, 456 (1980).
5. M. Stutzmann, W. B. Jackson and C. C. Tsai, Phys. Rev. B **32**, 23 (1985).
6. D. E. Carlson, Appl. Phys. A **41**, 305 (1986).
7. W. B. Jackson and J. Kakalios, Phys. Rev. B **37**, 1020 (1988).
8. J. Kakalios and R. A. Street, A.I.P. Conf. Proc. no. 157, ed. by B. L. Stafford and E. Sabisky, 179 (1987).
9. R. A. Street, C. C. Tsai, J. Kakalios and W. B. Jackson, Philos. Mag. B **56**, 305 (1987).
10. J. Shinar, R. Shinar, S. Mitra and J. Y. Kim, Phys. Rev. Lett. **62**, 2001 (1989).
11. X. M. Tang, J. Weber, Y. Baer and F. Finger, Phys. Rev. B **42**, 7277 (1990).
12. J. Shinar, R. Shinar, X.-L. Wu, S. Mitra and R. F. Girvan, Phys. Rev. B **43**,1631 (1991).
13. M. L. Albers, J. Shinar and H. R. Shanks, J. Appl. Phys. **64**, 1859 (1988).
14. A. Skumanich and N. Amer, J. Non-Cryst. Solids **59 & 60**, 249 (1983).
15. A. Rose, Concepts in Photoconductivity and Allied Problems, (R. E. Kreiger, New York) p. 33-53 (1978).
16. M. Cardona, Phys. Status Solidi B **118**, 463 (1983).
17. H. Coufal, Appl. Phys. Lett. **44**, 59 (1984); A. Mandelis, R. E. Wagner, K. Ghandi and R. Baltman, Phys. Rev. B **39**, 5254 (1989).
18. J. Fan and J. Kakalios, to be published.
19. R. Shinar, private communication.

# LIGHT-INDUCED CHANGES IN PHOTOCARRIER TRANSPORT IN MAGNETRON SPUTTERED a-Si:H.

J.R. Doyle, N. Maley, and J.R. Abelson
Coordinated Science Laboratory and Department of Materials Science,
University of Illinois, Urbana, IL 61801.

## ABSTRACT

The effect of light soaking on the steady-state reverse bias collection efficiency has been studied for hydrogenated amorphous silicon films produced by reactive magnetron sputtering. Films with optical gaps of 1.63 and 1.74 eV both showed considerable degradation in the collection efficiency, correlating with increases in sub-gap absorption and decreases in the spectral response quantum efficiency. The collection efficiency data have been fitted with the two-field Hecht expression, and effective mobility-lifetime products have been extracted. These results indicate that straightforward measurements on Schottky barriers can be utilized as sensitive monitors of light induced degradation in a-Si:H.

## INTRODUCTION

Changes in the opto-electronic properties of hydrogenated amorphous silicon with light soaking are typically characterized by photoconductivity, photosensitivity, and sub-bandgap absorption in single layer films using ohmic contacts in a coplanar geometry. However, it is not clear whether a unique correspondance exists between these measurements and the efficiency of a p-i-n solar cell, which is the ultimate technological criteria. Fabrication of p-i-n devices is a complex task, involving both the handling of dopant sources and optimization of very thin doped layers. Straightforward measurements on Schottky barrier structures may provide an attractive compromise between coplanar measurements and p-i-n structures because they provide information on transport of photocarriers in primary photoconduction and are much simpler to fabricate than p-i-n devices.

The interpretation of Schottky barrier photocarrier collection measurements is complex, generally requiring numerical modeling and knowledge of microscopic variables such as carrier mobility, defect distribution, capture cross sections, etc. More simplified analytic models have proposed[1], which may have utility for comparisons if the

248

model parameters are treated as phenomenological rather than fundamental.

Here we present a study of the effects of light induced degradation on the steady-state reverse bias photocarrier collection in Pd/a-Si:H Schottky barrier structures. The a-Si:H was grown by reactive magnetron sputtering, which has been shown to produce high quality material for optical gaps between 1.60 and 1.85 eV[2]. We will discuss the results in terms of a simple analytic model, the "two-field" Hecht relation. These results will be compared to subgap absorption and spectral response in the weakly absorbing regime, to acertain the relative sensitivity of these measurements to light soaking, an essential step in determining the suitability of these types of measurements to studies of degradation.

## EXPERIMENT

Films were deposited by reactive magnetron sputtering of silicon in argon + hydrogen. The Tauc optical gap was varied by changing the partial pressure of hydrogen. The substrates were glass/CTO/n+ a-Si:H supplied by Solarex. Companion samples were deposited on glass for constant photocurrent subgap absorption (CPM) measurements. Prior to deposition of the 1 μm a-Si:H layers the substrates were dipped for 10 sec in 20:1 methanol:HF. After deposition half of the sample was light soaked, and the other half was kept in the dark. Light soaking was for 7 hrs at 0.3 W/cm$^2$ red light (1.8 to 2.0 eV). Pd dots ( area 0.017 cm$^2$) were then deposited by evaporation following another methanol:HF dip.

The reverse bias collection efficiencies and the spectral response measurements were carried out using lock-in detection at 3 Hz. The samples were illuminated at 7 mW/cm$^2$ with red light bias, a value chosen as a compromise between actual solar cell conditions and minimizing degradation of the as-deposited samples during the measurements. Probe beam intensities were < 0.05 mW/cm$^2$.

## RESULTS AND DISCUSSION

In table 1 we list the results of the CPM-determined density of midgap states before and after light soaking  for 2 samples having Tauc optical gaps of 1.63 and 1.74 eV. Both films showed an increase in subgap absorption with light soaking, with the relative change larger for the higher gap sample.  This is consistent with previous results found in our laboratory[2].

Table 1

| Eopt (eV) | DOS (as dep) (cm$^{-3}$) | $\mu\tau$ (as dep) (cm$^2$/V) | DOS (lt skd) (cm$^{-3}$) | $\mu\tau$ (lt skd) (cm$^2$/V) |
|---|---|---|---|---|
| 1.74 | 6.0 x 10$^{15}$ | 2.0x 10$^{-8}$ | 2.2 x 10$^{16}$ | 1.8 x 10$^{-9}$ |
| 1.63 | 1.1 x 10$^{16}$ | 2.6 x 10$^{-9}$ | 2.1 x 10$^{16}$ | 7.0 x 10$^{-10}$ |

In figure 1 we show the reverse bias photocurrent collection efficiciency $\eta(V)$ for the two films, before and after light soaking.  $\eta(V)$ is defined by

$$\eta(V)=\frac{J(V)}{eFT(1-\exp(-\alpha d))}=CJ(V) \qquad (1)$$

where J is the measured current density, F is the probe beam light flux, T is the Pd layer transmittance, $\alpha$ is the absorption coefficient, and d is the sample thickness.  For all samples the wavelength was chosen such that $\alpha$ was approximately 3000 cm$^{-1}$, ensuring nearly uniform absorption.  The overall scale factor C is difficult to measure directly, requiring accurate knowledge of $\alpha$ and T, so it was treated here as a fitting parameter (see below).  In figure 1 we see that both films show drastic reduction in $\eta(V)$ with light soaking, as well as a qualitative change in the shape of the curve.

The light I-V characteristic is a complex function of the Schottky barrier material parameters and can only be adequately analyzed by numerical simulations.  However, it is useful to consider   simpler analytic models for  comparisons between samples and the effects of light soaking.  The so-called "two-field" Hecht expression is one such model that has been applied[1].  This model assumes that a single carrier lifetime-mobility product $\mu\tau$ (usually assumed to be that of the holes) determines the current-voltage relation.  It differs from the usual Hecht relation in that a uniform built-in electric field $F_i$ is assumed to exist over a depletion width $w_i$, in order to account for nonzero short-circuit current.  We have

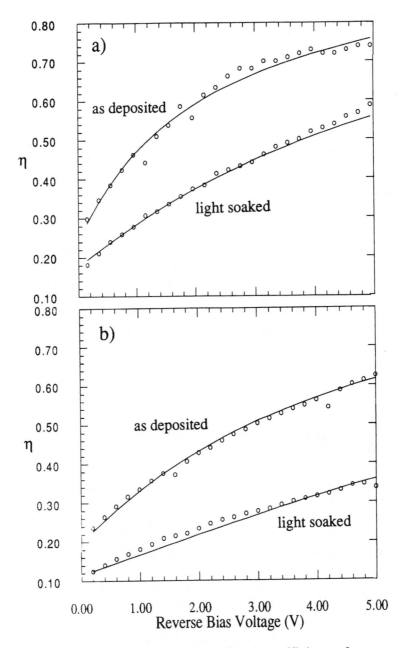

Figure 1. Reverse bias collection efficiency for
a) 1.74 eV and b) 1.63 eV optical gap.

$$\eta(V)=\frac{\mu\tau}{d}[(F+F_i)(1-\exp(\frac{-w_i}{\mu\tau(F+F_i)}))$$

$$+F\exp(\frac{-w_i}{\mu\tau(F+F_i)})(1-\exp(\frac{w_i-d}{\mu\tau F}))] \qquad (2)$$

where d is the sample thickness, F is V/d where V is the applied voltage, and $F_i$ is $V_i/w_i$ where $V_i$ is the diffusion potential for the barrier. The assumptions of uniform fields, a single carrier $\mu\tau$, and a depletion width that is independent of voltage are all difficult to justify a priori, even for optimum material. We also note that there are three adjustable parameters ($\mu\tau$, $V_i$, and $w_i$) in equation 2, in addition to the overall scale factor C, which would appear to underconstrain the fit. Nevertheless, this expression may provide a useful parameterization in terms of an "effective" $\mu\tau$ for primary photocurrent collection, if some additional constraints are imposed on the parameters. In particular, we will require that the normalization factor C is the same for both as-deposited and light soaked films (but is in general different for different films.) A less justifiable assumption is to take $V_i$ and $w_i$ as the same for all the films. This would be the case if the depletion width and the built-in field are determined by space charge residing in the band tails, i.e. independent of changes in the mid-gap defect density. Here, the bias light generation rate and the Eurbach slopes are similiar for all of the films.

The solid lines in figure 1 are fits of equation 2 to our data using these assumptions, where $V_i$ and $w_i$ were taken as 0.4 V and 0.25 μm respectively. We note that $\eta(V)$ approaches saturation only for the 1.74 eV film in the as-deposited state, and significant deviation from the fits begin to occur for all other samples at reverse bias voltages greater than 5 V, where the measured currents actually began to curve upward. This non-ideal behavoir was also observed for another set of low gap and high gap, and light soaked versus as-deposited films, and may be due to electron injection through the barrier at high reverse bias.

In table 1 we have listed the effective mobility-lifetime products $\mu\tau$ from the fits. For both films, the change in $\mu\tau$ correlates well with the CPM-determined density of mid-gap states, consistent with the results of Chu et al.[3] on (as-deposited) a-SiGe:H alloys, and Wronski et al. on glow discharge light soaked films[4]. Evidently then, this measurement and the simple fitting described above has good sensitivity for monitoring light soaking induced changes in primary photoconduction.  We note that $\mu\tau$ for the 1.63 eV gap film in the as-deposited state is close to that found for a similiar gap a-SiGe:H sample by Chu et al.[3], also using a two-field Hecht analysis.  We should point out, however, that the other fitting parameters and measurement conditions (in particular bias light intensity) were different in the two experiments, making direct comparison somewhat risky.

A complementary measurement to the reverse bias collection efficiency is the spectral response in the weakly absorbing wavelength region.  Here, wavelength is varied at short circuit and the current is measured.  The internal yield is defined as

$$Y(\lambda) = \frac{J(\lambda)}{eFT} \tag{3}$$

where F is the photon flux and T is the Pd transmission.  For absolute calibration, we used the method outlined in reference 5.   In figure 2 we have plotted $Y(\lambda)$ for the short-circuit case.  Again we find large changes with light soaking, which can be attributed to changes in the internal field, or changes in the effective $\mu\tau$ for collection. If we assume that the internal field is relatively unchanged, as in the two-field Hecht fit above, these results indicate  that  a considerable reduction in the effective $\mu\tau$ occurs on light soaking, consistent with the reverse bias collection efficiency results.

## CONCLUSIONS

We have presented a study of Schottky barrier photocarrier transport for 1.63 and 1.74 eV bandgap films produced by reactive magnetron sputtering.  The reverse bias collection efficiency showed large changes on light soaking, indicating a reduction in the "effective $\mu\tau$" for the material, for both films.  These changes correlate well with changes in the subgap absorption as measured by CPM, and with zero bias spectral response in the red. Ultimately, numerical modeling will be necessary to determine if simple analytic models, such as the two-field Hecht relation, can be reliably applied to these measurements.  Our results, however, indicate that straightforward measurements on Schottky barrier structures may be a sensitive monitor of light induced changes in a-Si:H.

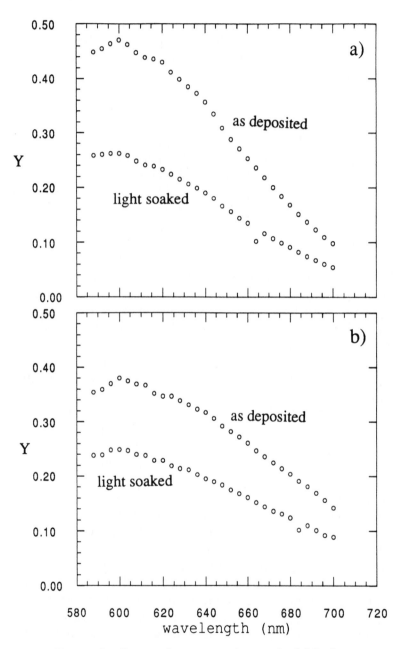

Figure 2.  Spectral response internal yields for
a) 1.74 eV and b) 1.63 eV optical gap.

## ACKNOWLEDGEMENTS

We would like to thank Murray Bennett of Solarex for providing the n+ substrates. This work was supported by the Thin Film Solar Cell Program of the Electric Power Research Institute under contract number RP 2824-1.

## REFERENCES

1.   V. Chu, J.P. Conde, D.S. Shen, and S. Wagner, Appl. Phys. Lett. **55**, 62, (1989).

2.   M. Pinarbasi, M.J. Kushner, and J.R. Abelson, J. Appl. Phys. **68**, 2255 (1990).

3.   V. Chu, S. Aljishi, J.P. Conde, Z E. Smith, D.S. Shen, D.Slobodin, J. Kolodzey, C. R. Wronski, and S. Wagner, Proceedings of the 19th IEEE Photovoltaics Specialists Conference (IEEE, New York, 1987), p.610

4.   C.R. Wronski, Z E. Smith, S. Aljishi, V. Chu, K. Shepard, D.S.Shen, R. Schwartz, D. Slobodin, and S. Wagner, in AIP Conference Proceedings No. 157, edited by B. Stafford and E. Sabisky (AIP, New York, 1987), p.70.

5.   C.R. Wronski, B. Abeles, and G.D. Cody, Solar Cells **2**, 245, (1980).

# PART V

# SOLAR CELL
# STUDIES

# Accelerated Light Soaking
## and Prediction of One-Sun Photostability
## in a-Si:H Solar Cells

T.Tonon, X.Li, A.E.Delahoy
Advanced Photovoltaic Systems, Inc., Princeton, NJ 08543

## ABSTRACT

A need exists for the rapid evaluation of photostability of a-Si:H solar cells. An accelerated method was developed that accurately reproduces for a-Si:H solar cells in a matter of minutes the extent of degradation (as characterized by I-V parameters) produced by weeks of continuous soaking at one-sun levels of irradiation. Such a method is suggested as a way to rapidly predict long term one-sun intrinsic photodegradation of a-Si:H solar cells fabricated by a variety of techniques.

## INTRODUCTION

In recent years, increased effort has been given to maximizing the photo-stabilized conversion efficiency of a-Si:H solar cells. There are various methods involving differences in deposition techniques and parameters, barrier layers, and post-deposition treatments conceived and presently experimented with in order to accomplish this maximization. Possible combinations of such methods thus become quite numerous. In addition, a necessary chore within such a maximization process is the photostability evaluation of each cell.

The most reliable method for the determination of photostability of a-Si:H solar cells is by direct experimentation, whereby cells are either placed in the field or placed under light irradiation and temperature conditions believed to be typical of what occurs in the field. Cells are thus "light soaked" until percent changes in their photovoltaic performance become relatively small on logarithmic time scales, at which point, equilibrium, or quasi-equilibrium in their material state is assumed. A major inconvenience of this method, however, is that, for completion, such experiments require relatively long times, ranging from weeks to months.

A reliable method to accelerate the photostabilization process would thus be a valuable tool in the design of solar cells. In this paper, we present such a method, which requires treatment times of the order of minutes. The method utilizes high cell irradiation levels achieved by means of a lens, a method first reported elsewhere[1]. Accelerated techniques involving the use of forward current injection may also have merit and may prove more convenient.

## EXPERIMENTAL APPROACH

If we limit ourselves to the Staebler-Wronski photodegradation widely observed in a-Si:H solar cells, the photodegradation extent of a given cell after a given time depends upon the soak conditions of irradiation and temperature. Evidence shows that photostabilization occurs when a balance is reached between defect formation, as induced by light (or more generally, by carrier recombination), and defect annealing, as induced by thermal energies.[2,3]

Thus, one might suppose that a useful accelerated light soaking (ALS) procedure would necessarily accelerate both the degradation and annealing processes. Hopefully, such a procedure would produce the same one-sun, long term extent of degradation in short times and thus provide a means to rank cells according to their susceptibility to photodegradation.

However, several mechanisms complicate the degradation and annealing processes; in particular, the light induced degradation mechanism appears to be thermally activated,[3,4] the metastable defects are formed with a distribution of activation energies for annealing,[3,4,5] there is evidence to show that the barrier to thermal annealing is reduced by irradiation,[6] and irradiation with UV light can lead to surface reconstruction.[7] The latter two effects appear to be caused by phonons emitted during photocarrier thermalization. In addition, the deepest soak state may be controlled by a limited number of potential defect sites.[8,9] With such complications in mind, it is not obvious at the outset whether a useful ALS technique exists.

Examples of ALS degradation are given in Figures 1 and 2. Figure 1 is a plot of normalized cell efficiency vs. soak time for three cells made by identical fabrication techniques. All soaks were conducted at open circuit and at 30 °C, but for three different irradiation levels of 10, 30, and 50 suns, without change of spectral content. If the parameter controlling degradation is written as $R^\gamma t$, where R is the carrier recombination rate (at open circuit, proportional to light intensity), and t is the light soak time, we find a horizontal intercept of the three curves at 30 % degradation yields $\gamma = 2.15$ (10 and 30 sun data) and $\gamma = 2.01$ (10 and 50 sun data). This result for $\gamma$ suggests that defect generation is a function of the parameter $R^2 t$, which is consistent with the result of Stutzmann et al.,[3] who predict that $N^3(t) - N^3(0) = kG^2 t$, where N is the density of recombination centers and G is the generation rate, based on the assumption that defects are created by a certain fraction of nonradiative, direct band-tail to band-tail transitions. It is also consistent with the Redfield and Bube model,[9] in which dispersive effects lead to a stretched exponential behavior for defect generation (or decay), which in turn may exhibit an apparent power law of the form $[R^2 t]^{0.3}$ over part of the range.

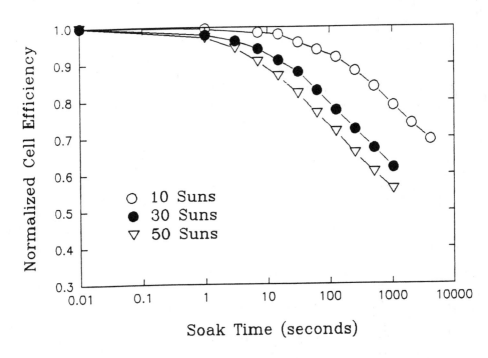

Figure 1. Data for three identical cells soaked at open circuit at different irradiation levels. Soaking and I-V measurements were performed at 30 °C.

Figure 2 is a plot of normalized cell efficiency vs. soak time for two identical cells soaked at open circuit and at 50 suns illumination, but with one cell soaked at 30 °C, and the other at 130 °C. The reader should note that the I-V data were obtained at the corresponding soak temperatures. Although we acknowledge that different I-V measurement temperatures contain information on the temperature coefficients of the cells, we remark that this figure appears to illustrate the increased degradation rate with temperature in the early portions of the soak. As the soak progresses, however, the curves cross, revealing the larger rate of thermal annealing at the higher temperature. We note that the $V_{oc}$ decline shown by 130 °C I-V measurements for the later soak times are not present in 30 °C measurements.

The method of this paper is phenomenological. The approach taken was to select a group of a-Si:H solar cells, deposited over a wide range of PECVD conditions, and compare their photostability behavior under varying soak conditions. From these observations, ALS conditions were selected such that the soak produced the same ranking in degradation behavior as occurred under one-sun soak conditions, as measured by I-V characteristics.

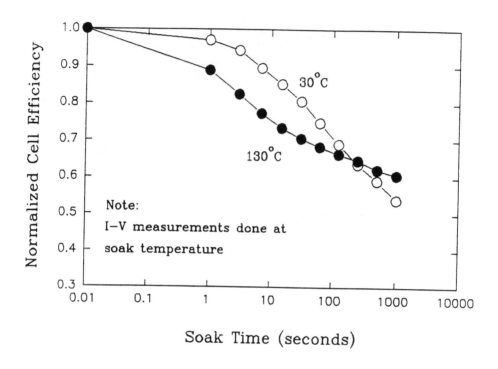

Figure 2. 50-sun soaking of two identical a-Si:H solar cells at open circuit and
at different temperatures

All cells were soaked at open circuit. We believe that power point soaking
affects chiefly the degradation rate and has little effect on long term stabilized
cell behavior. For the low light (about 0.9 suns), long term soaks, use was made
of a turntable and a multivapor lamp, with temperature controlled by means of
a fan. For ALS soaks, use was made of a specially-constructed apparatus that
enables accurate control of cell irradiation, temperature and soak time. The
apparatus contains a copper probe that contacts the entire area of the cell. This
probe provides electrical contact to the cell and ensures that the cell is
maintained within ±0.5 °C of specified temperatures. Cell $V_{oc}$ measurements
substantiate this claim. During the soak/measurement course of events, the cell,
probe and light source remain stationary. A motorized lens is admitted when
necessary to soak at elevated irradiation. The light source is a tungsten halogen
lamp, whose output is accurately maintained within ± 0.5 % by means of a
photocell monitor. Computerization accomplishes timing and sequencing, in
order to prompt lens motion, operate a shutter, and acquire I-V data.

In the following data, all cells with the same sample number were prepared
using adjacent areas of a-Si:H from a single substrate. Different sample numbers

correspond to different a-Si:H depositions. A total of eight depositions encompasses differences in i-layer thickness, substrate temperature, deposition rate, and boron tailing from the p-layer. Sample #1 is an a-Si:H/a-Si:H tandem-junction device having the structure, glass/SiO$_2$/SnO$_2$:F/p-i-n/p-i-n/Al, and all others are single junction devices having the structure, glass/SiO$_2$/SnO$_2$:F/p-i-n/Al. We acknowledge that there are differences in degradation behavior between tandem and single junction cells; however, we include both types of cells here since a reliable test for stability ranking ideally should distinguish between both types of cells.

## PRELIMINARY INVESTIGATION OF THE RELATIONSHIP BETWEEN PHOTOVOLTAIC DEGRADATION AND SOAK CONDITIONS

Table 1 presents data taken for a group of cells soaked under approximately 0.9 suns at 50 °C for 30 days, and an identical group of cells soaked at 50 suns (5.0 W/cm$^2$) and 50 °C for the times indicated. Numbers given are the percent changes of photovoltaic parameters experienced by the cells as a result of soaking. I-V data for the one-sun soaks were obtained from measurements made at one-sun (100 mW/cm$^2$) illumination in the temperature range 20 - 25 °C. I-V measurements made on these samples nine days before the 30-day measurement showed no more than a couple percent difference in cell efficiency. One-sun I-V data for the 50-sun soaks were taken at both 50 °C and 30 °C, as noted.

Data from Table 1 indicate firstly that the ALS conditions were able to drive the cells into a deeper soak state than the extended one-sun soak. However, the extent of degradation, as measured by I-V data and as left by the one-sun extended soak, was fairly closely duplicated by ALS when ALS was terminated at the specified intermediate times. The duplication was accurate enough to provide the same general ranking of photostability among the cells. Discrepancies exist for closely ranked cells, and these may be at least partly explainable because of the fact that the intermediate ALS data is from measurements made at a temperature significantly higher than that associated with the one-sun data. We note that the intermediate ALS and extended one-sun soak data show similar extents of degradation not only in cell efficiency, but also in $V_{oc}$, $J_{sc}$, and FF.

If we inquire whether the ALS data at a soak temperature of 50 °C could be used to predict one-sun ranking, we must conclude negative because of the fact that the ALS times necessary to produce the proper one-sun soak ranking differ widely among the samples. Thus, although we might conclude that ALS and extended one-sun soaking produce a similar nature and extent of degradation (as characterized by I-V data) when ALS is terminated at the proper time, we cannot say a priori what ALS times are necessary to produce the proper ranking.

Table 1. Percent Drops of I-V Parameters for One-Sun and 50-sun Soaks at 50°C

| SAMPLE | ONE-SUN SOAK, PERCENT CHANGES (20-25°C I-V measurement) | | | | | 50-SUN SOAK, PERCENT CHANGES (50°C I-V measurement, and 30°C measurement in parentheses) | | | | |
|---|---|---|---|---|---|---|---|---|---|---|
| | $V_{oc}$ (%) | $J_{sc}$ (%) | FF (%) | Eff (%) | S.Time (days) | $V_{oc}$ (%) | $J_{sc}$ (%) | FF (%) | Eff (%) | S.Time (sec.) |
| 1 | -1.9 | -0.5 | -14.8 | -16.8 | 30 | -4.6 | -0.1 | -13.7 | -17.7 | 1024 |
| 1 | | | | | | (-1.8) | (-0.7) | (-13.9) | (-16.2) | 1024 |
| 2 | -1.8 | -8.5 | -20.8 | -28.7 | 30 | 1.9 | -8.5 | -23.1 | -31.0 | 217 |
| 2 | | | | | | -3.0 | -15.0 | -28.1 | -40.8 | 1024 |
| 2 | | | | | | (-0.5) | (-15.7) | (-24.8) | (-36.7) | 1024 |
| 3 | -1.9 | -8.8 | -20.7 | -28.9 | 30 | -2.0 | -8.8 | -20.6 | -29.1 | 56 |
| 3 | | | | | | -5.0 | -25.1 | -27.9 | -48.7 | 1024 |
| 3 | | | | | | (-1.9) | (-24.8) | (-24.1) | (-44.2) | 1024 |
| 4 | -3.3 | -8.5 | -20.1 | -29.2 | 30 | -2.5 | -8.5 | -19.6 | -28.4 | 63 |
| 4 | | | | | | -5.5 | -20.9 | -28.7 | -46.7 | 1024 |
| 4 | | | | | | (-2.2) | (-22.1) | (-26.1) | (-43.6) | 1024 |
| 5 | -2.5 | -7.7 | -21.9 | -29.7 | 30 | -8.0 | -7.7 | -23.4 | -31.0 | 157 |
| 5 | | | | | | -4.5 | -14.6 | -29.0 | -42.1 | 1024 |
| 5 | | | | | | (-1.5) | (-16.4) | (-27.0) | (-39.5) | 1024 |
| 6 | -4.5 | -15.2 | -15.5 | -31.4 | 30 | -3.1 | -15.2 | -19.1 | -33.6 | 60 |
| 6 | | | | | | -6.8 | -30.9 | -27.1 | -53.0 | 1024 |
| 6 | | | | | | (-2.8) | (-32.9) | (-21.0) | (-48.8) | 1024 |
| 7 | -6.1 | -13.7 | -20.1 | -35.3 | 30 | -4.3 | -13.7 | -23.8 | -37.1 | 344 |
| 7 | | | | | | -6.0 | -17.0 | -27.0 | -43.1 | 1024 |
| 7 | | | | | | (-2.7) | (-18.5) | (-20.8) | (-36.8) | 1024 |
| 8 | -6.7 | -25.7 | -16.1 | -41.8 | 30 | -8.9 | -21.7 | -27.1 | -48.0 | 1024 |
| 8 | | | | | | (-4.4) | (-24.8) | (-22.3) | (-44.2) | 1024 |

## THE ACCELERATED METHOD

In searching for an accelerated method, one might suppose the reaction rates of the various mechanisms underlying degradation and annealing are such that an irradiation and temperature regime exists where the ranking produced by extended one-sun soaks becomes fairly insensitive to ALS times. A search was thus undertaken, and because of the flexibility of the research equipment, such a regime was indeed uncovered after relatively little experimentation.

The chief result of the investigation is that the one-sun ranking of cell degradation is fairly accurately duplicated by ALS performed at around 50 suns and 130 °C for 17 minutes. Data summarizing the results are presented in Table 2, which lists the one-sun soak data performed at 50 °C for 30 days under Condition (a), one-sun soak data performed at 30 °C for 20 days under Condition (b), and the ALS results performed at 130 °C for 17 minutes under Condition (c). All data of Table 2 were obtained by I-V measurements done at room temperature and one-sun illumination. We note anomalous behavior with Sample #6, which shows strikingly different ranking behavior under the one-sun 30 °C soak than it does under the other two soak conditions.

As can be seen from Table 2, the ALS procedure at the chosen conditions predicts, apart from Sample # 6, the same ranking as long term one-sun soaking, provided the one-sun ranking is for cells differing more than a few percent in relative efficiency drop. Similarly, one can observe from Table 2 that one-sun soaks performed at different temperatures also show, apart from Sample #6, slight discrepancies in ranking for cells with ranking within a few percent of relative efficiency drop. The anomalous behavior of Cell # 6 can perhaps be explained by an extraordinarily low level of annealing behavior at around 30 °C.

We note that, as can be seen from their extents of degradation, Samples 7 & 8 were deposited under conditions extremely unfavorable to photostability and are thus not of commercial interest. Commercially viable cells are deposited under conditions considerably more favorable to photodegradation. Samples 7 & 8 were included in this study primarily to illustrate the effectiveness of the ALS technique.

## SUMMARY

It was found that properly conducted accelerated light soaking (ALS) closely duplicates extended one-sun light soaking in a-Si:H solar cells deposited under a wide range of PECVD deposition parameters. In particular, ALS conditions of 50 suns and 130 °C for 17 minutes fairly accurately reproduces the ranking order of relative efficiency drops due to one-sun soaks conducted at 30 - 50 °C for 20 - 30 days. In addition, changes in I-V parameters ($V_{oc}$, $J_{sc}$, FF, and Eff)

Table 2. The Accelerated Method Used to Predict Long term One-Sun Ranking

| SAMPLE | SOAK CONDITIONS | PERCENT CHANGES | | | |
|---|---|---|---|---|---|
| | a) One sun, 50°C, 30 days, 20-25°C I-V<br>b) One sun, 30°C, 20 days, 30°C I-V<br>c) ALS: 50 suns, 130°C, 17 min, 30°C I-V | $V_{oc}$ | $J_{sc}$ | FF | Eff |
| 1 | a<br>b<br>c | -1.9<br>-3.7<br>-1.9 | -0.5<br>-0.3<br>3.4 | -14.8<br>-14.2<br>-12.5 | -16.8<br>-17.8<br>-11.4 |
| 2 | a<br>b<br>c | -1.8<br>-0.1<br>2.8 | -8.5<br>-7.6<br>-9.1 | -20.8<br>-22.8<br>-20.5 | -28.7<br>-29.1<br>-26.0 |
| 3 | a<br>b<br>c | -1.9<br>-2.5<br>-0.8 | -8.8<br>-10.7<br>-10.9 | -20.7<br>-25.0<br>-24.0 | -28.9<br>-34.9<br>-33.2 |
| 4 | a<br>b<br>c | -3.3<br>-3.1<br>-0.5 | -8.5<br>-10.1<br>-11.8 | -20.1<br>-24.5<br>-22.7 | -29.2<br>-34.3<br>-31.6 |
| 5 | a<br>b<br>c | -2.5<br>-3.4<br>-0.1 | -7.7<br>-9.3<br>-9.2 | -21.9<br>-22.5<br>-20.6 | -29.7<br>-32.0<br>-27.6 |
| 6 | a<br>b<br>c | -4.5<br>-3.9<br>-2.5 | -15.2<br>-20.8<br>-16.3 | -15.5<br>-21.7<br>-14.4 | -31.4<br>-40.0<br>-30.4 |
| 7 | a<br>b<br>c | -6.1<br>-3.9<br>-1.5 | -13.7<br>-12.9<br>-14.7 | -20.1<br>-22.9<br>-23.0 | -35.3<br>-35.6<br>-35.7 |
| 8 | a<br>b<br>c | -6.7<br>-5.0<br>-4.6 | -25.7<br>-20.1<br>-26.0 | -16.1<br>-25.2<br>-21.8 | -41.8<br>-43.7<br>-44.6 |

for long term one-sun soaking is fairly accurately duplicated by ALS under the stated conditions.

Although the reliability of this ALS approach requires further testing and adjustment in specific features might prove necessary, the technique holds promise as a valuable tool in rapidly evaluating intrinsic photostability of a-Si:H solar cells made under a wide range of deposition conditions.

## REFERENCES

1. H.Volltrauer, S.C.Gau, F.J.Kampas, Z.Kiss, and L.Michalski, Light-Induced Changes in a-Si:H at High Illumination; Inverse Staebler-Wronski Effect, Proc. 18th IEEE Photovoltaic Specialists Conf., Las Vegas, Nevada, Oct. 21-25, 1985, ISSN: 0160-8371, p. 1760.

2. A.E.Delahoy, J.Kalina, C.Kothandaraman, and T.Tonon, Advanced Technology Amorphous Silicon Photovoltaic Modules, Proc. 9th European Photovoltaic Solar Energy Conf. (W.Palz, G.T.Wrixon, P.Helms, Eds.). Kluwer, Dordrecht, 1989, p. 599.

3. M.Stutzmann, W.B.Jackson, and C.C.Tsai, Light-Induced Metastable Defects in Hydrogenated Amorphous Silicon: A Systematic Study, Phys. Rev. B 32, 23 (1985).

4. R.S.Crandall, Metastable Defects in Hydrogenated Amorphous Silicon, Phys. Rev. B 36, 2645 (1987).

5. A.E.Delahoy, T.Tonon, J.A.Cambridge, M.Johnson, L.Michalski, and F.J.Kampas, Light Soaking Studies on Amorphous Silicon Photovoltaic Devices and Modules, Proc. 8th European Photovoltaic Solar Energy Conference (Kluwer, Dordrecht, 1988; I.Solomon, B.Equer, P.Helm, Eds.) pp. 646-652.

6. A.E.Delahoy and T.Tonon, Light-Induced Recovery in a-Si:H Solar Cells, American Institute of Physics Conference Proceedings, No. 157, p 263 (1987).

7. J.Dutta and G.Ganguly, Optically Induced Restructuring of a Hydrogenated Amorphous Silicon Thin-Film Surface, Appl. Phys. Lett. 57, 1227 (1990).

8. H.R.Park, J.Z.Liu, and S.Wagner, Saturation of the Light-Induced Defect Density in Hydrogenated Amorphous Silicon, Appl. Phys. Lett. 55, 2658 (1989).

9. D.Redfield and R.H.Bube, Reinterpretation of Degradation Kinetics of Amorphous Silicon, Appl. Phys. Lett. 54, 1037 (1989).

# T-INDUCED DEGRADATION IN a-Si ALLOY
## LAR CELLS AT INTENSE ILLUMINATION

A. Banerjee, S. Guha, A. Pawlikiewicz, D. Wolf, and J. Yang
United Solar Systems Corp., 1100 W. Maple Rd., Troy, Michigan 48084

## ABSTRACT

Light-induced degradation has been investigated in a-Si alloy p-i-n solar cell structures as a function of cell deposition temperature and light intensity. Cells are deposited at temperatures ranging between $200°C$ to $300°C$; degradation has been carried out at intensities up to 50 times AM1.5 illumination at $35°C$. The cell characteristics have been measured under AM1.5, blue and red illuminations. The degradation is found to have a power law dependence on the product of square of generation rate and light-soaking time. Most cells show saturation in degradation under 50 times AM1.5 illumination beyond 1000 sec, which is equivalent to approximately 800 hours under AM1.5 intensity. However, some cells showed continued degradation at the high intensity up to $6 \times 10^4$ sec without any saturation; the cell properties could be restored to their original values after annealing. Computer simulation studies have been carried out to analyze the result on the basis of existing theories.

## INTRODUCTION

Light-induced degradation in a-Si alloy (a-Si:H) solar cells is one of the most challenging problems confronting the photovoltaic industry. The degradation, also know as the Staebler-Wronski effect,[1] is attributed to the generation of defects in the intrinsic layer of a p-i-n solar cell. There is a wealth of information describing various aspects of the effect. Basically, the defects are created by the recombination of electrons and holes[2] which result in an enhancement of the density, $N_s$, of deep localized states. The higher value of $N_s$ leads to the degradation of the transport properties of the material. Various workers have reported the degradation behavior of cells under both outdoor and simulated illumination conditions.[3]

Three important aspects of light-induced degradation have been identified. First is the saturation behavior. It has recently been shown[4,5] that, irrespective of the intensity of illumination, the light-induced defect density attains a steady state value, $N_{sat}$. Saturation may occur either due to the attainment of steady state between generation and annealing (light-induced or thermal) or depletion of available defect sites. The saturation behavior, therefore, dictates the final quality of the degraded material. The typical value of $N_{sat}$ has been shown[5,6] to lie in the range $5 \times 10^{16}$ to $2 \times 10^{17}$ cm$^{-3}$ for a large number of samples prepared under different deposition conditions. Second is the dependence of the time, $t_{sat}$, to reach the onset of saturation on the carrier generation rate, G, which in turn is proportional to the intensity, I, of illumination. The approximate relationship $t_{sat} \propto 1/G^2$ has been explained by different models.[7,8] Third is the dependence of

$N_{sat}$ on the initial deposition conditions. Recent work[5,6] has suggested that the value of $N_{sat}$ increases with increasing bandgap. Since higher bandgap material contains more hydrogen, this could be explained on the basis of the metastable defects to be hydrogen-related.

The correlation of the above aspects with cell performance is of considerable interest. It is generally observed that under AM 1.5 illumination the cell efficiency initially degrades fast and then tends to saturate. The saturation phenomenon in cell performance is, however, not clearly established. Furthermore, the value of saturation time and its dependence on light intensity has not been investigated. Such information can provide valuable clues regarding how long after installation it will take a module to attain its degraded saturation output. In this paper, we investigate the saturation behavior of a-Si alloy single junction p-i-n solar cells at different intensities and correlate the behavior with fundamental material properties. In order to investigate the role of hydrogen, cells have been deposited at two different temperatures so as to have different hydrogen contents.

## EXPERIMENTAL DETAILS

p-i-n cells were grown by glow discharge technique on stainless steel substrates. The back reflector used was a double layer ZnO/Ag coating. The top contact was ITO. Details of the cell fabrication technique are given elsewhere.[9] Two sets of samples were prepared at nominal substrate temperatures of $200^{\circ}$C (LT) and $300^{\circ}$C (HT). All other deposition conditions for the two sets were kept the same. The thickness of the i-layers of the two samples was similar ~ 3400-3700 Å.

Separate i-layer samples were prepared under the same conditions on glass and Si wafer substrates for optical bandgap and infrared absorption studies. The infrared absorption curves were used to calculate the hydrogen content in the films. The I-V characteristics were measured under AM 1.5 global illumination. I-V characteristics under red and blue illumination were obtained by incorporating appropriate filters. The cells were exposed to 50 AM1.5 and 5 AM1.5 white light intensity for accelerated light-induced degradation. 50 AM1.5 (HI) and 5 AM1.5 (LI) intensities are defined as the intensities at which the Jsc of the cells are equal to 50 and 5 times, respectively, that of the corresponding value obtained under AM1.5 conditions. The illumination time ranged from 10s to $10^4$s. The concentrated illumination was obtained using an incandescent lamp with suitable focusing lenses and neutral density filters. The temperature of the cells was maintained at near room temperature ($\sim 35^{\circ}$C) during the illumination with the help of cooling fans. The I-V characteristics under AM1.5, red, and blue illumination were obtained after each exposure. After each exposure (HI or LI) was over, the cells were annealed and re-measured to ensure that the I-V characteristics recovered to the initial values. Cells which did not recover on annealing were rejected. The exposure sequence on each sample (LT or HT) was LI exposure - anneal - HI exposure - anneal. Both exposures were done on the same sample.

## RESULTS AND DISCUSSION

The nominal values of the Tauc optical bandgap were 1.87 eV and 1.81 eV for the LT and HT samples, respectively. The ir absorption curves showed the evidence of only Si-H bonds and no Si-H$_2$ bonds in both samples. The hydrogen content in the LT and HT samples were 18 and 12 at %, respectively. The higher deposition temperature of the HT material accounts for the lower hydrogen incorporation in the films and consequently the lower bandgap.

The various cell parameters measured initially before degradation are shown in Table I. Results for light-induced degradation for both LI and HI illumination are shown in Figs. 1 to 4. Since the light-induced defect states affect mostly the carrier collection efficiency, which in turn affects the fill factor, only this parameter of the cells has been plotted as a function of illumination time. Referring to Table I, the initial efficiency for the LT and HT samples are, respectively, 9.4% and 8.7%. The open-circuit voltage of the LT sample is higher than that of the HT sample. This is understandable in view of the higher optical gap of the LT sample. The short-circuit current density for this sample is also lower for the same reason. The FF for the HT sample is poorer than that of the LT sample. It appears that, at this substrate temperature, the material quality has already started getting poorer.

Table I.   Initial cell performance for the LT and HT samples under global AM 1.5 illumination.

| CELL TYPE | $V_{oc}$ (V) | $J_{sc}$ (mA/cm$^2$) | FF | $\eta$ (%) |
|-----------|--------------|----------------------|-----|------------|
| LT | 0.965 | 14.5 | 0.67 | 9.4% |
| HT | 0.943 | 15.4 | 0.60 | 8.7% |

The onset of saturation for the FF values occurs at $\geq 10^2$s for the HI case (Fig. 2). However, there is no clear evidence of saturation for the LI case (see Fig. 1). Instead, there are some signs of saturation for the LI sample after $10^4$s since the AM1.5 FF is ~ 0.57, which is the same as the FF$_{sat}$ for the HI case. The $10^4$s degraded values of the red and blue FF for the LI case are also similar to the corresponding saturated values for the HI case. Assuming a $t_{sat} \propto 1/G^2$ law, $10^4$s of LI degradation is equivalent to $10^2$s of HI degradation. It thus appears that the onset of saturation occurs at $10^4$ of LI exposure and the $1/G^2$ law is obeyed.

The significance of the red and blue FF measurements needs some discussion. The effect of introducing any filter during the measurement is firstly to reduce the intensity of the light. The lower intensity would increase the value of the measured FF since series resistance is smaller, and, also, the electric field is stronger due to the smaller space charge. Second, the color of the filter governs which section of the i-layer absorbs the light. With a blue filter, most of the light is absorbed in the top section (close to the p-layer), and the value of the measured FF is dominated by the electron transport mechanism. Incorporation of a red filter leads to more uniform absorption throughout the thickness of the i-layer, and the FF gives a measure of the hole transport properties. As shown in Figs. 1 and 2, the degradation of the red and blue FF are qualitatively similar (the curves are almost parallel), thereby showing that both the electron and hole transport properties suffer similarly, and finally attain saturation values.

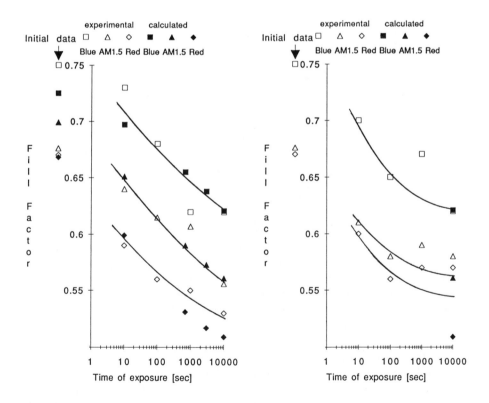

Fig. 1. Experimental and theoretical values of AM1.5, red, and blue FF as a function of exposure time for the LI case and LT sample.

Fig. 2 Experimental and theoretical values of AM1.5, red, and blue FF as a function of exposure time for the HI case and LT sample.

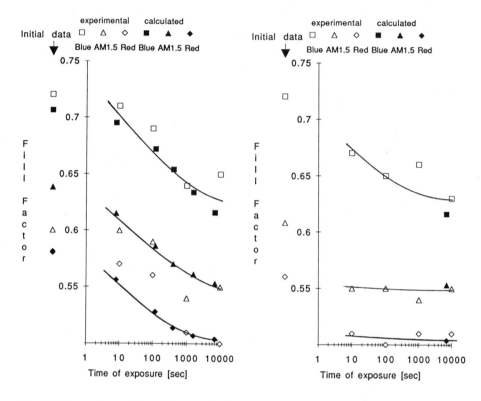

Fig. 3. Experimental and theoretical values of AM1.5, red, and blue FF as a function of exposure time for the LI case and HT sample.

Fig. 4. Experimental and theoretical values of AM1.5, red, and blue FF as a function of exposure time for the HI case and HT sample.

The degradation of the FF for the HT samples as a function of light exposure time is shown in Figs. 3 and 4, corresponding to the LI and HI cases, respectively. As can be observed from Figs. 3 and 4, the onset of saturation occurs at ~10s and ~$10^3$s for the HI and LI cases, respectively. Thus, the $1/G^2$ dependence for $t_{sat}$ is clearly obeyed in this sample. The value of $FF_{sat}$ is ~0.55, and is lower than the corresponding value for the LT sample. The red and blue FF degrade in a similar manner implying that both the electron and hole transport properties deteriorate similarly.

We have also investigated the degradation behavior of the cells using the numerical simulation model developed by us earlier.[10,11] The device transport parameters were chosen so that the calculated initial I-V characteristics agree with the experimental data. The parameters chosen for the LT samples are $\mu_n$ = 15 cm$^2$ V$^{-1}$ sec$^{-1}$, $\mu_p$ = 4 cm$^2$ V$^{-1}$ sec$^{-1}$, minimum density of states = 2 x $10^{16}$cm$^{-3}$ eV$^{-1}$ and neutral trap capture cross section $\sigma_n$ = 1.2 x $10^{-16}$ cm$^2$. The transport parameters for the HT samples were chosen to be somewhat poorer, namely, $\mu_n$ = 15 cm$^2$ V$^{-1}$ sec$^{-1}$, $\mu_p$ = 2 cm$^2$ V$^{-1}$ sec$^{-1}$, minimum density of states = 3 x 10$^{16}$cm$^{-3}$ eV$^{-1}$ and $\sigma_n$ = 1.2 x 10$^{-16}$ cm$^{2.}$ The optical bandgap of the LT sample was assumed to be 40 mV higher than that of the HT sample. All other transport parameters were kept the same as described in our earlier publications.[10,11] Exposure to light is assumed to change only the minimum density of states as shown in Fig. 5 with saturation taking place at high-intensity, long-time exposure. The results show good agreement with experiments (Figs. 1-4) demonstrating the usefulness of the model.

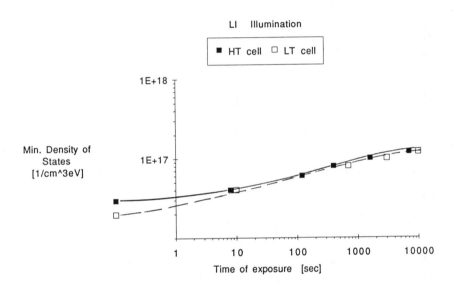

Fig. 5.   Calculated minimum density of states as a function of light exposure for the two samples.

We thus find that the degradation behavior for the two samples is qualitatively similar. Both samples exhibit saturation under concentrated light exposure. However, the quantitative results are different. The degradation in the FF is 15% and 8% and in $\eta$ is 21% and 17% for the LT and HT samples, respectively. It is known that both high temperature deposited materials and cells degrade less than their low temperature counterparts. The relevant issue is not the extent of degradation but the performance/quality of the saturated cell/material. In the narrow deposition temperature range (200-300°C) investigated in this work, the cell with the lower deposition temperature possesses a higher value of $\eta_{sat}$ ~ 7.4% and FF$_{sat}$ ~ 0.57. This is an interesting result since it is generally believed that a material deposited at higher substrate temperature (having lower hydrogen content) has lower saturation defect density.

We should, however, mention that earlier results showing poorer stability with higher hydrogen content usually covered materials with much higher hydrogen content. In that regime, one has poorer morphology of the material and hydrogen incorporated in di- and poly-hydride configurations. In the range of temperature investigated by us, even the LT sample has only 18% hydrogen and did not show presence of di-hydride. In fact, the initial quality of the material was superior to that of the HT material as demonstrated by the higher initial fill factor. One can therefore speculate that the deleterious effect of hydrogen on stability takes place only when hydrogen exceeds a certain concentration; even the initial quality of the material is affected in that case.

It is noteworthy that some cells, the results of which are not reported here, exhibited anomalous behavior. For example, some cells did not show any saturation in the FF after high intensity exposure up to 6 x $10^4$ seconds even though they recovered completely on annealing. Some cells exhibited degradation which could not be recovered on annealing. Hence, there are several aspects of light-induced degradation which we have not addressed in this paper. Further work will be necessary to understand these problems.

## ACKNOWLEDGEMENTS

We thank V. Trudeau for preparation of the manuscript. The work was supported in part by Solar Energy Research Institute under Subcontract Number ZM-1-19033-2.

## REFERENCES

1.   D. L. Staebler and C. R. Wronski, Appl. Phys. Lett. 31, 292 (1977).
2.   S. Guha, J. Yang, W. Czubatyj, S. J. Hudgens, and M. Hack, Appl. Phys. Lett. 42, 588 (1983).
3.   For a recent review, see C. R. Wronski in Proc. 21st IEEE PV Specialists Conf. (IEEE, 1990) p. 1487.
4.   H. R. Park, J. Z. Liu, and S. Wagner, Appl. Phys. Lett. 55, 2658 (1989).
5.   H. R. Park, J. Z. Liu, P. Roca i Cabarrocas, A. Maruyama, M. Isomura, S. Wagner, J. R. Abelson, and F. Finger, Materials Res. Soc. 192, 751 (1990).
6.   H. R. Park, J. Z. Liu, P. Roca i Cabarrocas, A. Maruyama, M. Isomura, S. Wagner, J. R. Abelson, and F. Finger, Appl. Phys. Lett. 57, 1440 (1990).
7.   M. Stutzmann, W. Jackson, and C. Tsai, Phys. Rev. B32, 23 (1985).
8.   D. Redfield and R. H. Bube, Appl. Phys. Lett. 54, 1037 (1989).
9.   Final Annual Subcontract Program Report, SERI, Subcontract No. ZB-7-06003-4 (Feb. 1990).
10.  M. Hack and M. Shur, J. Appl. Phys. 54(10) 5858 (1983).
11.  A. H. Pawlikiewicz and S. Guha, Proc. 20th IEEE PV Specialists Conf., (IEEE, 1988) p. 251.

# Accelerated Light Degradation of a-Si:H Solar Cells and its Intensity and Temperature Dependence

L. Yang, L. Chen and A. Catalano
Solarex Thin Film Division, 826 Newtown-Yardley Rd., Newtown, PA 18940

## ABSTRACT

The long term kinetics of the light induced degradation of single junction a-Si:H solar cells was studied by an accelerated test. A simple scaling law for cell degradation was found between the light intensity (I) and the exposure time (t), i.e. $I^{1.8}t = $ const. No apparent saturation was observed near the room temperature, even though an AM1.5 equivalent exposure time of over 25,000 hours has been reached using the high intensity light of ~14 $W/cm^2$. Saturation, however, did occur at higher temperatures and the level of saturation was found to depend on the temperature and the light intensity. These results suggest that the long term stability of a-Si:H devices is governed by two competing effects, namely the light induced degradation and the thermal annealing.

## INTRODUCTION

Despite many years of effort, understanding and eliminating the light induced degradation in amorphous semiconductors and devices remains one of the greatest challenges. The lack of understanding of the effect and the difficulty of conducting long term tests under normal, AM1.5, operating light intensity has severely limited the ability to design and fabricate solar cells with optimized long term efficiency. In fact, the degradation behavior of solar cells beyond 10 years of normal operating period has hardly ever been explored. The kinetics of long term exposure and the question whether saturation occurs after prolonged light soaking and, if it does, what causes it are of much practical as well as scientific interest.[1,2] An accelerated degradation test is extremely useful in assessing the long term stability and getting quick feedback for further device optimization. Moreover, an accelerated test is necessary to acquire the large amount of experimental data needed to determine the degradation kinetics of various materials and devices, which is the first step towards a complete understanding of the light induced metastability in amorphous semiconductors.

In this paper, we report our first accelerated degradation study of a-Si:H single junction solar cells. The light intensity dependence near the room temperature is investigated to determine the scaling relationship between the light intensity and the exposure time for solar cell degradation. The temperature dependence, on the other hand, is studied to provide detailed information on the mechanism which governs the long term stability of the devices.

## EXPERIMENTAL

The accelerated degradation measurements were done using a 150 W Xenon arc lamp as the light source. The high intensity white light, which is collimated to a beam diameter of ~4 cm when it exits the lamp housing, is further concentrated by focussing the beam to an area ~0.5 cm$^2$. Filters are used to cut down the strong peaks in the infrared region to reduce heating and shape the spectrum of the Xenon lamp to that of the sun. The light intensity, up to ~14 W/cm$^2$ (or 140 times AM1.5), was calibrated by integrating the actual light spectrum measured using a Li-cor spectrophotometer. Neutral density filters were applied to reduce the intensity by a known factor.

Measuring and controlling the temperature of the sample being illuminated by such an intense light are non-trivial tasks. Two criteria have to be met in order to accurately determine the sample temperature using a thermocouple. One is that there is no temperature gradient near the thermocouple junction to ensure good thermal equilibrium with the sample. The other is that the thermal conductivity of the thermocouple wires must be sufficiently low so that the contact by the thermocouple does not interfere with the thermal equilibrium of the sample. We have used an ultra thin Chromel/Alumel thermocouple (0.0005") to fulfill these criteria and accomplish stable and consistent temperature measurements. The thermocouple was carefully attached to the Ag back contact of the solar cells. An electrically switched shutter was used to control the light exposure time. Simple calculations show that upon illumination the sample should reach its thermal equilibrium almost instantaneously (<1 sec) and the temperature gradient within the sample should be negligible due to the extremely small sample thickness.

In the absence of active cooling, the solar cell exposed to ~14 W/cm$^2$ illumination reached a temperature as high as ~240° C. A constant room temperature nitrogen flow reduced the cell temperature to ~160° C. Lower cell temperatures was obtained using a flow of nitrogen cooled by a liquid nitrogen bath.

The single junction a-Si:H solar cells used in the degradation study were all made in one deposition run and have a simple p-i-n structure. The i-layer thickness, which can strongly affect the cell degradation rate, was ~5000Å. The initial cell efficiencies were around 9%. Degradation was performed under open circuit conditions.

## RESULTS AND DISCUSSION

The light intensity dependence of the degradation is of prime importance because a simple scaling law between the light intensity and the exposure time, if found, can greatly enhance the practical usefulness of the accelerated degradation test. Furthermore, the intensity dependence is also an essential part of the kinetics governing the degradation process. Plotted in Fig. 1(a) are the degradation curves of normalized efficiencies versus exposure time on an logarithmic scale for five different light intensities. The cell temperature was maintained at 50° C during light soaking.

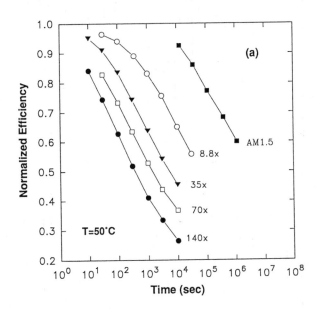

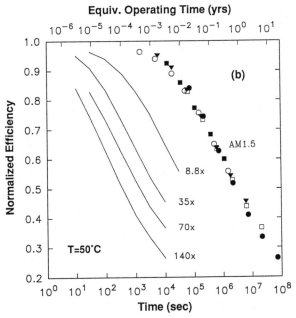

**Fig. 1** Normalized cell efficiency versus light exposure time at various intensities. The temperature was maintained at 50° C during light exposure. Data is renormalized to AM1.5 intensity in (b) using Eq. 1. Curves are guides to the eye.

The degradation curve with the lowest light intensity, i.e. 100 mW/cm$^2$ (AM1.5), was taken using a calibrated Na$^+$ vapor lamp which is our standard light source for normal degradation tests. The other curves were all measured using the focused Xenon arc lamp with neutral density filters. As has been commonly observed[3,4], the normalized cell efficiencies degraded linearly with log time over many orders of magnitude near the room temperature. The slope of the linear section is almost independent of the light intensity and the displacement of the curves on the logarithmic time scale is roughly proportional to the light intensity. These observations suggest that the light intensity dependence of cell degradation bears a simple scaling relationship,

$$I^\alpha t = const \qquad\qquad (1)$$

where I is the light intensity, t is the exposure time, and $\alpha$ is a constant. Figure 1(b) illustrates the accuracy of this scaling law by normalizing all the data in Fig. 1(a) to the AM1.5 intensity with the exponent $\alpha$=1.8. The scaling law, Eq. 1, and the exponent are in excellent agreement with earlier studies on the degradation of film properties including spin density, sub-band gap absorption and transport.[5]

Another striking feature of the data shown in Fig.1 is that no apparent saturation was observed at this temperature, even though an AM1.5 equivalent light exposure time of over 25,000 hours, neglecting the effect of thermal annealing, has been reached at the highest intensity. Recently there has been some controversy on whether the light induced degradation will reach saturation at low temperature.[1,2,5] This is interesting because it would shed much light on the basic mechanism which governs the long term stability of the material. If saturation does occur in the absence of thermal annealing, one can then argue that the total number of convertible sites in the material may be the limiting factor determining the long term stability. Recent light soaking experiments on mid-gap defect density in a-Si:H films appeared to support this mechanism because saturation was observed after ~1000 hours of AM1.5 equivalent light soaking and the saturation level was found to be almost independent of the light intensity. This result is in sharp contrast with our cell degradation data shown in Fig. 1. It is worth pointing out that the heavily degraded cells with efficiencies as low as only ~25% of the original value can all be completely restored by subsequent thermal annealing.

While there is no evidence from our measurements that saturation would occur due to the exhaustion of convertible sites, it did occur when light soaking experiments were done at higher temperatures. Figure 2 shows the degradation curves measured at four different temperatures, i.e. 50° C, 100° C, 150° C, 190° C. The maximum light intensity of the Xenon lamp, ~14 W/cm$^2$, was used for all measurements. It is clear that all degradation curves shown in Fig. 2 begin to bend towards a saturation value after a certain period of light exposure. This period, however, decreases with increasing temperature and complete saturation was apparently reached at 150° C and 190° C within the total duration of the experiment. These results suggest that the degradation process is mainly governed by two competing effects: the light induced defect generation and the thermal annealing which heals the defects. At low temperature and high intensity, the light induced defect generation dominates the degradation process as shown by the data in Fig. 1. With increasing temperature, the annealing effect becomes more important and eventually balances the degradation to reach a saturation. The saturation level thus increases with increasing temperature, as shown in Fig. 2. It is interesting to notice that in the early stages of light exposure, cells actually degraded faster at high temperature than at low. This effect has been observed previously in both film and device studies and was attributed to a weak temperature dependence of the defect generation.[5,6]

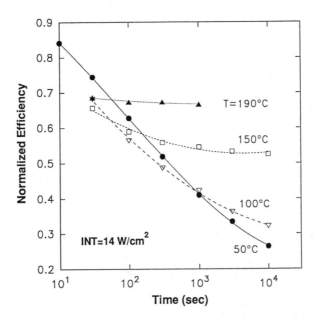

**Fig. 2** Normalized efficiency versus exposure time measured at various temperatures as indicated in the figure. The light intensity used was 140 times AM1.5. Curves are guides to the eye.

The fact that the saturation is caused by the balance between the light induced degradation and thermal annealing can be further illustrated by examining the saturation level as a function of light intensity at a relatively high temperature. It is expected that the saturation level would decrease with increasing light intensity at a given temperature. Figure 3 shows the degradation curves measured at three different intensities at 100° C. The saturation behavior is evident in these curves and the level of saturation indeed decreases monotonically with increased light intensity.

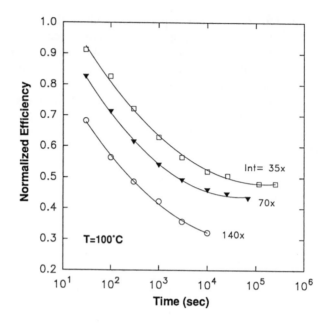

**Fig. 3** Normalized cell efficiency versus exposure time at three intensities. The degradation temperature was 100° C. Curves are guides to the eye.

A number of models for the kinetics of light induced degradation and annealing have been proposed.[1,5,7] However, these models can only be used to describe the light induced degradation effect on the defect density of the material. A relationship between the defect density and the cell parameter, e.g. the efficiency, is required in order to compare the kinetic data of solar cell degradation with the theories. Based on the experimental observation that over a wide range of exposure time the cell efficiency degrades linearly with log time and the defect density in the meantime follows a power law with time[5], we propose the following relationship,

$$\eta \sim - A \log N \qquad (2)$$

where $\eta$ and $N$ are the cell efficiency and the i-layer defect density, respectively, and A is a constant depending on the structure of the solar cell, i.e. the i-layer thickness, etc. Using the above empirical relationship, the kinetic data presented in this paper was compared to the model proposed by Stutzmann, Jackson and Tsai[5] and to the more recent model by Redfield and Bube.[1] While both models can fit the data quite satisfactorily, self consistency of the parameters used in fitting both the intensity and the temperature dependence can be only established using the Redfield and Bube model. Details of this comparison with the theories will be published elsewhere.[8]

## CONCLUSION

Light intensity and temperature dependence of the light induced degradation of a-Si:H solar cells have been studied using an accelerated test. A simple scaling law was demonstrated between the light intensity and the exposure time near the room temperature. It was concluded that the defect saturation is a result of a detailed balance between the light induced defect generation and the thermal annealing. Therefore, it is important to realize that the measurements done at high intensity underestimate the effect of annealing on cells being degraded under normal operating condition. In fact, our model suggests that at $50°$ C and AM1.5 intensity the saturation would occur in about 15000 hours.[8]

## ACKNOWLEDGEMENT

The authors would like to thank many colleagues at Solarex for helpful discussion and technical assistance. This work is partially supported by the Solar Energy Research Institute under subcontract No. ZM-0-19033-1.

## REFERENCES

1. D. Redfield and R.H. Bube, Appl. Phys. Lett 54, 1037 (1989).
2. H.R. Park, J.Z. Liu, and S. Wagner, Appl. Phys. Lett. 55, 2658 (1989).
3. M.S. Bennett and K. Rajan, Proc. of the 20th IEEE Photovoltaic Specialists Conf., Las Vegas, Nevada (1988) 67.
4. A. Catalano, Proc. of the 21th IEEE Photovoltaic Specialists Conference, Dissimimee, Florida (1990) 36.
5. M. Stutzmann, W. Jackson, and C. Tsai, Phys. Rev. B32, 23 (1985).
6. M. Bennett, private communication.
7. W.B. Jackson, Mat. Res. Soc. Symp. Proc., Vol. 149, 571 (1989).
8. L. Yang and L. Chen, to be published.

# SELECTING THE BAND GAP FOR BEST LONG-TERM PERFORMANCE OF a-Si:H SOLAR CELLS

X. Xu,  M. Kotharay,  N. Hata,  J. Bullock and S. Wagner
Department of Electrical Engineering, Princeton University
Princeton, NJ 08544

## ABSTRACT

We calculate the light-soaking time dependence of the a-Si:H solar cell efficiency with a simple analytical model that expresses the efficiency as a function of the deep-level defect density in the i-layer. We use the model to predict the cell stability as a function of the optical gap of the i-layer. The band gap for best long-term performance lies in the vicinity of 1.6 eV.

## INTRODUCTION

Recent theoretical[1] and experimental[2,3] work has demonstrated that a correlation exists between the saturated light-induced defect density and the rate of defect buildup in a-Si:H. Because the saturated defect density can be reached and measured within a few hours, this correlation provides a tool for evaluating the long-term light-soaking behavior of a-Si:H *material* from short-term data.  We have now begun to exploit this approach for the modelling of the *solar cell* performance under light-soaking, i.e., for inferring the long-term light-soaking behavior of cells from short, high-intensity, soaking experiments.  We expect that such modelling eventually will enable us to infer the entire light-soaking history of solar cells from their characteristics in the defect-saturated state.  The underlying assumption is that the exposure-dependence of cell efficiency can be described entirely in terms of the growth of light-induced deep-level defects in the i-layer.

In this paper we concentrate on an analytical, semi-quantitative, formulation of the time-dependent cell efficiency. We also will use one early result of first-principles, quantitative, numerical, modelling, which we have used so far only for computing the initial cell efficiency.

The model infers the long-term cell performance from a measurement of the saturated defect density $N_{sat}$. We apply the model by combining it with an empirical correlation between $N_{sat}$ and the optical gap $E_{opt}$.[4] This combination provides a prediction of cell stability as a function of the i-layer bandgap. While the $N_{sat}$-$E_{opt}$ correlation is purely heuristic, its application to cell stability furnishes an example for connecting cell to material parameters.

## THE MODEL

To set up the analytical model we relate:
- $N_{sat}$ to the defect density during light-soaking $N_S(t)$, using the Redfield-Bube model for a two level system:[1]

$$N_S(t) = N_{sat} - (N_{sat} - N_{s0})\exp(-Kt^b).\qquad(1)$$

Here, $N_{s0}$ is the annealed-state defect density, b is a dispersion parameter, $b = T(^\circ K)/605$, and K is the rate constant which is proportional to the carrier generation rate G. The values of K and G used in this work are $4 \times 10^{-4}s^{-1}$ and $5 \times 10^{20}cm^{-3}s^{-1}$, respectively.
- $N_S(t)$ to the carrier collection length $l_{c0}(t)$ for zero applied voltage using a correlation established by Faughnan and Crandall:[5]

$$l_{c0}(t) = \frac{\mu\ V_{bi}}{r\ d\ N_S(t)}\qquad(2)$$

where $\mu$ is the ambipolar mobility, $V_{bi}$ the built-in voltage, r the recombination rate constant, and d the i-layer thickness. $d = 0.5\ \mu m$ and $\frac{\mu}{r} = 5 \times 10^{8}cm^{-3}v^{-1}$ [ref.6]. For $V_{bi}$ see eq.(8).
- $l_{c0}(t)$ to the fill factor FF(t) with a relation established by Smith, Wagner and Faughnan[7]

$$FF(t) = FF_1 + k_1\log(l_{c0}(t)/d).\qquad(3)$$

Here $FF_1$ (=0.39) and $k_1$ (=0.30) are two constants obtained by fitting to experimental results.[6,7] The link between $N_{sat}$, $N_S(t)$, $l_{c0}(t)$ and FF(t) allows to compute the time-dependence of the fill factor from the saturated defect density.

The time dependence of the short-circuit current density $J_{sc}(t)$ is introduced as a function of the collection length $l_{c0}(t)$ via the Hecht equation:

$$J_{sc}/J_p = l_{c0}(t)/d\ [1-\exp(-d/l_{c0}(t))]\qquad(4)$$

where $J_p$ is the maximum value of the photocurrent.

Prompted by published results for the long-term light-soaking of solar cells[8] we assume that the open-circuit voltage $V_{oc}$ remains constant during light-soaking.

The cell efficiency is calculated as,

$$\eta(t) = \eta(0)\ [J_{sc}(t)/J_{sc}(0)]\ [FF(t)/FF(0)].\qquad(5)$$

We used a first-principles numerical model to compute the initial cell efficiency $\eta(t)$.[9]

For an example of the application of this model we calculate the bandgap-dependence of $\eta(t)$. We take advantage of the observation that an apparent lower bound exists to $N_{sat}$ as a function of $E_{opt}$, which is given by[3]

$$\log[N_{sat}(cm^{-3})] = 17.0 + 3.1[E_{opt}(eV) - 1.70]. \qquad (6)$$

In this way we make a first attempt at predicting cell stability from a material parameter. The $N_{sat}$-values were determined at ~35°C, which is the temperature for the efficiency model.

When the optical gap $E_{opt}$ is introduced as a parameter, the numerical calculation of the initial cell efficiency requires as an input the $E_{opt}$-dependent initial, annealed-state, defect density $N_{s0}(E_{opt})$. A generic function for $N_{s0}(E_{opt})$ could be taken from the equilibrium theory,[10] but instead we take an empirical relation taken on the same samples that provided eq.(6):

$$\log[N_{s0}(cm^{-3})] = 15.7 - 6.0[E_{opt}(eV) - 1.70]. \qquad (7)$$

The bandgap dependences of the initial, annealed-state, defect density $N_{s0}(E_{opt})$ and the saturated defect density $N_{sat}(E_{opt})$ that we use for the model are shown in Fig.1.

$E_{opt}$ also affects the built-in voltage which enters eq.(2), and we assume that

$$V_{bi}(V) = E_{opt}(eV)/q - 0.5. \qquad (8)$$

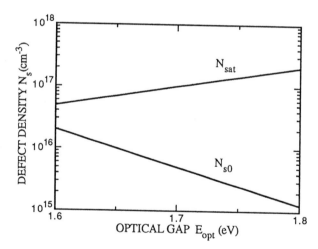

Fig.1 Optical gap dependence of defect densities in the annealed (initial) and saturated states.

MODELLING THE BANDGAP DEPENDENCE OF CELL STABILITY

Experiments on the saturation of the light-induced defect density suggest that the saturated density $N_{sat}$ increases with increasing bandgap. It is not clear whether this gap-dependence of $N_{sat}$ reflects a fundamental energy-dependence of $N_{sat}$, or whether the correlation is observed because both $N_{sat}$ and $E_{opt}$ are tied to another parameter that determines stability. We couple the observed, empirical correlation of $N_{sat}$ with $E_{opt}$ with the analytical model explained above. The result provides an estimate of the bandgap-dependence of the cell efficiency. For the $N_{sat}$-$E_{opt}$ relation we use eq.6, which represents our current estimate for the lower limit of $N_{sat}= f(E_{opt})$. The analytical model was our early vehicle to estimating the effect of material parameters on the cell stability. We plan to employ the first-principles numerical model, which we used for computing the initial cell efficiencies in the present work, for modelling the entire light-soaking process.

Results on the fill factor (eq.3) and the short-circuit current (eq.4) are discussed first. Fig.2 shows the dependence of the fill factor on continuous light-soaking time for Tauc gaps from 1.60 to 1.80 eV. Note that the initial fill factor is 0.75 except for the i-layer with $E_{opt} = 1.60$ eV, which has the highest initial defect density $N_{s0}$ (Fig.1). FF begins to drop after approximately one day of continuous illumination ($G=5\times10^{20}cm^{-3}s^{-1}$) and saturates – or, reaches a steady-state – within one or two years. The final value of FF is seen to rise noticeably with decreasing $E_{opt}$, reflecting the concurrent drop in $N_{sat}$. The initial short circuit current $J_{sc}$ again drops slightly with decreasing gap (Fig.3). $J_{sc}$ begins to level off after about one

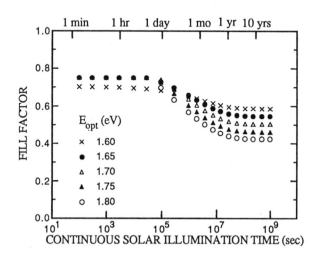

Fig.2 Fill factor of a-Si:H cells with optical gaps from 1.6 to 1.8 eV as a function of continuous light-soaking time.

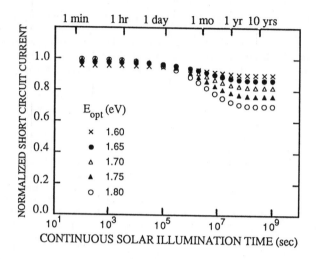

Fig.3 Normalized short circuit current versus light-soaking time.

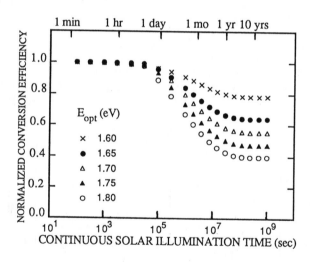

Fig.4 Normalized cell efficiency versus light-soaking time.

year, and the final $J_{sc}$ lies substantially higher for low $E_{opt}$ than for high $E_{opt}$. The product of FF(t) and $J_{sc}$(t) is plotted in Fig.4, where it is normalized to an initial value of 1.0 to highlight the differences in the final value. It is clear that the drop of $N_{sat}$ with $E_{opt}$ translates to substantially higher long-term stability for cells with lower-gap i-layer. Although the analytical model represented by eqns.1 to 5 is useful for predicting trends, it is not reliable enough for the calculation of absolute cell efficiencies.   We computed the initial efficiencies $\eta(0)$ with a quantitative model, and then derived changes of efficiency with illumination by multiplying with the data of Fig.4.   The parameters employed in the numerical modeling are listed in Table I.   The initial and final, long-term, efficiencies so determined are shown in Fig.5.   While the initial cell efficiency is highest at high values of $E_{opt}$, the final, long-term value is highest for the lowest gap.

The results of the model suggest that the best long-term performance of a-Si:H cells may be achieved with low-gap material. This conclusion is plausible, but not firm.   On the one hand, the $E_{opt}$-dependence of $N_{sat}$ given by eq.6 is an estimate which has not yet been confirmed on i-layer samples with gaps around 1.6 eV.   On the other hand, very recent experiments show that $N_{sat}$ can be reduced to values less than given by eq.6,[11] which would translate to a higher stability than the one we calculated here. Furthermore, $N_{sat}$ (or, rather, the steady-state defect density) under solar illumination intensity is quite sensitive to temperature,[12] and is likely to lie lower than the value determined under high-intensity illumination, so that the long-term cell stability would be raised by raising T.

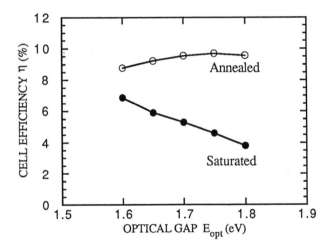

Fig.5 Optical gap dependence of the initial  cell efficiency calculated   numerically and the final cell efficiency obtained by combining this numerical  result with the normalized long-term  efficiency of Fig.4.

Such reservations notwithstanding, it is clear that models that use data from high-intensity light-soaking can produce stimulating information about the long-term stability of a-Si:H solar cells.

## SUMMARY

We have calculated the long-term stability of a-Si:H solar cells by combining a semi-quantitative analytical model with experimental data on the defect saturation in i-layer material. The model expresses changes in the cell efficiency as function of changes in the density of deep-level defects. These changes are calculated from the experimentally determined initial and saturated defect densities, which in turn are expressed as functions of the optical gap. The values of Tauc gap that we can cover with some confidence range from 1.6 to 1.8 eV. The model predicts that the rate of efficiency degradation is lowest, and the final efficiency is highest for cells with 1.6 eV i-layers.

## ACKNOWLEDGEMENTS

We thank Prof. S. J. Fonash, Dr. J. K. Arch and J. Y. Hou for making their program, "Amorphous Modelling at Penn State", available and for their help with the numerical model. We also thank M. Isomura, M. Ch. Lux-Steiner and N. W. Wang for useful discussions. This work is supported by the Thin-Film solar Cell Program of the Electric Power Research Institute.

### Table I. Principal input parameters

Front contact barrier height (eV)   $1.25(E_{opt}=1.6)-1.45(E_{opt}=1.8)$

Back contact barrier height (eV)    $0.21$

|  | p-layer | i-layer | n-layer |
|---|---|---|---|
| Thickness (Å) | 150 | 4650 | 200 |
| Acceptor or donor doping (cm$^{-3}$) | $1\times10^{19}$ | 0 | $1\times10^{19}$ |
| Electron mobility (cm$^2$v$^{-1}$sec$^{-1}$) | 20 | 20 | 20 |
| Hole mobility (cm$^2$v$^{-1}$sec$^{-1}$) | 2 | 2 | 2 |
| Char.valence band tail energy (meV)* | 60 | 45-55 | 60 |
| Char.conduction band tail energy (meV)* | 40 | 27-30 | 40 |
| Midgap density of states (cm$^{-3}$ev$^{-1}$) | $5\times10^{18}$ | $2\times10^{15}$-$2\times10^{16}$ | $5\times10^{18}$ |
| Tail state density at band edges (cm$^{-3}$ev$^{-1}$) | $4\times10^{21}$ | $4\times10^{21}$ | $4\times10^{21}$ |
| Optical gap (eV) | 1.6-1.8 | 1.6-1.8 | 1.6-1.8 |

* We assume that the characteristic energies of both valence band tail states, $E_v$, and conduction band tail states, $E_c$, vary with $E_{opt}$ as $E_v$(meV) $= 50 - o.5/[E_{opt}(eV) - 1.70]$ and $E_c$ (meV) $= 28.5 - 0.15/[E_{opt}(eV) - 1.70]$.

REFERENCES

1) D. Redfield and R. H. Bube, Appl. Phys. Lett. 54, 1037 (1989).
2) M. Isomura, H. R. Park, N. Hata, A. Maruyama, P. Roca i
   Cabarrocas, S. Wagner, J. R. Abelson and F. Finger, Record of
   the 5th International Photovoltaic Science and Engineering
   Conference, Kyoto, Japan, Nov.26-30, 1990, p.71.
3) M. Isomura, X. Xu and S. Wagner, Solar Cells, to be published.
4) H. R. Park, J. Z. Liu and S. Wagner, Appl. Phys. Lett. 55, 2658
   (1989).
5) B.W. Faughnan and R. S. Crandall, Appl. Phys. Lett. 44, 537
   (1984).
6) Z E. Smith, S. Aljishi, V. Chu, J. Conde and S.Wagner, Record
   of the 3rd International Photovoltaic Science and Engineering
   Conference, Tokyo, Japan, Nov.3-6, 1987, p.589 .
7) Z E. Smith, S. Wagner and B. W. Faughnan, Appl. Phys. Lett.
   46,1078 (1985).
8) Y. Uchida, M. Nishiura, H. Sakai and H. Haruki, Solar Cells 9,
   3 (1983).
9) P. J. McElheny, J. K. Arch and S. J. Fonash, Appl. Phys. Lett.
   51, 1611 (1987); J. K. Arch, F. A. Rubinelli, J. Y. Hou and
   S. J. Fonash, J. Appl. Phys., in press.
10) Z E. Smith and S. Wagner, Phys. Rev. Lett. 59, 688 (1987).
11) M. Isomura and S. Wagner, this volume.
12) N. Hata and S. Wagner, this volume.

ENHANCED SURFACE RECOMBINATION IN a-Si:H SOLAR CELLS CAUSED BY LIGHT STRESS

W. Kusian and H.Pfleiderer
Siemens AG, Corporate Research and Development, D-W-8000 München

ABSTRACT

The change of the spectral photocurrent characteristics of amorphous silicon pin solar cells with light induced degradation is compared with the effect of slightly doping the "i-layer". Both treatments yield similar results. Light stress lets the primary photocurrent, measured with blue light, decrease and the secondary photocurrent, measured with red light, increase. The similar change occurs when a slight n-doping of the "i-layer" is replaced by a slight p-doping. A simple interpretation in terms of uniform fields and preponderant surface recombination is possible and will be outlined.

We additionally resort to numerical simulations. Degradation is to be simulated by the introduction of stronger recombination. The recombination rate will be distributed in space. We indeed find that enhanced surface recombination plays the key role in guiding the simulations towards our experiment.

INTRODUCTION

Amorphous silicon pin solar cells degrade under "prolonged illumination" ("aging" by "light soaking"). Light stress is known to create additional defects within the amorphous network and increases the recombination rate of excess carriers, in particular of photocarriers. The recombination rate may be inhomogeneously distributed through the pin structure, and also the additional recombination   process developing during light stress. In a rough way, "surface" recombination near the p/i and i/n junctions may be distinguished from "bulk" recombination across the main part of the i-layer. Light stress will enhance both surface and bulk recombination, but perhaps one of them predominantly. A widespread view today is that essentially the bulk degrades[1,2]. We shall collect a number of arguments here in favour of the opposite possibility, namely that the surface degradation outbalances the bulk degradation.

We analyse own spectral response measurements by analytical and numerical modeling.

EXPERIMENTAL TECHNIQUES

The observed diodes are standard pin solar cells and others, where the "i-layer" was slightly doped with small amounts of phosphorus or boron[3]. We call them psn diodes. Our sample set comprises a diode with an undoped s-layer (proper i-layer) and doped diodes (obtained by adding 3 ppm $PH_3$ or 3 ppm $B_2H_6$ to the $SiH_4$ gas during plasma CVD deposition of the s-layer). Spectral small-signal photocurrent characteristics were measured by means of chopped light in connection with standard lock-in technique used to suppress the dc-current under forward voltage bias. We did not apply bias light during the spectral response measurements. The light soaking was made by 64h of AM1-illumination under open-circuit. A part of the experiments presented here was already published previously[3].

DIODE STRUCTURES pνn AND pπn

The external collection efficiency Q of a solar cell represents the ratio of the short-circuit current under a monochromatic illumination of wavelength $\lambda$ over the incoming photon flux $\Phi$. Amorphous silicon (a-Si:H) cells are structured as pin diodes and $Q(\lambda)$ for them depends on the applied voltage U. Thus the full function $Q(\lambda,U)$ should be considered. In "slightly" doping the i-layer, a pin diode transforms into a "psn" diode. We distinguish between $s = \nu$ (P doped "i-layer")

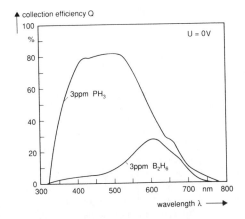

Fig. 1. External collection effiency Q versus wavelength $\lambda$ of psn diodes doped with 3 ppm $PH_3$ and 3 ppm $B_2H_6$. Bias voltage $U = 0$ (short-circuit)

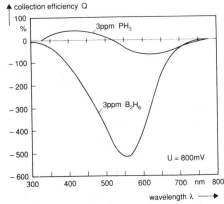

Fig. 2. As Fig. 1, but bias voltage $U = 0.8$ V

and $s = \pi$ (B doped "i-layer"). Any s-layer provides a certain $Q(\lambda,U)$ picture. Fig. 1 shows the short-circuit efficiency $Q(\lambda,0)$ of a p$\nu$n and a p$\pi$n diode, both for illumination through the p-contact. Increasing voltage U shifts $Q(\lambda)$ downwards and also below the line $Q = 0$. Fig. 2 presents $Q(\lambda)$ for the relatively high voltage $U = 0.8$ V. The sign reversal of Q indicates a reversal of the photocurrent direction. While primary photocurrents are limited to the closed interval $0 < Q < 1$, secondary photocurrents $Q < 0$ can easily go below the line $Q = -1$. The Q efficiency of a pin diode would lie within the frame defined by the p$\nu$n and p$\pi$n diodes. The comparison between Figs. 1 and 2 reveals a certain asymmetry. A more elaborate description of this asymmetry can be found in our previous publication[3].

Another spectral-response representation offers the internal collection efficiency $q(\lambda,U) = -Q(\lambda,U)/Q_s(\lambda)$, with saturation value $Q_s(\lambda)$ of $Q(\lambda,U)$, attainable under a sufficiently high reverse voltage $U < 0$[4]. This representation has the advantage that it excludes optical losses like reflection at the substrate and absorption in the TCO-layer, and only involves recombination losses within the pin semiconductor. The minus sign in the definition of q is arbitrary. It assures that q has the same sign as the photocurrent. The primary photocurrent interval becomes $-1 < q < 0$. In Fig. 3 we present a set of characteristics $q(U,\lambda)$ for both of the diodes shown in Figs. 1 and 2. The already mentioned asymmetry now appears in another fashion.

The following conceptual discussion will be based on the simple arguments already introduced earlier[3]. An idea of how space charges and recombination rates are distributed

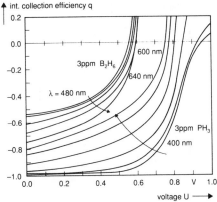

Fig. 3. Internal collection efficiency q versus voltage U of psn diodes doped with 3 ppm $PH_3$ and 3 ppm $B_2H_6$. Different wavelengths $\lambda$

through the s-layer is required from the start. The electric field through the s-layer is more or less uniform, with spikes at the p/s and s/n boundaries. The assumption of a "uniform field" is justifiable at least with respect to a pin diode[5]. The uniform field will be controlled by the voltage U. A weak illumination, to be used for the measurements, will not alter the field. Besides a "bulk" recombination of the photocarriers, an extra "surface" recombination is to be expected, because photoelectrons (photoholes) find plenty of holes (electrons) to recombine with around the p/s junction (s/n junction). The field spikes push the photocarriers out of the regions with an extra strong recombination strength, called "surfaces". Thus a high (low) field spike will hinder (promote) surface recombination. The doping of the "i-layer" is a means to vary the p/i and i/n junctions and the associated field spikes. The p/$\nu$ and $\pi$/n junctions are high, and the p/$\pi$ and $\nu$/n junctions are low. Surface recombination, especially at the frontal p-side, weakens primary photocurrents. Therefore, the primary efficiency q of a p$\nu$n diode (p$\pi$n) diode is high (low), in accordance with the experiment shown in Fig. 3. Moreover, surface recombination feeds secondary photocurrents. Therefore, a p$\nu$n diode (p$\pi$n diode) allows low (high) secondary photocurrents. The low (high) surface recombination velocity at the front (back) side of a p$\nu$n diode explains also the splitting of the q(U) curves for the different values of the parameter $\lambda$. The photocarriers generated by the deeply penetrating red light (high $\lambda$) "see" the strongly recombining back side. As a consequence, the primary efficiency becomes better towards the blue end of the spectrum (with lower $\lambda$). Just the opposite conditions prevail for p$\pi$n diodes!

## DIODES pin IN STATES A AND B

The annealed state of an a-Si:H pin cell is termed A as usual. Light-soaking leads to state B, and annealing again back to A. The A→B transition will be accompanied by a deformation of Q($\lambda$,U) and q($\lambda$,U). Figs. 4 and 5 show Q($\lambda$) of a pin diode in states A and B for U = 0 and U = 0.8 V. The comparison with Figs. 1 and 2 reveals some similarity. The transition $\nu$→$\pi$ (Figs. 1,2) corresponds to A→B (Figs. 4,5). This observation suggests an identical cause in both cases, i.e. essentially an increase of surface recombination.

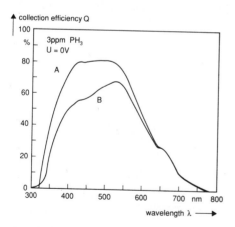

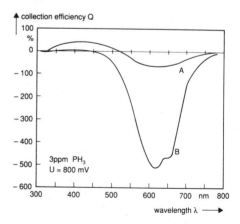

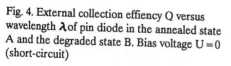

Fig. 4. External collection effiency Q versus wavelength $\lambda$ of pin diode in the annealed state A and the degraded state B. Bias voltage U = 0 (short-circuit)

Fig. 5. As Fig. 4, but bias voltage U = 0.8 V

Fig. 6 gives the q picture of a pin diode in states A and B, shown already in a previous work[4]. A similarity with Fig. 3 is recognizable, but only a faint one. The q(U) curves of a pin diode show a conspicuous feature, a compromise between the $p\nu n$ and $p\pi n$ cases. The q(U) curves for different $\lambda$ cross through a point with coordinates $q_P$, $U_P$. In the voltage regime $U > U_P$ ($U < U_P$), the sequence of the q(U) with varying $\lambda$ follows the proper sequence of a $p\nu n$ ($p\pi n$) diode, as shown in Fig. 3. This sequence just reverses at $U = U_P$. In other words, a pin diode in states A and B shows two faces, directed towards a $p\nu n$ and $p\pi n$ diode. It should be possible to gradually shift the cross point to and fro by incorporating very small amounts of doping atoms (P or B) into the i-layer. Then A→B could become equivalent to $\nu$→$\pi$ on a quite small scale. This vague consideration calls for further support.

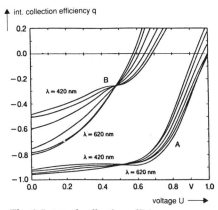

Fig. 6. Internal collection efficiency q versus voltage U of pin diode in the undegraded state A and the degraded state B. Different wavelengths $\lambda$

Our conceptual arguments stem from simple assumptions (uniform field, surface recombination). It is straightforward to build with them an analytical photocurrent collection model[4,6]. The uniform field substitutes for Poisson's equation. The neglection of bulk recombination (for photocarriers) decouples the continuity equations of electrons and holes. Therefore, a drift-diffusion equation remains for both photocarriers. The linkage between the two equations comes in by the optical generation rate being common for electrons and holes. The p- and n-layers are considered through boundary conditions involving surface recombination velocities. The model parameters are mobilities and surface recombination velocities for electrons and holes, and a flat-band voltage $U_F$, providing a vanishing field. The q(U) characteristics of a symmetrical diode (identical electron and hole parameters) for different $\lambda$ cross through a point at $U = U_F = U_P$. The A→B transition of Fig. 6 can be modeled using a symmetrical diode by increasing the surface recombination velocity (on both sides)[4]. A real pin diode will be slightly asymmetrical. The p/i junction is higher than the i/n junction. The surface recombination velocity, therefore, of electrons (holes) at the p/i and (i/n) junction has to be lower (higher). Then the calculated q(U) curves for the longer $\lambda$ miss the crossing at $U_P < U_F$ of the lower $\lambda$[6]. If the diode is made even more asymmetrical, we can arrive at the q(U, $\lambda$) family of $p\nu n$ in Fig. 3, and reversing the asymmetry, also for $p\pi n$. Hence the function q(U,$\lambda$) of psn diodes can be simulated correctly with the uniform-field model in a qualitative sense. The best agreement between model and experiment is possible with a pin diode, where the field distribution is most uniform. Anyway, all of our uniform field arguments can be elaborated more quantitatively by uniform field modeling. In addition, the q(U) crossing can be modeled in this way.

### NUMERICAL SIMULATION OF PIN DIODE

Simple arguments and modeling can identify the meaning of Fig. 6. Deviations from standard conditions (psn diodes, light soaking) helped to probe our understanding. A more realistic description of pin diodes serving as solar cells can be given by numerical simulations that solve the three semiconductor equations in one dimension. We now apply this possibility in order to justify our hitherto gained knowledge. The calculations to be presented here will be limited in scope. In connection with numerical simulations the physical assumptions to be worked out and the experiments to be covered become crucial. The q(U) characteristics due to the wavelengths $\lambda = 400$ nm and $\lambda = 600$ nm were simulated for a pin diode of thickness 0.5 $\mu$m and are

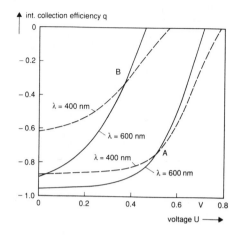

Fig. 7. Calculated internal collection efficiency q versus voltage U. Wavelengths $\lambda = 400$ nm and 600 nm. States A and B

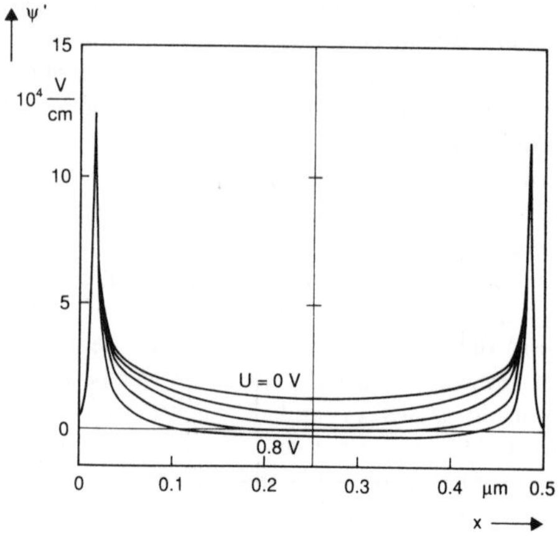

Fig. 8. Calculated derivative of potential $\psi'$ versus coordinate x through pin diode in the dark. Voltages U. Annealed state A

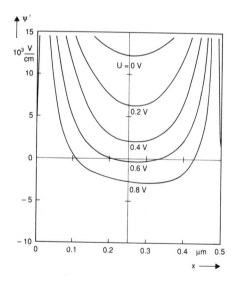

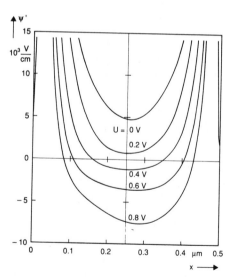

Fig. 9. Like Fig. 8. Another scale

Fig. 10. Like Fig. 9. But degraded state B

presented in Fig. 7. The input parameters were manipulated in order to obtain examples for states A and B. Quantitative agreement between theory and experiment is out of question. What counts is the inherent similarity of Figs. 6 and 7. The physical model behind the simulation is modest. The internal collection efficiency q was calculated by relating the photocurrent to the integral generation rate introduced by monochromatic illumination with the low photon flux $\Phi = 10^{10}$/cm$^2$s. A density of gap states (DGS) distribution, and laws for trapping and recombination, were assumed of a kind introduced earlier[7]. The band gap was assumed to be constant through the pin structure (no C-buffer layer), as well as band mobilities ($\mu_n = 15$ cm$^2$/Vs for electrons and $\mu_p = 5$ cm$^2$/Vs for holes). The dark characteristic behind Fig. 7 show diode factors n = 1.3 (n = 1.6) and the saturation current $I_0 = 2.44*10^{-10}$ A/cm$^2$ ($I_0 = 8.27*10^{-9}$ A/cm$^2$) in state A (B). The derivative of the potential (negative electric field) through the simulated pin diode is plotted in Fig. 8 for voltages U = 0 to U = 0.8 V in the case of state A. The field spikes mark the p/i and i/n junctions. Magnified field pictures are given in Figs. 9 and 10 for states A and B.

We now come to a quick discussion of Fig. 7 in order to disclose the physical mechanisms inherent to our simulation. The q(U) characteristics for state A have horizontal and ascending branches. A horizontal branch only appears under two previsions: First, the bulk recombination in the i-layer is (sufficiently) weak; second, the electric field in the middle part of the i-layer remains negative (and therefore drives a primary photocurrent). The characteristics bend upwards as soon as the electric field approaches zero, somewhere in the i-layer, and then becomes positive (and drives a secondary photocurrent), see Fig. 9. In order to be definite we introduce the flat-band voltage $U_F$, the lowest voltage allowing a vanishing field. We furthermore define the dispersion $d(U) = q(U, \lambda_b) - q(U, \lambda_r)$ with $\lambda_b = 400$ nm and $\lambda_r = 600$ nm indicating blue and red illumination. A special value is $d(U_p) = 0$. The A→B transition (simulation of light soaking) was effected by two manipulations. First, the frontal surface recombination was enhanced; second, the bulk recombination in the i-layer was enhanced too, together with the DGS. We at once give the relevant data. One of our recombination parameters is the factor $S = v_t \sigma$, with $v_t$ thermal velocity of carriers and $\sigma$ capture cross section of neutral traps. Enhanced "surface" recombination was allowed in lifting S in a thin region near the p/i junction from $5*10^{-10}$ cm$^3$/s to $5*10^{-7}$ cm$^3$/s. Enhanced bulk recombination and bulk trapping was allowed by lifting S and DGS in midgap, S from $1*10^{-9}$ cm$^3$/s to $5*10^{-9}$ cm$^3$/s, and DGS from $1*10^{16}$/cm$^3$eV to $5*10^{16}$/cm$^3$eV. So A→B was achieved by a massive increase of "surface" recombination and a moderate increase of bulk recombination and trapping. (The rear surface recombination near the i/n junction remained untouched. In reality, this back-side recombination may increase too with light soaking.) Enhanced surface recombination broadens the short-circuit dispersion d(0), because it shifts mostly the $\lambda_b$ characteristic. Enhanced bulk recombination and trapping reduce $U_F$ and $U_p$. In fact, enhancing the frontal surface recombination leads (besides broadening d(0) also) to a later characteristics' crossing, i.e. to a higher voltage $U_p$, in contradiction to the experiment, Fig. 6. In order to bring $U_p$ down then, we had to decrease $U_F$ and that can be done by increasing bulk recombination and trapping. The result of $U_p \approx U_F$ in state A and $U_p > U_F$ in state B may be of minor importance, and perhaps reflects an inconvenience of our (by no means perfect) simulation. In any case, in order to launch a simulation of A→B, we had to start with enhancing the surface recombination in order to broaden d(0). Blue light generates more carriers near the "surface" than red light. So a reduction of the blue efficiency indicates enhanced surface recombination. This obvious notion is compatible with our simulation. We finally can conclude that enhanced surface (bulk) recombination is of major (minor) importance with degradation under exposure to light.

In order to visualize the simulation we add plots of the recombination rate r depending on the coordinate x through the pin structure. Fig. 11 presents the function r(x) in the dark of state A. We consider the recombination rate in excess of thermal equilibrium. (Hence r = 0 for U = 0.) The "structures" to the left and to the right reflect special assumptions for S nearby. According to them, an extra surface recombination is apparent already in the dark. An earlier paper exploited this circumstance[8]. Now we are more interested in the additional recombination caused by illumination, the photorecombination rate δr. Where r is high in the dark, δr induced by weak illumination may stay insignificant. The relative photorecombination $\mathfrak{s} = \delta r/r$ offers a way out. Figs. 12 and 13 present functions $\mathfrak{s}(x)$ for states A and B, calculated for blue and red illumination ($\lambda_b$ and $\lambda_r$). Because $\mathfrak{s}$ depends on the illumination strength (the photon flux assumed here is $\Phi = 10^{10}/cm^2 s$) and on the voltage U, only the functional form $\mathfrak{s}(x)$ is significant. Increasing parameters U let r increase (Fig. 11) and hence $\mathfrak{s}$ decrease. (The limit U→0 means also r→0, and $\mathfrak{s}$ tends to infinity.) The lower limit for the plots is $\mathfrak{s} = 10^{-4}$, corresponding to our (chosen) numerical accuracy. It suffices then to look at $\mathfrak{s}(x)$ for U = 0.2 V in Figs. 12 and 13. We see a steep pile-up of $\mathfrak{s}(x)$ against the outer boundaries of the pin diode. This picture stresses an extra surface recombination of photocarriers. The surface recombination, seen through the plots in Figs. 12 and 13, naturally increases with the blue shift $\lambda_r \to \lambda_b$, and also with A→B, as it should.

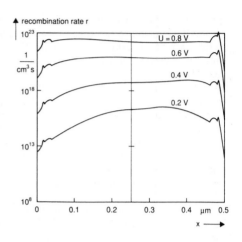

Fig. 11. Calculated recombination rate r versus coordinate x. Annealed state A

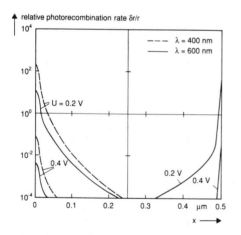

Fig. 12. Calculated relative photorecombination rate $\mathfrak{s} = \delta r/r$ versus coordinate x. Wavelengths $\lambda = 400$ nm and 600 nm. Annealed state A

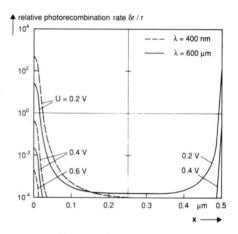

Fig. 13. Like Fig. 12, but degraded state B

ACKNOWLEDGEMENT

This work was supported under the technological program of the Federal Ministry of Research and Development.

REFERENCES

1. C.M. Fortmann, et al., J. Appl. Phys. Vol. 64 No. 8 (1988) 4219
2. P. Chatterje, et al., J. Appl. Phys. Vol. 67 No. 8 (1990) 3803
3. W. Kusian, et al., J. Appl. Phys. Vol. 64 No. 10 (1988) 5220
4. W. Kusian, et al., Mat. Res. Soc. Symp. Vol. 118 (1988) 183
5. J. Furlan, et al., Proc. of the PVSEC-5 (1990) 347
6. H. Pfleiderer, et al., Proc. of the 20th IEEE PV Spec. Conf. (1988) 180
7. M. Hack, et al., J. Appl. Phys. Vol. 58 No. 2 (1985) 997
8. H. Pfleiderer, Journal of Non-Cryst. Solids 115 (1989) 57

Design Considerations for Stable Amorphous Silicon Solar Cells

Vikram L. Dalal, B. Moradi, and G. Baldwin
Department of Electrical and Computer Engineering
and Microelectronics Research Center
Iowa State University
Ames, IA 50011

## ABSTRACT

We examine the effects of light-induced degradation in material properties of a-Si:H upon the device physics of an a-Si:H $p^+in^+$ cell. It is shown that the increase in defect density upon degradation can lead to a significant decrease in electric field over middle regions of the i layer. This decrease is particularly severe under forward bias (near the maximum power point), and can explain the degradation in fill factor of the cell upon light soaking. We show that by using an innovative cell design, with graded gap i layer, we can increase field in the middle regions, and reduce the degradation in fill factor.

## INTRODUCTION

Light-induced degradation in the performance of a-Si:H solar cells is a major technological problem.[1,2] The degradation arises because of changes in the electronic properties of i layers of p-i-n cells upon exposure to light. These changes include an increase in defect density, a shift in Fermi level, and increases in recombination rates of electrons and holes. To address the problems of degradation, both scientific and technological approaches are being tried. The scientific approaches address the questions of origins of light-induced degradation and how to alter structure and bonding to reduce degradation.[3,4] The technological approaches address the questions of improving the purity of the material,[5,6] and cell design. In particular, tandem-cell designs, with 3-stack tandem cells, seem to have found favor as an ad-hoc design approach for improving stability.[7,8] In this paper, we examine the cell design considerations for improving stability by examining the basic device physics of p-i-n junction solar cells. In particular, the concepts of bandgap grading and deliberate changes in material properties are examined to improve stability. The simulations clearly show that appropriate bandgap grading can improve stability, both because of the additional field that can be provided, and because of the increase in absorption that can result when the bandgap is reduced.

## REVIEW OF PHYSICS OF a-Si SOLAR CELLS

At this stage, it is useful to review the basic physics of a-Si p-i-n junction solar cells[9,10] A schematic band diagram of the $p^+$-i-$n^+$ solar cell under short-circuit is shown in Fig. 1. It is well known that a-Si solar cells have two junctions--the front $p^+i$ junction, and the back $i-n^+$ junction. The total built-in voltage between $p^+$ and $n^+$ layers is distributed unequally between the front and the back junctions, with the division of the voltage depending on the original positions of Fermi levels in the $p^+$,i and $n^+$ materials. Typically, out of a total built-in voltage of $\sim 1.0$ V, about 0.6 V is in the front $p^+i$ junction and 0.4 in the rear, assuming an $(E_c - E_f)$ of $\sim 0.75$ eV in the i material.

298

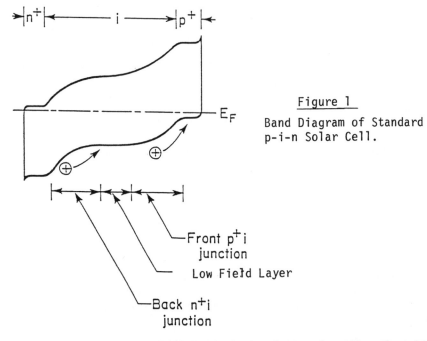

Figure 1

Band Diagram of Standard
p-i-n Solar Cell.

—Front $p^+$i
junction

Low Field Layer

—Back $n^+$i
junction

Typically, the i layer is thin enough that the two junctions overlap, with a substantial field in the middle portion of the i layer. It is useful to remember that the field is non-uniform in the i layer, with very high fields ($\sim 10^5$ V/cm) near the $p^+$ and $n^+$ contacts, and low fields ($\sim 10^4$ V/cm) in the middle region. The high degree of non-uniformity of the field is a consequence of the fact that there is a distribution of defect states g(E), throughout the bandgap of the material, with high defect densities near the valence and conduction band edges, due to Urbach tails. As a result of this continuous and non-uniform g(E), the regions of the i layer nearest the $p^+$ and $n^+$ junction layers in a $p^+in^+$ cell contribute more space charge to the depletion regions than the middle regions of the i layer. Therefore, as a consequence of Poisson's equation, the regions of i layer nearest $p^+$ and $n^+$ junctions have the highest fields.

A consequence of the non-uniformity of space charge is that when the junction goes under forward bias, i.e., when the solar cell is operating near maximum power point, the fields near the $p^+$ and $n^+$ junctions remain high, but the field in the middle portion of the i layer decreases significantly. Thus, a quasi-neutral or low field region develops in the middle portion of the i layer, and carrier collection efficiency can be significantly reduced if the physical extent of the low-field region is less than the diffusion length or field-assisted range of the minority carrier.

These simple considerations hold good even when more complicated phenomena, such as carrier trapping and non-uniform lifetimes are taken into account. From first principles, hole concentrations, and hence trapped hole concentrations, are going to be high near the $p^+$ junction, and low near the $n^+$ junction. Thus, positive space charge due to trapped holes adds to the positive space charge due to defect states in the $p^+$i junction, and correspondingly, the negative space charge due to trapped electrons adds to the negative space charge of defect states in the $n^+i$ junction region. Thus, carrier trapping may be considered as contributing to further narrowing of the high field regions, thereby reducing the field in the middle.

At this time, it is appropriate to discuss carrier collection. As shown earlier by a number of workers,[9,11] except in a narrow region of i layer adjacent to the $p^+$ layer, holes are the limiting, or minority carriers in the standard $p^+in^+$ solar cell. This is a consequence of preferential hole trapping and a lower $(\mu\tau)$ product for holes (where $\mu$ is mobility and $\tau$ is lifetime) than for electrons. Typically, in good a-Si:H, $(\mu\tau)$ for holes is $\sim 5 \times 10^{-8}$ cm$^2$/V, and $5 \times 10^{-7}$ to $1 \times 10^{-6}$ cm$^2$/V for electrons. In a thin (0.6μm) undegraded a-Si:H cell, the minimum field of $\sim 10^4$ V/cm assures a minimum range of $\sim 5$ μm for holes, thereby assuring a high collection efficiency for all photo-generated holes. [Collection efficiency is $e^{-x/\mu\tau\epsilon}$, where x is distance to the $p^+$ layer, and $\epsilon$ is electric field.] The field-free diffusion length for holes in undegraded material is 0.36 μm.

## LIGHT-INDUCED CHANGES IN MATERIAL PROPERTIES AND THEIR INFLUENCE ON DEVICE PHYSICS

Upon exposure to light, changes occur in material properties such as defect densities, and electron and hole $(\mu\tau)$ products. In particular, mid-gap defect densities increase from $1\text{-}2 \times 10^{15}$/cm$^3$ in virgin material to $4\text{-}8 \times 10^{16}$/cm$^3$ (or more) upon prolonged light soaking.[12] This increase in mid-gap defect densities reduces electron and hole lifetimes and $(\mu\tau)$ products.

The increase in defect densities has several consequences.

1.  Since the majority of defects generated by light are in the lower half of the gap, (1.0-1.4 eV below $E_c$), Fermi level $(E_f)$ in the i layer moves down such that the number of occupied and unoccupied states is equal. $E_f$ may move down by 0.08-0.10 eV. This movement of $E_f$ has important consequences for redistribution of junction voltages in a $p^+in^+$ cell. In particular, the built-in voltage of the front junction $(p^+i)$ is reduced, and of the back junction $(in^+)$ is increased, by the amount of shift of $E_f$. This phenomenon will redistribute the field profiles, making them more symmetrical than in the virgin state.

2.  The increase in defect density, to the mid $10^{16}$/cm$^3$ level, can have a significant impact on field distribution in the device. In particular, the fields will become less uniform (more peaked), and the mid-gap region will see a decrease in the field. When the decrease in the field is coupled with the tenfold decrease in hole $(\mu\tau)$, the decrease in range, $\mu\tau\epsilon$, can be catastrophic for efficient collection of photo-generated carriers. In particular, the collection efficiency of holes generated by long wavelength photons can be significantly reduced. This phenomenon is particularly troublesome under forward bias, i.e., at maximum power point, where the increased defect density may lead to a middle region of i layer which is nearly devoid of field. Hence the carriers have to travel by diffusion alone, with a reduced diffusion length, $L_p$, of 0.114 μm. We will show in the next section that this is indeed what happens, with a resultant significant loss in fill factor of the cell.

## ANALYTICAL MODELS AND RESULTS

For purposes of design and understanding, we shall use a simple model which approximates the field profiles quite well. The model assumes that the defect density increases exponentially with distance from the middle of the i layer towards the junctions. Such an increase can approximate for the increasing space charge density found nearest the junctions. Thus, the defect density $N_D(x)$, at any point is given by $N_o e^{ax}$, where a is a parameter and x is measured from the point where the depletion-region field is zero. The maximum mid-gap defect density, which is found nearest the tail-states, i.e., nearest the $p^+$ and $n^+$ regions, is determined by the value of a, and is typically (5-10)x $N_o$. Of course, more complicated models, such as models using exponential increases in g(E) away from $E_f$, can also be used,[13] but they give results qualitatively similar to the simple model used here.

The results of calculations of field profiles, $\varepsilon(x)$, in the bulk of the i layer (except for the very narrow regions nearest $p^+$ and $n^+$ contacts where the tail states determine the profiles) are shown in Fig. 2 and 3 for a standard, 0.5 μm thick i layer $p^+in^+$ a-Si:H solar cell. Fig. 2 shows $\varepsilon(x)$ curves in virgin and degraded states for short-circuit conditions, and Fig. 3 for a forward bias of +0.6 V. The a-Si:H had a starting total defect density of $2.5 \times 10^{15}/cm^3$, $(N_o = 1 \times 10^{15}/cm^3)$ and a defect density of $2.5 \times 10^{16}/cm^3$ after degradation. The hole ($\mu\tau$) products were $5 \times 10^{-8} cm^2/V$ in the virgin (or annealed) state, and $5 \times 10^{-9} cm^2/V$ in the degraded case.

From Fig. 2, we see that under short-circuit conditions, the field in the middle regions of the i layer is reasonably high ($10^4$ V/cm) after degradation. This implies a range of 0.5 μm in the middle, giving a reasonable collection efficiency for holes generated in the middle and back (in$^+$) regions of the cell by light incident from the $p^+$ side. The collection efficiency does suffer some loss for short-circuit conditions after degradation, a result that agrees with experimental results.

However, as is clear from Fig. 3, the field collapses in the middle regions of the cell upon forward bias of 0.6 V after degradation. As a result, the holes generated in the middle and back regions of the cell can only travel to the front by field-free diffusion, with a diffusion length of 0.114 μm, significantly smaller than the 0.5 μm field-assisted range obtained under short circuit conditions. Hence, collection efficiency (QE) suffers a significant loss, implying a loss in fill factor, in agreement with experimental results.

## DESIGN OF A MORE STABLE CELL

The results of the preceding section allow us to design a more stable cell. In particular, we can take advantage of the experimental fact that bandgap of a-Si:H can be controlled within limits by changing H dilution in the gas phase and growth temperature.

Using these means, it is possible to vary bandgaps between 1.75 and 1.65 eV, and still maintain reasonable quality of films, with mid-gap ESR defects in the ~$10^{15}/cm^3$ range, electron $\mu\tau$ of $10^{-6} cm^2/V$, and hole $\mu\tau$ of $3–5 \times 10^{-8} cm^2/V$. Further reductions in bandgaps, to ~1.6 eV, can be achieved by adding small amounts of Ge to Si. Electronic qualities of these high gap alloys are still good.[14]

The small variations in bandgap can be used to advantage in designing a more stable cell with a graded-gap i layer.

## Electric Field in Standard Cell

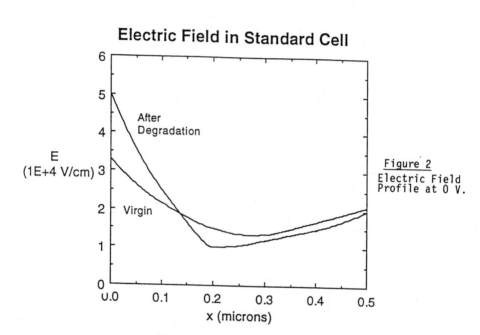

E
(1E+4 V/cm)

After Degradation

Virgin

x (microns)

Figure 2
Electric Field
Profile at 0 V.

## Field Profile:  V=+0.6V
## (Standard Cell)

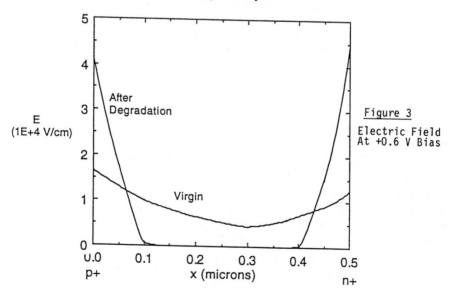

E
(1E+4 V/cm)

After Degradation

Virgin

p+

x (microns)

n+

Figure 3
Electric Field
At +0.6 V Bias

A graded-gap cell structure is shown in Fig. 4. It consists of a $p^+$ a-(Si,C):H layer ($\sim E_g$ of 2.0 eV), a thin (200A°) buffer i layer with $E_g$ varying from 2.0 to 1.65 eV, a low-gap region ($\sim 0.15$ μm thick) of 1.65 eV, a graded-gap region where the gap changes from 1.65 to 1.75 eV, over $\sim 0.15$ μm, and a final 0.15 μm thick region of 1.75 eV, followed by a thin $n^+$ layer and reflecting contact. The light is incident from $p^+$ side.

Such a structure can be expected to be more stable because of two considerations.

1.  By using a graded gap in the middle, we are achieving a grading of the valence band, thereby creating an electric field in a region where there was low field before. This field is in a direction to assist the holes. Typically, out of 0.1 eV change in bandgap, perhaps 70 meV may show up as a change in valence band position, leading to a field of $5 \times 10^3$ V/cm in the middle. The slight retarding field ($2$-$3 \times 10^3$ V/cm) faced by electrons will not deter them since their ($\mu\tau$) is so much higher. Thus, we restore field-assisted transport in the middle, which persists even under $+0.6$ V operation.

2.  By using lower gap region in the front of the cell, we increase light absorption in the front regions, thereby increasing the fill factor.

These considerations are borne out by the numerical simulation results shown in Fig. 5, where we compare the QE of graded cell under forward bias after degradation with QE of standard cell, also under forward bias and degradation. The substantial improvement in QE of graded cell is obvious from the figure. Fig. 6 shows the improvement in electric field due to grading. At this stage, it is appropriate to examine the role of the front buffer layer. Its purpose is to reduce recombination at the $p^+$i interface, by driving electrons away from the interface, thereby achieving both high $V_{oc}$ and high QE in the blue region of the spectrum.

To improve the cell stability further, one may wish to further lower bandgap of low-gap region to the 1.6 eV range, thereby achieving higher absorption and valence band grading. However, we would have to be careful about the quality of 1.6 eV material, since a significant decrease in electronic properties and an increase in defect density may defeat the expected improvements.

Note that the design considerations outlined here are particularly appropriate for a 2-junction tandem cell, where we want a lower gap in the second cell. Using these design considerations, one may be able to achieve both higher efficiency and higher stability in a 2-junction cell, by using only a-Si:H, or a-Si-H with small amounts of Ge. The advantages of using a 2-junction tandem cell over a 3-junction cell are obvious from considerations of manufacturing and robustness against spectral-content variations.

## CONCLUSIONS

In conclusion, we have shown that upon light-induced degradation, there is a catastrophic collapse of electric field in the p-i-n device under forward bias (maximum power point). This leads to poor collection efficiency for holes generated in the middle and back regions of the i layer, leading to a poor fill factor. The fill-factor degradation can be reduced by using bandgap grading of i layer, using only conservative changes in bandgaps. Such designs may be particularly applicable to 2-cell tandem-junction structures, made from high gap (1.8 eV) a-Si:H as the first cell, and a graded gap (1.6-1.8 eV) a-Si:H or a-(Si,Ge):H as the second cell. It may even be possible to design such cells without Ge, using techniques which can produce good quality low gap a-Si:H. The development of high quality, low gap a-Si:H alloys should

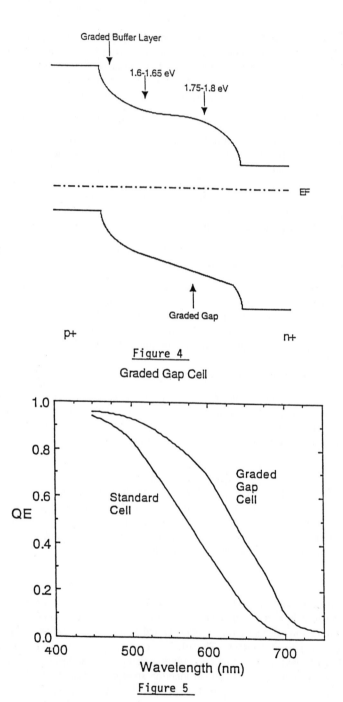

Figure 4

Graded Gap Cell

Figure 5

Quantum Efficiency of Graded Gap Cell at +0.6 V

be a high priority for future research programs.

### REFERENCES

[1]  C. R. Wronski, *Proc. AIP Conf on Stability of a-Si Alloy Materials and Devices,* AIP Conf., Vol. 157, 1 (1987).

[2]  K. Asaoka, et al., *J. Non-Cryst. Solids,* 115, 24 (1989).

[3]  R. Biswas, Proceedings of this Conference.

[4]  V. L. Dalal, et al., Proceedings of this Conference.

[5]  V. L. Dalal, *Proc. AIP Conf.,* Vol. 157, 257 (1987).

[6]  T. Unold and J. D. Cohen, Proc. *Material Res. Society,* 192, 719 (1990).

[7]  J. Yang, et al., *Proc. IEEE Photovolt. Spec. Conf.,* 19, 241 (1988).

[8]  A. W. Catalano, et al., *J. Non-Cryst. Solids,* 115, 14 (1989).

[9]  V. L. Dalal, *Solar Cells,* 2, (1980).

[10]  M. Hack and M. Shur, *J. Appl. Phys.,* 58, 997 (1985).

[11]  M. Hack and M. Shur, *J. Appl. Phys.,* 58, 1656 (1985).

[12]  H. R. Park, et al., Proc. *Material. Res. Society,* 192, 751 (1990).

[13]  P. Sichangurist, M. Konagai and K. Takahashi, *J. Appl. Phys.,* 55, 1155 (1984).

[14]  V. L. Dalal, J. Booker and M. Leonard, *Proc. IEEE Photovolt. Spec. Conf.,* 18, 1500 (1985).

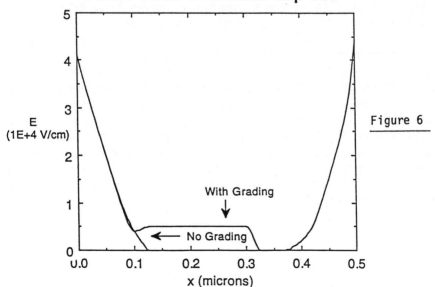

## Field Profile In Graded-Gap Cell

Figure 6

THERMAL ANNEALING OF PHOTODEGRADED a-SiGe:H SOLAR CELLS

M. Bennett and K. Rajan, Solarex Corporation,
Thin Film Division, Newtown, PA  18940

## ABSTRACT

Heterojunction p-i-n solar cells were made with a-SiGe:H i-layers. The bandgaps of these i-layers were 1.55 eV for one set of cells and 1.65 eV for another. Both sets of cells were photodegraded, then thermally annealed at various temperatures between 90°C and 150°C. The cells could be fully annealed. The characteristic annealing curves (fractional recovery of the fill factor vs. time) were the same as those previously found for a-Si cells[1]. However, the activation energy for annealing was higher, about 1.6 eV, compared with 1.2 eV for a-Si:H i-layer cells.

## INTRODUCTION

Amorphous silicon and the amorphous silicon based alloys suffer from photodegradation (the Staebler-Wronski effect)[2]. In order to achieve optimal solar cell performance it is necessary to study the processes involved so that either this degradation can be eliminated or the cells can be designed in such a way as to minimize its impact. One strategy to minimize the impact of photodegradation is to use three junction devices[3]. These combine high initial efficiency with good resistance to photodegradation but introduce a further complication, namely, that amorphous SiGe:H must be used. In order to maximize the cell output, both the initial properties of this alloy and the effect on these of exposure to light must be known.

Previous studies[4] have shown that the higher bandgap alloy, a-SiC:H, is much more susceptible to light-induced degradation than a-Si:H, however there are indications, at least from device measurements, that a-SiGe:H may be less susceptible than a-Si:H[5]. Certainly as the bandgap is lowered the relative loss in efficiency decreases. The lowered degradation seen in a-SiGe:H cells could be the result of lowered susceptibility to defect creation, it might be the result of an increased rate of thermal annealing, or it may be simply be a reflection of the poorer initial efficiency of a-SiGe:H cells.

The objective of this study is to determine the effect of thermal annealing on the improved resistance to photodegradation of a-SiGe:H solar cells. We will not address the possibility of an inherently lower generation rate in a-SiGe:H nor the more complicated question regarding a possible relationship between rate of degradation and the initial efficiency.

## EXPERIMENTAL PROCEDURE

Heterojunction p-i-n solar cells were deposited on conductive tin oxide (CTO) coated glass and had the structure glass/CTO/p-SiC:H/i-SiGe:H/n-Si:H/ metal, where the metal contact was a layer of indium tin oxide overcoated with Ag. The i-SiGe:H layers had bandgaps and thicknesses as shown in Table I. Because the structure incorporates no grading of the intrinsic layer, it is not optimally efficient[6]. However the use of a homogeneous i-layer makes the assignment of a activation energy to a particular alloy meaningful.

Table I

| Eg | Thickness | Efficiency | | Fill Factor | |
|---|---|---|---|---|---|
| (eV) | (A) | Initial | After Deg. | Initial | After Deg. |
| 1.55 | 2000 | 4.6% | 3.2% | .50 | .41 |
| 1.65 | 2000 | 4.8% | 3.5% | .55 | .44 |

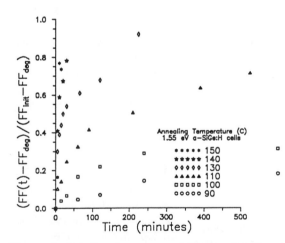

**Figure 1.** Fractional recovery of FF as a function of annealing time for various annealing temperatures. Cells contain 1.55 eV a-SiGe:H i-layers.

These experiments were carried out using procedures identical to that used for a-Si studies[1]. The cells were degraded under 100 mW/cm$^2$ white light for about 1000 hours. During this period the cells were held at $V_{oc}$ and at 45°C. Temperature was maintained using fans, the intensity was set using a suitably filtered and calibrated crystalline Si solar cell. Degradation proceeded until the cells had reached the efficiencies shown in Table I. The fill factor (FF) was monitored as in ref. 1 because it changes more than the other photovoltaic parameters (short circuit current ($J_{sc}$) and open circuit voltage ($V_{oc}$)) and because it is less affected by variations in measurement temperature and intensity. I-V measurements were done on the two source simulator described in ref 7.

## EXPERIMENTAL RESULTS

Figure 1 shows the fractional recovery of the FF as a function of annealing time for the 1.55 eV a-SiGe:H cells at various temperatures. The quantities in the quotient which labels the vertical axis are the fill factor as a function of annealing time (FF(t)), the fill factor in the degraded state, prior to commencement of annealing (FFdeg), and the fill factor after annealing is complete (FFinit). The cells can be completely annealed although the curves in Figure 1 follow only about the first 75% of the recovery. It is obvious that the annealing process is strongly influenced by temperature.

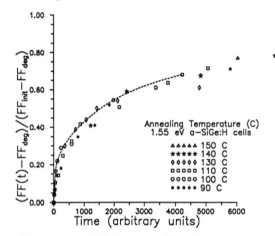

**Figure 2.** Curves of Figure 1 with time axes shifted so that the curves coincide.

The procedure which was used to analyze the data and which will be outlined below is the same as that described in ref 1. Although it is not obvious, the curves in Figure 1 all have the same functional dependence. This is shown in Figure 2 for cells with 1.55eV a-SiGe:H. For each curve in Figure 1 the time axis has been compressed or expanded, as required, to cause all six curves to fall on top of one another. It is possible to make the 6 curves fall along a common line to a very good approximation, thus demonstrating the assertion made above that the curves all have the same shape. Figures 3 and 4 show the corresponding curves for 1.55 eV cells. The dashed lines in Figures 2 and 4 are cubic spline fits which will be referred to later in the discussion.

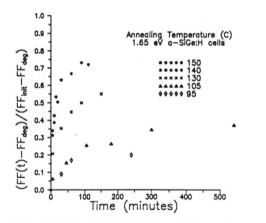

**Figure 3.** Fractional recovery of FF as a function of annealing time for various annealing temperatures. Cells contain 1.65 eV a-SiGe:H i-layers.

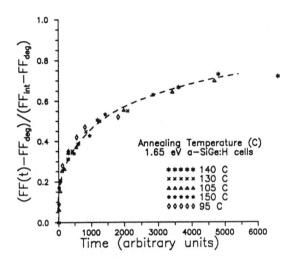

**Figure 4.** Curves of Figure 3 with time axes shifted so that the curves coincide.

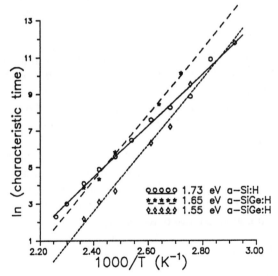

**Figure 5.** Natural log of characteristic annealing time vs. inverse temperature.

Since the annealing curves for different temperatures all have the same shape and since temperature has such a strong influence on annealing rate, it is natural to inquire as to whether the process is thermally activated. We arbitrarily define a characteristic time to be that which is required for a 70% recovery of fill factor to take place. Plotting the natural log of this characteristic time against the inverse of temperature, we obtain an Arrhenius relationship (Figure 5) for each of the a-SiGe:H sets of data, where for convenient comparison the corresponding curve for Si has been included. As shown in reference 1 and as would be guessed from Figures 2 and 4 the definition of characteristic time is not critical. Defining the characteristic time as that required for 20% or 50% recovery, for example, would lead to the same conclusions.

Taking slopes in Figure 5, we find that the activation energy for annealing is 1.54 eV for a-SiGe:H having a bandgap of 1.65 eV and 1.56 for a-SiGe:H having a 1.55 eV bandgap. It is difficult to assign uncertainties to these values. Referring to Figure 5 however, it is clear that the Arrhenius relations corresponding to the a-SiGe:H cells have a different slope than that corresponding to a-Si:H cells. It is less certain whether or not there is a real difference between the slopes of the a-SiGe:H lines. One can conclude, however, that the annealing activa-

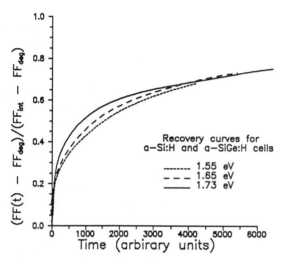

**Figure 6**. Comparison of the functional dependence on time of the fractional recovery of FF for solar cells with 1.73 eV a-Si:H i-layers and cells with 1.65 eV and 1.55 eV a-SiGe:H i-layers.

tion energy is greater for the a-SiGe:H cells.

### DISCUSSION

The similarity between the annealing kinetics of 1.55 eV a-SiGe:H and 1.73 eV a-Si:H is striking. Figure 6 compares the recovery curves for Si:H with 1.55 eV and 1.65 eV a-SiGe:H, where the a-SiGe:H curves are the fits taken from Figures 2 and 4 above and the a-Si:H curve is taken from the corresponding Figure in ref 1. It was found for a-Si:H cells that the initial part of the recovery was slower if the photodegradation prior to annealing was longer in duration. The slower initial recovery of the a-SiGe:H cells may result simply from the fact that they were photodegraded for a much longer period than the a-Si:H cells had been. In any case, the similarity of the curves is striking. This indicates that the annealing is governed by the same process in both cases. The annealing process is largely independent of material, although the rates are somewhat different.

Previously we had associated the activation energy for annealing with the activation energy for H diffusion in a-Si:H. We have not measured the activation energy for H diffusion in a-SiGe:H and do not know of any previous measurements. That the improved stability of a-SiGe:H is not due to increased annealing is evident. It can be seen from Figure 5 that, while at high temperatures a-SiGe:H anneals much faster than Si, at temperatures under 100°C or so the rate of annealing for a-SiGe:H cells is less than that of a-Si:H cells so that the apparent improvements in stability seen in a-SiGe:H devices is not due to an enhanced rate of annealing.

### ACKNOWLEDGEMENTS

This work was partially supported by SERI contract #ZM-0-19033-1.

### REFERENCES

1.   M.S. Bennett, J.L. Newton, K. Rajan and A. Rothwarf, J. Appl. Phys. 62, 3968 (1987).

2.    C.R. Wronski, Stability of Amorphous Silicon Alloy Materials and Devices, edited by B.L. Stafford and E. Sabisky (American Institute of Physics Conference Proceedings 157, New York, 1987) p. 1.

3.    A. Catalano, 21st IEEE Photovoltaic Specialists Conference - 1990 (Institute for Electrical and Electronic Engineers, New York, 1990) p. 36.

4.    M.S. Bennett, S. Wiedeman and K. Rajan, Amorphous Silicon Technology - 1989, edited by A. Madua, M.J. Thompson, P.C. Taylor, Y. Hamakawa and P.G. Lecomber (Mater. Res. Soc. Proc. 149, Pittsburgh, PA 1989) p. 577.

5.    M.S. Bennett, A. Catalano, K. Rajan, and R.R. Arya, 21st IEEE Photovoltaic Specialists Conference - 1990 (Institute for Electrical and Electronic Engineers, New York, 1990) p. 1653.

6.    R.R. Arya, M.S. Bennett, K. Rajan, and A. Catalano, 9th E.C. Photovoltaic Solar Energy Conference edited by W. Palz, G.T. Wrixon, and P. Helm (Kluwer Academic Publishers, Dordrecht, 1989) p. 251.

7.    M.S. Bennett, and R. Podlesny, 21st IEEE Photovoltaic Specialists Conference - 1990 (Institute for Electrical and Electronic Engineers, New York, 1990) p. 1438.

# STABILITY OF AMORPHOUS SILICON SOLAR CELLS

P.K. Bhat*, D.S. Shen+ and R.E. Hollingsworth*
Glasstech Solar, Inc., 6800 Joyce Street, Golden, CO 80403
* Materials Research Group, Inc., 12441 West 49th Avenue, Wheat Ridge,
CO 80033
+Electrical Engineering Dept., University of Alabama, Huntsville,
Huntsville, AL 35899

## ABSTRACT

We report results on optimization of "stable" amorphous silicon solar cells. Approaches used for optimization were based on hydrogen content variation in the intrinsic layer of p-i-n solar cells. Hydrogen content was varied in two ways: by changing the substrate temperature and by using hydrogen diluted silane during intrinsic layer deposition. Single junction solar cells of efficiencies greater than 7% after 600 hours of continuous light soaking were obtained. Single junction solar cells having microcrystalline intrinsic layers were prepared for the first time and show promise for use in the top cell of tandem modules.

## INTRODUCTION

Hydrogenated amorphous silicon (a-Si:H) has received much attention because of its potential in many types of semiconductor devices. For solar cell applications, much of the recent effort has been geared towards producing "stable" solar modules. GSI has been working on similar band gap silicon-silicon multijunction modules to produce a module with stable efficiencies greater than 10%. In this regard we have looked into optimization of process parameters to produce a stable cell rather than optimization for initial high efficiencies.

Ideally, a high band gap i-layer is preferred for the top cell (to gain high $V_{OC}$) and a low band gap i-layer is preferred for the bottom cell (to gain in red current) of a tandem solar cell. However, without alloying (e.g., with carbon or germanium), the band gap is adjusted by changing the hydrogen concentration of the material. Two problems have to be addressed here: initial electronic properties and the stability. In general, the initial electronic properties will degrade when the hydrogen concentration deviates from the optimized value. Furthermore, for low-hydrogen material fabricated at high temperatures, impurity diffusion is an additional concern, while for high-hydrogen intrinsic a-Si:H material, light stability is a concern. Materials work was directed towards producing intrinsic thin films of amorphous silicon with band gaps of 1.9 eV and 1.7 eV. Stability of intrinsic layers and solar cell devices was studied in detail.

Two well known methods of varying the hydrogen content were used in this study: substrate temperature variations and hydrogen dilution of the source gas. Increasing the deposition temperature reduces the hydrogen content monotonically. Hydrogen dilution effects on the hydrogen content of the film are not so simple. There is competition between an increased hydrogen arrival rate which would tend to increase the hydrogen content in the film and etching by atomic hydrogen which reduces the content in the film. At high dilution ratios, etching wins out and microcrystalline films are deposited with hydrogen contents below 5%.

## EXPERIMENT

The deposition of a-Si:H films and devices was carried out in a commercially available GSI multichamber PECVD system. RF power was supplied to the 130 cm$^2$

312

lower electrode by a variable frequency oscillator and a wide-band power amplifier. Temperature dependent experiments were performed using the standard 13.56 MHz frequency while hydrogen dilution experiments were done using 110 MHz plasmas. Materials were characterized by measurements of the dark and light conductivities, intensity dependence of the photoconductivity, dark conductivity activation energy, and the optical absorption. The intensity dependence of the photoconductivity is proportional to $G^\gamma$, where G is the photon flux and the exponent $\gamma$ is used to characterize the carrier recombination. Standard devices were fabricated using glass/SnO$_2$:F/p-i-n/silver structures with 1 cm$^2$ opaque top contacts. Additional devices were fabricated with ITO/Ag back reflectors and exhibited no difference or slightly better stability under light soaking compared to devices with simple silver contacts. The doped layers were deposited using 13.56 MHz discharges. The p layer used here was a wide band gap boron doped a-SiC:H layer. Device performance was measured under global AM1.5 (100 mW/cm$^2$) illumination conditions. Stability of devices was studied during 600 hours of light exposure (AM1.5 intensity) at about 40°C.

## RESULTS AND DISCUSSION

Figure 1 shows the photoconductivity (AM1.5 100 mW/cm$^2$ and with a 600 nm bandpass filter) and dark conductivity vs. the deposition temperature. Figure 2 shows the optical gap Eg, absorption coefficient at 600 nm wavelength, and the power index of photoconductivity vs. light intensity. The photoconductivity remained constant until the substrate temperature dropped below 150°C. The dark conductivity increased by about one order of magnitude while the band gap decreased from 1.83 eV to 1.70 eV as the temperature was increased. The photosensitivity is of the order of $10^6$. The band gap decreased and the red absorption coefficient increased linearly as the deposition temperature increased. The activation energy of the dark conductivity was also measured to estimate the position of the Fermi level. For an i-layer with Tauc's gap of 1.70 eV, the activation energy was 0.85 eV. This activation energy and good electronic properties as shown in Figures 1 and 2 suggest the 'impurity' level in the intrinsic layer was low.

Intrinsic layers and companion devices were fabricated to clarify the relationship between the degradation and hydrogen concentration which is assumed to decrease as the deposition temperature increases. Intrinsic layers and p-i-n solar cells were made with intrinsic layer substrate temperatures from 120-330°C. Figure 3 shows the degradation of the AM1.5 photoconductivity of the intrinsic layers with light soaking time. All of the layers show nearly an order of magnitude decrease in photoconductivity, except for the 120°C layer which started very low. The intensity dependence of the photoconductivity as characterized by the exponent $\gamma$ remains essentially constant with light soaking time. The photoconductivity is controlled by the total defect states between the dark Fermi level and the quasi Fermi level. Since $\gamma$'s were measured under weak light, they reflect the density of states near midgap while the AM1.5 photoconductivity will reflect the total states from midgap into the tails. Therefore, the fact that $\gamma$'s (weak light) do not change but $\sigma$'s (strong light) change significantly suggests that the light induced states are close to the tail states.

Figure 4 shows the absolute photovoltaic parameters of devices made with different intrinsic layer temperatures. The device deposited at 330°C has a very low efficiency because of boron diffusion through the i layer from the p layer as found with SIMS depth profiling. Differences in initial efficiency of the other devices is largely due to the bandgap variation giving different short circuit currents. The devices deposited at 140-

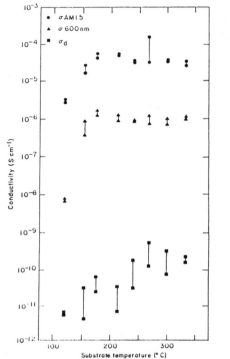

Figure 1. Dark and photoconductivity of intrinsic layers made at different substrate temperatures.

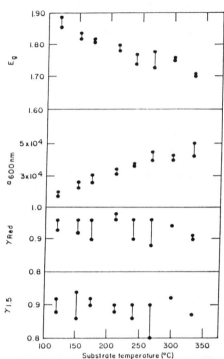

Figure 2. Optical gap, absorption coefficient at 600 nm, and power index of photoconductivity of intrinsic layers made at different substrate temperatures.

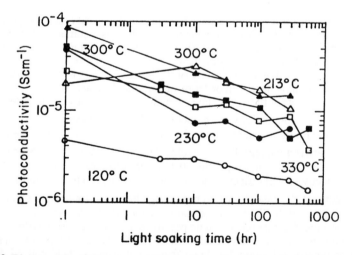

Figure 3. Photoconductivity versus light soaking time for intrinsic layers prepared using different substrate temperatures.

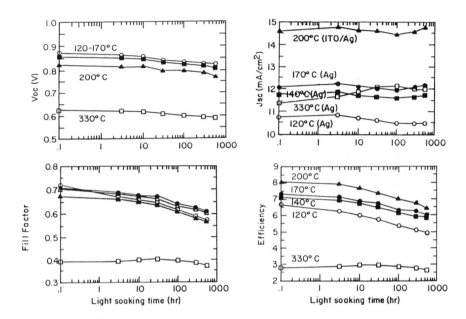

Figure 4. Photovoltaic parameters versus light soaking time for devices prepared with intrinsic layers at different temperatures.

200°C degrade at essentially the same rate while the 120°C device degrades slightly faster. This suggests a weak dependence of stability on hydrogen concentration with the concentration changed only by means of the deposition temperature.

As an alternative method of varying the hydrogen content in the film, hydrogen dilution of the silane source gas was used. Figure 5 shows the dark and photo conductivities of intrinsic layers as a function of the $H_2/SiH_4$ mass flow ratio. The sharp increase in dark conductivity at a flow ratio of 19 is due to a transition to microcrystalline growth which has been confirmed by RAMAN scattering. The Tauc band gap shows a smoother increase from 1.8 to 2.1 eV, contradictory to the anticipation of a lower band gap as the hydrogen concentration is reduced.

Solar cells have been produced with intrinsic layers made using a wide range of hydrogen dilutions and 110 MHz plasmas. The dilution ratio for the microcrystalline transition might be different when grown on TCO due to the importance of nucleation on the growth of microcrystalline material. The pin solar cells fabricated all used the same a-SiC p layer with a graded a-SiC buffer layer and a standard a-Si n layer. The intrinsic layers were approximately 2000Å thick. The initial photovoltaic properties are shown in Figure 6 as a function of $H_2/SiH_4$ mass flow ratio. The best performance in this series occurs for a flow ratio of 1, being slightly better than pure silane. All of the photovoltaic properties show a monotonic decrease with increasing hydrogen content above a ratio of 1. The decrease in short circuit current is easily explained by the lower absorption at higher hydrogen dilutions. Interestingly, the open circuit voltage decreases even though the intrinsic layer bandgap is increasing. The blue current increases slightly as the hydrogen dilution increases, suggesting that some etching of the p/buffer layer occurs during the initial period of the intrinsic plasma. Additionally, the blue fill factor is relatively low for all devices, which generally indicates a poor p-i interface. The red fill factor drops in the same way as the AM1.5 fill factor.

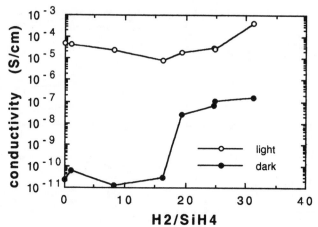

Figure 5. Dark and photo conductivities of intrinsic layers made using high frequency (110 MHz) plasmas.

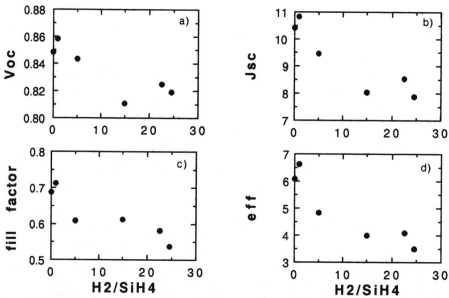

Figure 6. Initial photovoltaic parameters of devices made with hydrogen diluted intrinsic layers.

The absolute and normalized efficiencies vs. light soaking time are shown in Figure 7 for devices with different hydrogen dilutions. The devices with intrinsic layers made with high hydrogen dilution show the best stability, having as little as 9% degradation after 844 hours of light soaking. (High bandgap amorphous cells of comparable thickness made by reducing the deposition temperature show 25% degradation after 600 hours of light soaking.) The devices made from pure silane and with a dilution ratio of 1 exhibit a monotonic decrease in open circuit voltage with light soaking while the hydrogen diluted devices ($H_2/SiH_4 \geq 5$) show an initial increase in $V_{OC}$ of as much as 40 mV during the first 10 hours and a constant value after that. There is no consistent variation of the final $V_{OC}$ with hydrogen dilution. The short circuit current of all devices decreases slightly, with the decrease being almost entirely in the blue end of the spectrum. The fill factors of all devices converge to about 0.57, except for the one with the highest dilution.

The hydrogen diluted material has some potential advantages for use in the top cell of a tandem junction device. Taking into account the reduced red current in the top cell compared to a single junction cell of the same thickness, dilution ratios around 10-15 with top cell thickness of about 1500Å would provide a current matched tandem cell having an a-Si bottom cell. Using a standard amorphous intrinsic layer in the top cell would require a thickness of only about 600Å for current matching. The increased thickness of the hydrogen diluted cells provides distinct advantages in terms of production yield from the point of view of a reduced number of shorted cells. The stable efficiency of the existing hydrogen diluted cells is close to that required for the top cell of a 10% stable tandem.

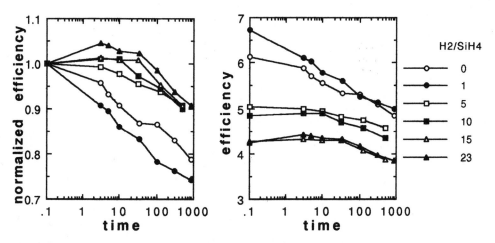

Figure 7. Absolute and normalized efficiency versus light soaking time for devices having intrinsic layers made with various hydrogen dilutions.

## CONCLUSIONS

The effects of intrinsic layer process variations which change the hydrogen content of the deposited film have been examined as methods of improving the stability of amorphous silicon p-i-n solar cells. Substrate temperature variations and hydrogen dilution of the source gas at high rf frequencies were examined. The photoconductivity of intrinsic layers made at temperatures from 140-330°C dropped nearly an order of magnitude after 600 hours of light soaking, essentially independent of the deposition temperature. The optimum deposition temperature for single junction solar cells was approximately 200°C.

The efficiency dropped at lower temperatures due to an increased band gap and correspondingly lower short circuit current. The efficiency dropped at higher temperatures because of boron diffusion from the p layer into the intrinsic layer. The light induced degradation rate was nearly independent of deposition temperature in the range 140-200°C.

Intrinsic layers made with hydrogen dilution and high frequency (110 MHz) plasmas showed a sharp transition to microcrystalline growth at an $H_2/SiH_4$ ratio of 19. Tauc band gaps increased from 1.8 to 2.1 eV as the dilution ratio was increased. Solar cells were made using dilution ratios in the range which give both amorphous and microcrystalline intrinsic layers. Devices made with dilution ratios $\geq 5$ exhibited an increase in open circuit voltage of as much as 40 mV during the first few hours of light soaking and stable voltages after that. Devices with lower dilution ratios showed a monotonic decrease in open circuit voltage of 20-30 mV during light soaking. Devices made with dilution ratios $\geq 5$ exhibited an efficiency drop of less than 10% after 600 hours of light soaking.

## ACKNOWLEDGEMENTS

We would like to thank Cheryl Matovich, Clare Marshall and Clay DeHart for preparing the samples and performing some of the measurements. Al Campaan of the University of Toledo performed the RAMAN scattering measurements. This work was supported by the Solar Energy Research Institute under subcontract #ZM-0-19033-3.

# PART VI

# SUMMARY AND
# CLOSING REMARKS

# PANEL ON METASTABILITY MODELING

W. Fuhs, Chairman

Panelists:  H. Branz, W. Jackson, D. Redfield, B. Street, and M. Stutzmann

## SUMMARY

All the panelists presented papers at this meeting that provide details on their models of a-Si:H metastability[1-5].  The discerning researcher should read the panelists' and other researchers' original contributions in this proceeding.  The discussion started with brief statements of the panelists in which the microscopic models were characterized and the differences between them were emphasized.  The statements and the discussion concentrated on

- hydrogen related models
- weak-bond breaking model
- charge trapping in pre-existing defects
- rehybridization two-site center (RTS) model

The models each seem to be able to describe the main experimental observations.  It seems obvious that kinetic studies will not be able to decide clearly in favor of one or the other model.  The stretched-exponential behavior observed quite generally for defect creation and annealing is an inherent property of all these models and results for instance from a broad distribution of defect creation energies or annealing energies.  As a decisive experiment, the ESR and light-induced ESR results on compensated films have been stressed.  These data show that with increasing doping level of phosphorus and boron the recombination path shifts from bandtail to phosphorous donor states.  This is accompanied by a reduced production of defects by light soaking and therefore this behavior is believed to strongly support the charge induced weak-bond breaking model.  Some room was given to the discussion of the role of impurities.  The panelists agreed that the present data allow to exclude nitrogen as a possible RTS-center and that it's unlikely that oxygen could be involved.  In case of carbon the situation still seems to be unclear.  Another matter of debate was the nature of the involved states.  The panelists agreed that Si-dangling bonds are formed but principle problems were mentioned with the weak-bond concept and in particular with the assignment of these states to bandtail states.  It is still a surprise that the weak-bond concept allows an almost quantitative description of equilibration of defects and dopants in a-Si:H.

The primary areas of agreement:

-   dangling bonds are created
-   hydrogen may be involved either directly by forming unstable hydrogen sites or indirectly by stabilizing defects or determining electronic structure of the films
-   kinetic studies will not allow a decision between the microscopic models.
-   the film stability might be improved beyond today's device quality materials. Each of the panelists arguing from their model's viewpoint made suggestions: modification of the hydrogen bonding, reduction of the number of weak bonds, reduction of potential fluctuations and of the concentration of impurity atoms.

## REFERENCES

1.   Howard M. Branz, Richard S. Crandall, and Marvin Silver, this proceeding.

2.   W.B. Jackson, S.B. Zhang, C.C. Tsai, and C. Doland, this proceeding.

3.   David Redfield and Richard H. Bube, this proceeding.

4.   R.A. Street, this proceeding.

5.   Martin S. Brandt and Martin Stutzmann, this proceeding.

# STABILITY OF a-Si:H MATERIALS AND SOLAR CELLS - CLOSING REMARKS

P.G. LeComber
University of Dundee, Dundee DD1 4HN, Scotland, U.K.

## ABSTRACT

This paper summarises the important stability issues which were identified at this meeting and reviews some of the progress made in attempting to address these.

## INTRODUCTION

It is over a decade since Staebler and Wronski reported[1] the observation of a light induced change in the dark conductivity of amorphous silicon (a-Si:H) films produced by the glow discharge of silane. In their original paper the authors were optimistic that this effect might not degrade the performance of a-Si:H solar cells. Unfortunately, this optimism was not substantiated by subsequent work and the effect has been the subject of a great deal of attention ever since. Many people concerned with photovoltaic materials in general now believe, rightly or wrongly, that unless this effect can be understood and eliminated then the use of a-Si:H solar cells is likely to be restricted to consumer products rather than large-scale power generation. This meeting on "Stability of a-Si:H Materials and Solar Cells" is a consequence of this belief.

In attempting to make a few remarks to close the meeting, I have an impossible task. Twenty-one papers were given orally, about another twenty were presented as posters and we had two excellent discussions - all in two and a half days! Suffice to say that I shall be forced to omit mention of some excellent papers and I would strongly encourage the discerning reader to study the proceedings and form their own opinions rather than rely on my inevitably distorted summary!

The meeting attracted about eighty participants from a range of countries and I am sure I am speaking on their behalf when I congratulate the organisers on arranging a most interesting meeting. If there was one fault with the programme then it was the shortage of papers on solar cells themselves. We can hardly blame the organisers for that - it probably reflects the difficulty that the academic community, at least, has in fabricating state-of-the-art cells with all their technical refinements.

323

The motivation for the meeting was clearly understood.   In the words of Jack Stone in his opening address "....... there are good engineering solutions to stability problems but is the material (viz *a*-Si:H) *inherently* stable?"   The Department of Energy in the USA, in common with a number of other similar organisations elsewhere, are investing significant sums of money into solar cell research and development, and material stability has become the central focus of their *a*-Si:H programme.   The near-term goal$^2$ for *a*-Si:H solar cells is a *stabilised* efficiency $\geq 10\%$ at a cost of $\leq \$90/m^2$ with a 30 year cell lifetime in order to reach the target of 12 cents/(kW hr) at 1986 prices.

Three important issues were identified at the meeting:

(i)    to understand the mechanism underlying light induced degradation in a-Si:H and to determine if it is an *inherent* property of the material;

(ii)   to identify material parameters which can be *demonstrated* to correlate with solar cell efficiency upon light soaking;   and

(iii)  to impress upon the community that it is essential to optimise the solar cell material for the best *stabilised* efficiency, which may not necessarily be the same as the best initial efficiency.

## Models of Light Induced Instability

Five probable causes of metastability have been identified. These are:

- hydrogen

- impurities

- nanoscale inhomogeneities

- strained bonds

- charged defects

By far the majority of the work presented at the meeting concentrated on hydrogen as a possible cause.

From a correlation of material properties and solar cell parameters with hydrogen content $c_H$, Wronski[3] reported that a consortium from Universities, Industries and EPRI believed that further reduction of $c_H$ is important. At this point of the conference I was reminded of a diagram I first showed in public in 1983 at the Tokyo Amorphous and Liquid Semiconductor Conference and this is reproduced[4] as Fig.1. The vertical axis shows the logarithm of the ratio of the room temperature conductivities $\sigma$ in the annealed state A to that in the light degraded state B, normalised by the factor $b$ to its value after exposure for one hour at an intensity of 100 mW/cm$^2$ (where the normalisation has been done both assuming a linear dependence on exposure time $t$ and photon flux, $N_p$, or the $t^{1/3} N_p^{2/3}$ behaviour suggested in the literature). With a few simplying assumptions, the quantity $b\ln(\sigma_A/\sigma_B)$ can easily be shown to be a measure of the density of Staebler-Wronski sites divided by the density of electronic states at the Fermi energy. The values of this parameter for samples from a number of laboratories are plotted as a function of the hydrogen content $c_H$. Clearly the strong correlation between the magnitude of the Staebler-Wronski metastability and $c_H$ has been evident for some

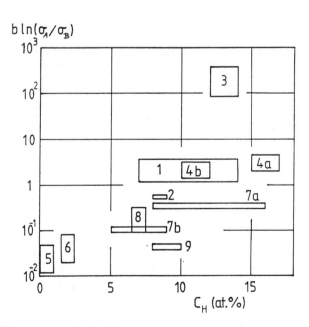

Fig. 1.    Stability of undoped a-Si:H films, as measured by the ratio of the conductivities before and after illumination with 100 mW/cm$^2$, plotted as a function of hydrogen content [After Ref. 4]

time.   However, nobody is suggesting that the *total* hydrogen content is the determining factor and recent work has shown that it is possible to deposit films with the same $c_H$ but with different stabilities.   However, the number of Si-H$_2$ bonds, the density of weak bonds, the details of the Si-Si network, perhaps even the impurity density, are all likely to change with $c_H$ and the challenge is to identify which of these underlies the metastability mechanism. At this meeting we were reminded by Jackson in a discussion session that hydrogen generally lowers the energy of any defect in a-Si:H and is therefore likely to play an important role in *any* model of instability.

The instability was discussed by Street[5] in terms of the hydrogen density of states.   He argued that it can be related to the defect density, the diffusion of H, the metastability effects and to the structure during growth.   In subsequent discussion it became apparent that it is not clear if the hydrogen density of states controls the electronic density of states or vice versa - a problem reminiscent of the chicken and egg debate!   Nevertheless there seemed general agreement that it is important to obtain more information on the hydrogen density of states and more ideas are needed as to how this could be measured.   Possibly the best information presently available comes from the measurements of Jackson et al[6] reported at this meeting.

Cohen and Leen[7] have measured the light induced changes in the density of dangling bonds by transient photocapacitance spectroscopy and the changes in the density of electrons in band tails in n-type specimens.   Their aim was to eliminate certain defect reactions by correlating their results with the predicted behaviour of these quantities.   In principle, this should provide some important information but during the subsequent discussion Stutzmann pointed out that the authors appeared to be measuring significantly smaller changes than those reported by other groups.   If the techniques used by the authors are not able to measure *all* the states then their conclusions may be in doubt and clearly some *independent* measurements of the changes in the dangling bond densities of their samples is required.

Branz et al[17] discussed the possible role of charged defects on the stability of a-Si:H films.   They reminded us that there are many examples of metastable charged defects known to exist in crystalline materials so that their existence in amorphous films would not be surprising.   Recent work on the "Defect-Pool" model by Schumm and Bauer[18] predicts a 2 to 10 times higher density of charged defects compared to neutral defects in thermal equilibrium.   The evidence for this is not as yet unambiguous.   For example, measurements[19] of the temperature dependence of the electron mobility-lifetime product for an undoped a-Si:H sample with $\varepsilon_c$-$\varepsilon_f \simeq 0.75$ eV indicate capture of the electrons into neutral centres. This would be very surprising even if the density of positively charged centres were only equal to the density of neutral centres, let alone greater as predicted by the model. Nevertheless, the charged-defect model can correctly explain recent bias-annealing data[20] for a-Si:H thin-film transistors and clearly further work is required in this area.

The possible role of hydrogen in the Staebler-Wronski effect has been studied using molecular dynamics[8] by using a local "hot-spot" to incorporate the excess energy required by the metastability and then calculating what happens as this propagates through the model. In the models studied to date, two cases have been identified where weak Si-Si bonds can be broken. Both of these involved dihydride sites and no bond breaking was observed at monohydride sites. However, more work is needed in the way the hot-spot is incorporated into the dynamics before more definite conclusions can be drawn from this work.

The major competition to the hydrogen mechanism of the instability arose from the work of Redfield and Bube[9] who proposed a rehybridised two-site (RTS) model involving an "impurity" atom. A major advantage of this mechanism, if it were correct, would be that it would not necessarily be an inherent limitation of a-Si:H!

There was considerable debate of the two mechanisms at a lively Panel discussion. Although the majority of the panel favoured a hydrogen related mechanism, Redfield was not convinced. All panel members agreed that, whatever the mechanism, it involved carriers, barriers to reconfiguration, dangling bonds and states stable against H! On a more serious note, a number of experiments which did appear to offer the hope of distinguishing between a H mechanism and the RTS impurity model were proposed at the meeting. The first of these has already been attempted by a number of groups but with conflicting results! Basically it involves seeing if the number of light induced defects really saturates with light exposure. Although this seems a relatively simple experiment there are a number of difficulties and careful experimentation is required. If the density does not saturate, but can be shown to increase above a particular impurity concentration, then clearly this can be used to exclude that impurity from playing a role in an RTS impurity mechanism. In this way, it has been possible to exclude the dopants P and B as well as N. Some experiments were quoted[10] which would also make C unlikely. However, that still leaves O as well as H and the ubiquitous weak bonds!

The second experiment that could help distinguish between the two mechanisms is one in which the average photon flux during light exposure from a pulsed laser is kept constant but the peak power is varied[11]. In this way the authors were able to demonstrate that the light induced defect density could be made to "saturate" at various densities and furthermore could be increased to values in excess of $10^{18}$ cm$^{-3}$ without any sign of saturation. The experiments on the effect of light soaking on the reverse bias current in p-i-n diodes would also appear to indicate that saturation can be observed at various levels[12]. It may be difficult to reconcile these experiments with the RTS impurity model but no doubt we shall hear more on this!

Many detailed calculations of the metastability make use of the valence band tail states to model weak bonds and the question was asked if there is a good fundamental justification for this or even if a simple correlation is known to exist between the two. Somewhat surprisingly the evidence for this did not appear to be overwhelming and appeared to need further investigation.

If the majority are correct and the presence of hydrogen is responsible for the instabilities, then we have to face the consequence that it is likely to remain an inherent property, at least to some extent, of a-Si:H films. In spite of this the meeting was generally optimistic about the potential for further advances in the stability of high efficiency a-Si:H solar cells. In the last decade we have seen (Fig.2) an increase in stability resulting from a reduction in impurities and a reduction in hydrogen content. We may even need some reduction in the amount of disorder but it is likely we shall stop significantly short of zero disorder!

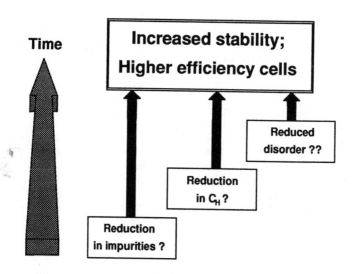

Fig. 2.    Possible routes to increased stability of a-Si:H
solar cells

## Novel Techniques for Reduction in Hydrogen Content

If the correlation between solar cell stability and low $c_H$ is indeed correct, then are there ways we can reduce $c_H$ still further and still produce high efficiency material?   To my knowledge nobody has reported *amorphous* Si-H films with $c_H$ less than about 4 or 5 at.% when made by PECVD of silane.   Is this a fundamental limitation or will the material always be microcrystalline at very low $c_H$?   In the early papers on a-Si:H:F produced from PECVD of $SiF_4$ and $H_2$, amorphous films with $c_H \sim 1$ at.% were reported.   Are films produced in this way more stable than a-Si:H films for the *same* solar cell efficiency?   This question was answered in a discussion session by Guha who stated that although this may have been true initially, the a-Si:H films have now improved so that no difference is apparent!

At the meeting three techniques for producing a-Si:H films with lower $c_H$ were reported.

The first of these[13], Remote Plasma CVD, was first reported some years ago but suffers from the serious disadvantage of very low deposition rates.   Nevertheless, if it can be used to demonstrate low $c_H$ high-efficiency stable cells, then it will have made an important contribution.

Shirai et al[14] reported some new results for samples prepared by the Chemical Annealing technique.   Since this uses PECVD for the actual deposition it could in principle be scaled up for large areas if the annealing gases excited in the microwave discharge can be injected over large areas.

There was considerable interest in the results presented by Mahan and Vanecek[13] on films deposited at up to 15 Å/sec with $c_H$ from 0.1 to 1.0 at.% by a hot wire technique.   However, this is still a relatively new technique requiring considerably more study and among the questions to be addressed the following were identified:

- are the films really different from CVD films?

- can both n- and p-type doping be incorporated into the technique?

- can films be deposited uniformly over large areas?

- can lower substate temperatures be used without degradation of film properties?

If any of the above techniques look promising, there is still the need to demonstrate that *stable high-efficiency* solar cells can be deposited over large areas.

## CONCLUSIONS

What progress was made at the meeting in the issues raised at the start? With regard to understanding light induced degradation, good progress was made in that some candidates were excluded and some experiments were identified to determine the role of "impurities".

I was less impressed with the progress in identifying material parameters which correlate with solar cell efficiency under all conditions. Fabricating and light soaking high-efficiency solar cells in order to test all changes in material parameters is both expensive and time consuming and it would be a major step forward if we could identify simple material parameters which can be shown to *always* correlate with solar cell efficiency. That this will not be easy was demonstrated by von Roedern[16] but, in my view, it is sufficiently important that we have to find a way round this problem.

A number of papers were reported at the meeting on accelerated light soaking. If we are to prepare solar cells under a range of conditions and test these for periods significantly longer than $10^4$ hours, then in order to have realistic feedback to the material growers it is imperative that an accepted method of accelerated testing is forthcoming. Some progress was reported but a number of questions remain - particularly as to whether or not the material is the same in all respects after accelerated and normal testing, even though a particular property may be identical. Again it may come down to identifying which material property is the best measure of ultimate solar cell performance.

How near are we to the goal of 10% efficient stabilised cells? Clearly this has not yet been demonstated but there was general optimism that this was not far away.   Can we achieve 10% stabilised efficiency   with   the   present   material   by   clever   engineering solutions?   Can we optimise other solar cell parameters so that even after light exposure we still have a 10% cell?   If we develop an understanding of the metastability mechanism, can we make   use of this to produce more stable high-efficiency material or will there still be a material limitation?   Unfortunately we do not yet know all the answers but I found this conference made an important contribution in clarifying in my mind what are the important questions, what we presently know and where we need further work.

## REFERENCES

1. D.L. Staebler and C.R. Wronski, Appl. Phys. Letters **31**, 292 (1977).

2. W. Luft, B. Stafford and B. von Roedern, these proceedings.

3. C.R. Wronski and N. Maley, these proceedings.

4. P.G. LeComber, presented orally at 10th ICALS, Tokyo (1983) and J. Non-Crystal. Solids **90**, 219 (1987).

5. R.A. Street, these proceedings.

6. W.B. Jackson, S.B. Zhang, C.C. Tsai and C. Doland, these proceedings.

7. J.D. Cohen and T.M. Leen, these proceedings.

8. R. Biswas and I. Kwon, these proceedings.

9. D. Redfield and R.H. Bube, these proceedings.

10. S. Muramatsu, N. Nakamura, S. Matsubara, H. Itoh and T. Shimada, Jap. J. Appl. Phys. **24**, L744 (1985).

11. M. Stutzmann, J. Nunnenkamp, M.S. Brandt and A. Asano, to be published.

12. R.A. Street, to be published.

13. M.J. Williams, C. Wang and G. Lucovsky, these proceedings.

14. H. Shirai, J. Hanna and I. Shimizu, these proceedings.

15. A.H. Mahan and M. Vanecek, these proceedings.

16. B. von Roedern, these proceedings.

17. H.M. Branz, R.S. Crandall and M. Silver, these proceedings.

18. G. Schumm and G.H. Bauer, Phil. Mag. B (1991) in press.

19. W.E. Spear and P.G. LeComber, Phil. Mag. B. **52**, 247 (1985).

20. M.J. Powell, S.C. Deane, I.D. French, J.R. Hughes and W.I. Milne, Phil. Mag. B. **63**, 325 (1991).

# Author Index